THE PAST DECADE IN PARTICLE THEORY

THE PAST DECADE IN PARTICLE THEORY

Edited by

E. C. G. SUDARSHAN
Y. NE'EMAN

GORDON AND BREACH
London   New York   Paris

Copyright © 1973 by

    Gordon and Breach, Science Publishers Ltd.
    42 William IV Street
    London W. C. 2.

Editorial office for the United States of America

    Gordon and Breach, Science Publishers Inc.
    1 Park Avenue
    New York, N.Y. 10016

Editorial office for France

    Gordon and Breach
    7 - 9 rue Emile Dubois
    Paris 14e

Library of Congress catalog card number 71-181810. ISBN 0 677 12010 9 All rights reserved. No part of this book may be reproduced or utilized in any form or by any means, electronic or mechanical, including photocpoying, recording, or by any information storage or retrieval system, without permission in writing from the publishers. Printed in Great Britain.

FOREWORD

A symposium on "The Past Decade in Particle Theory" was held April 14-17, 1970, at The University of Texas at Austin. The symposium brought together about one hundred theoreticians and experimentalists in particle physics to review the results achieved in this field during the decade of the sixties and to discuss the prospects for the future. Some twenty lectures and a round table discussion provided the opportunity of making a fairly thorough assessment of the achievements, hopes, and frustrations of the status of particle physics.

Our symposium, in spite of an exhausting program of lectures and discussions, was eminently successful and even turned out to be a festival of sorts. The success of the symposium was assured by the distinguished members of the scientific advisory committee (E. P. Wigner, Chairman, R.H. Dalitz, R. E. Marshak, A. Salam, and W. Thirring), and a remarkable group of invited lecturers and participants. Their presence and work at the symposium made it a historic and memorable event.

On behalf of my colleagues and myself at the Center for Particle Theory, I take great pleasure in presenting the proceedings of our symposium to a wider audience of fellow particle physicists. We hope they will find the proceedings useful.

E. C. G. Sudarshan

The papers in these proceedings were first published in the journal Fields and Quanta Volume 1 Numbers 1 - 4 (January - July 1971); Volume 2 Numbers 1 - 4 (October 1971 - April 1972).

ACKNOWLEDGMENTS

In the organization of our symposium on "The Past Decade in Particle Theory," we were very fortunate in receiving support from numerous sources. We are grateful to the Chancellor and the President of The University of Texas at Austin whose financial support made it possible to hold the symposium in Austin. We thank the Atomic Energy Commission for its interest in the symposium and financial support toward the editorial expenses of the proceedings. Several distinguished physicists, acting as scientific secretaries, assisted us in the preparation of manuscripts for publication in the proceedings, and we are very grateful to them. The entire staff of the Center for Particle Theory provided invaluable assistance in the successful organization of the symposium.

CONTENTS

Foreword ... v

Breaking of Conformal Symmetry and Renormalization of Non-Polynomial Theories
    A. Salam, Imperial College London, England and International Centre for Theoretical Physics, Miramare, Italy ... 1

Symmetry Breakdown and Small Mass Bosons
    Y. Nambu, The Enrico Fermi Institute and Department of Physics, The University of Chicago ... 33

Advances in the Application of Symmetry
    Y. Ne'eman, Department of Physics and Astronomy Tel Aviv University, Tel Aviv, Israel Center for Particle Theory University of Texas, Austin, Texas ... 55

Particle Spectroscopy
    N.P. Samios, Brookhaven National Laboratory Upton, New York 11973 ... 93

Exact Results in the Analytic S-Matrix Theory of Strong Interactions for High Energies
    Virenda Singh, Tata Institute for Fundamental Research, Bombay, India ... 151

Phenomenology of High-Energy Scattering
    Loyal Durand, III, University of Wisconsin Department of Physics ... 171

Behavior of Inelastic Nucleon-Nucleus Cross Sections from 10^2 to 10^5 GeV
    G. B. Yodh, J. R. Wayland, and Yash Pal, University of Maryland ... 233

Experimental Verification of Dispersion Theory and High Energy Scattering
    S.J. Lindenbaum, Brookhaven National Laboratory Upton, New York ... 255

Dispersion Theory and the Complex J-Plane
    Stanley Mandelstam, University of California, Berkeley ... 295

The Construction of Crossing-Symmetric Unitary
Scattering Amplitudes
    Robert L. Warnock, Illinois Institute of Technology,
              Chicago and Argonne National
              Laboratory, Argonne, Illinois   345

Discussion of Professor Warnock's Talk
    Robert L. Warnock   392

Development of Quantum Field Theory
    I. Segal, Mathematics Department
        Massachusetts Institute of Technology
        Cambridge, Massachusetts   395

Nonleptonic Decay and the Current x Current Theory
    S. Pakvasa, University of Hawaii, Honolulu
    S.P. Rosen, Purdue University, Lafayette,
            Indiana   437

Experimental Verification of Weak Interactions
and the Muon Puzzle
    V. Telegdi, Physics Department
         University of Chicago, Illinois   481

Action-At-A-Distance
    E. C. G. Sudarshan, University of Texas,
                   Austin, Texas   519

Headlights of the Last Decade in Field Theory
    John R. Klauder, Bell Telephone Laboratories,
                Murray Hill, New Jersey   561

Gravitation and Particle Physics
    S. Deser, Brandeis University,
           Waltham, Massachusetts   599

Weak Interactions
    Jan Nilsson, Institute of Theoretical Physics,
            Goteborg, Sweden   637

CP Violation
    L. Wolfenstein, Department of Physics
               Carnegie-Mellon University
               Pittsburgh, Pennsylvania   719

Relativity and Quantum Mechanics
    P. A. M. Dirac, Physics Department
                     Florida State University
                     Tallahassee, Florida   747

Partons
    R. P. Feynman, Department of Physics
                    California Institute of Technology
                    Pasadena, California   773

Subject Index   815

BREAKING OF CONFORMAL SYMMETRY AND RENORMALIZATION OF NON-POLYNOMIAL THEORIES

A. Salam

Imperial College
London, England

and

International Centre for Theoretical Physics
Miramare, Italy

Abstract: This paper is based on a talk delivered by Prof. A. Salam at the symposium "The Past Decade in Particle Theory" held at The University of Texas at Austin on April 14-17, 1970. The talk discusses the non-linear realizations of the conformal group and the group of general relativity. In the second part of the talk the general problem of non-polynomial Lagrangians is reviewed. In particular the non-polynomial, non-linear Lagrangian Theories do not appear to suffer from the ultraviolet infinities of the usual polynomial Lagrangian Theories. The relativistic theory of gravitation is included in this discussion of non-linear non-polynomial theories since, when this theory is taken into account as a quantum theory, it appears possible that all infinities still surviving in renoxmalized theories like quantum electrodynamics will disappear. Explicit calculations with models do show this result. A discussion of this talk follows.

In the first part of my talk I shall take up the subject of conformal symmetries from where Professor Nambu left off in his elegant talk on non-linear realizations.* I shall consider non-linear realizations of the conformal group and the group of general relativity. In the second part of the talk I shall consider non-polynomial Lagrangians in general. These include non-linear realizations. I shall be concerned with the problem of what sort of field theories such non-linear, non-polynomial Lagrangians define. In particular I wish to show that in spite of the complicated look of non-polynomial, non-linear Lagrangians these theories do not appear to suffer from the ultra-violet infinities of usual polynomial Lagrangian theories. I included the general relativistic theory of gravitation in my list of non-polynomial, non-linear theories for the special reason that I believe when we take general theory of relativity into consideration as a quantized theory, using the calculational techniques which I shall exhibit, it explicitly appears that all infinities still surviving in renormalized theories like quantum electrodynamics disappear. This has been conjectured before, by Landau, Klein, and Deser and others. What I wish to do is to show that explicit calculations do give this result.

As we heard from Professor Nambu, the introduction of non-linear realizations provides us with a conceptual advance in the understanding of intrinsic symmetry breaking. Intrinsic symmetry breaking is one of the important concepts of present day physics. The non-linear realization method provides an exceedingly elegant and natural formulation for the concept. Just to remind you of what Professor Nambu showed us, consider chiral symmetry, $SU(2) \times SU(2)$. Particle-multiplets do not correspond to the linear realizations of this group. One reason for this is that for spin

---

* See Professor Nambu's talk this issue

1/2 cases all these linear realizations must correspond to massless particles. If the baryons correspond to linear representations of SU(2) × SU(2), they must be massless. The non-linear realization method avoids this difficulty. We start by assuming that physical particles--massive nucleons for example--correspond to linear representations of the isotopic subgroup SU(2) of the chiral group and only to <u>non-linear</u> realizations of the full chiral group SU(2) × SU(2). To accomplish this the method--which goes back to Wigner--specifies that we introduce a set of <u>preferred</u> fields (in the case of the group SU(2) × SU(2) these are the three π-mesons) and replace every derivative ∂_μ by what may be called "covariant derivative". For example,

$$\partial_\mu \rightarrow \partial_\mu - \frac{i\underline{\pi} \times \partial_\mu \underline{\pi}}{1+\pi^2} \qquad (1)$$

(This covariant derivative is a Christoffel symbol for a metric which one can associate with the group space. The metric is parameterized in terms of the preferred fields, and one proceeds in the time-honored methods of differential geometry). For my purposes the important point is that the total number of preferred fields one needs is equal to the number of generators of the higher symmetry group, (the group SU(2) × SU(2) in the present case) <u>minus</u> the number of the generators of the subgroup whose linear representations give the physical particle multiplets (SU(2)). To check, there are six generators of SU(2) × SU(2) and three of SU(2) so that there have to be 6-3=3 pion fields. The differential geometry tells you what the precise form of the Lagrangian for these preferred fields will be. It also tells you that these particles can have no mass and they are in fact Goldstone-Nambu particles. The existence of these zero mass objects means that the vacuum is degenerate and even though the full Lagrangian is invariant under the (non-linear) transformations of fields of the higher symmetry group SU(2) × SU(2), this symmetry is intrinsically broken. All this is well known for the internal symmetry groups SU(2) × SU(2) or SU(3) × SU(3).

Now there is no reason why one should not do all this for higher space-time groups which contain Poincare or Lorentz groups as subgroups. Physical particles would correspond to linear realizations of Poincare group and non-linear ones of these higher groups. This is what we have started to do in Trieste with J. Strathdee and C. J. Isham. As a first example consider the conformal group--Dirac's gift to Physics. There are 15 generators altogether. Ten of these are the Poincare generators, translations and the rotations P_μ, $J_{\mu\nu}$. In addition there are five generators which correspond to dilations D and special conformal transformations K_μ where D has the form

$$D\psi = -(x^a \partial_a)\psi + \ell\psi. \tag{2}$$

The dilatation operator associates with every field a number ℓ which is related to the dimensionality of the field in a rough sense. For free fields one can show that $\ell = -1$ for (massless) Bose particles and $-3/2$ for (massless) Fermi particles. Physical particles, however, do not correspond to linear representations of the conformal group. The difficulty here is the same which we met for the chiral group case. All linear representations of the conformal group correspond to massless particles. And this time this is true not only for all nucleons but also for all mesons. In hadron physics, the use of linear realizations of the conformal symmetry (even if enough particles could be found to make up complete multiplets) would give a very approximate description of nature. Conformal symmetry could be postulated as a symmetry of nature and its consequences examined for lepton physics in this linear formulation but the results are rather meager and hard to test. From analogy with the case of chiral symmetry one gets the idea that perhaps conformal symmetry is an intrinsically broken symmetry of hadron physics, except for the fact that hadrons transform non-linearly under it. The non-linear Lagrangian would then be invariant for the full conformal group but the vacuum would not be, and the symmetry thus broken intrinsically. If this view is adopted there is no reason to take the masses of the particles to be zero. We must, however, introduce preferred fields. The number of preferred fields in this case--the objects which correspond to the π mesons is 5--you arrive at this number by subtracting from the number of conformal generators (15) the number of Poincaré generators (10). These five particles arrange themselves as a multiplet of vector mesons, four 1^- particles which I shall call C_μ, plus one scalar particle 0^+ which I shall call χ. These particles must be massless from the considerations of Goldstone and Nambu, but they are the only massless particles in the theory. All other particles are massive. The ℓ-value for these particles equals -1.

Consider the χ-field first. Group theory gives us definite rules for writing down the covariant derivatives corresponding to the conformal metric. One can easily show that for this case the metric tensor is given by $g_{\mu\nu} = \eta_{\mu\nu}\chi^2$ where $\eta_{\mu\nu}$ is the Minkowski metric tensor and χ is the 0^+ preferred field. The rules for writing conformal-invariant Lagrangians are simple. Write <u>any</u> Lorentz-invariant Lagrangian. Replace the ordinary derivatives by the "covariant" derivative

$$\partial_\mu \psi \to \partial_\mu \psi + [(\ell \eta_{\mu\nu} - iS_{\mu\nu})\partial_\nu \chi/\chi]^\psi, \tag{3}$$

where $S_{\mu\nu}$ is the spin-matrix for the field ψ; multiply all terms of the Lagrangian by an appropriate number of powers of χ such that the resulting Lagrangian has an ℓ value of -4. (For example, the $f\bar{\psi}\psi$ term appears in the form $f\bar{\psi}\psi\chi$). One can show that one consequence of intrinsic symmetry-breaking is that χ must have a non-zero vacuum expectation value $<\chi>_o \neq 0$. The bare mass term associated with the field ψ is then given by $f\bar{\psi}\psi<\chi>_o$. We still have to add on a (conformal-invariant) Lagrangian for the χ field. For this there is one unique form.

$$\frac{1}{2}(\partial \chi)^2 + K\chi^4 \tag{4}$$

One can easily show that the χ field obeys an exact equation of motion irrespective of what the total Lagrangian was. This equation is given by

$$\partial_\mu D_\mu = \frac{1}{2}\partial^2(\chi^2) + \theta_{\mu\mu} = 0 \tag{5}$$

where D_μ is the dilatation current associated with the dilatation operator D and $\theta_{\mu\mu}$ is the trace of the symmetric energy-momentum tensor.

To establish the existence of this intrinsically broken conformal symmetry of nature, we need to show that a field χ exists. There are two possible suggestions, not mutually exclusive. The first is to associate the massless χ-particle with scalar gravity. It is amusing that Eq. 5 is the field equation postulated previously by Dicke, Gürsey and others for scalar gravity from different considerations. Accepting this, we would then make no distinction between leptons and hadrons; <u>all</u> particles correspond to non-linear realizations of the conformal symmetry and of our Lagrangians broken spontaneously by the degeneracy of the vacuum which arises from the existence of the massless scalar-graviton χ. There is, however, a second point of view possible. We may assume that the χ field, (the "dilaton") couples only to hadrons. Only the hadrons correspond to non-linear realizations of the conformal group while (zero-mass) leptons are linearly realized. Now Professor Nambu raised the problem already for the SU(2) × SU(2) case; how do the preferred fields themselves acquire mass in a non-linear formulation? We face the same problem here. We know of a number of (broad) 0^+ objects which might correspond to the dilaton. But all have mass. How do we arrange to give χ a mass?

Professor Nambu gave no answer to the question he posed for SU(2) × SU(2) except the rather inelegant answer; at the present stage of our understanding of the theory we simply add an explicit symmetry breaking to the Lagrangian which among other possible effects will also give the Goldstone particles a mass. For the conformal case we shall follow exactly the same procedure, except that we shall try to economize. Can we think of just one term which breaks both conformal and chiral symmetries? (Implicit in this is of course the statement that if this one term did not exist, one could write Lagrangians which are both conformally and chirally symmetric). The answer to our question is: Yes. We can construct a number of different models. For example, we may take the well-known Gell-Mann - Lévy model for SU(2) × SU(2) where instead of the usual condition $\sigma^2 + \pi^2 = 1$, we write it in the form $\sigma^2 + \pi^2 = \chi^2$, with $<\chi>_0 = 1$ as the self consistency condition. If we do this, we find that the only explicit symmetry breaking term one needs to introduce to break both chiral and conformal symmetries is $L_1 = A\sigma$. One can easily show that in this model the χ-particle mass and pion mass are simply related-- $m_\chi^2 = 3m_\pi^2$. One could take a different model, for example,

SU(3) × SU(3) with a symmetry breaking term of the type $(3,\bar{3}) + (\bar{3},3)$. Here one finds $8m_\chi^2 = 3m_\pi^2 + 6m_K^2$ so that m_χ would be 440 Mev. One could also take a third model in which the symmetry breaking is of an (8,8) variety; here one finds $3m_\chi^2 = 2m_\pi^2 + 4m_K^2$ i.e., m_χ = 590 Mev. As you know, a number of 0^+ broad-width objects have been reported from time to time; whether any one of them corresponds to the χ-particle, I shall not discuss. I shall finally mention in passing a nice way of writing the relation (5) when the Lagrangian contains a symmetry-breaking term $L_1 = A\chi$. Coleman, Gross and Jackiw have recently suggested that the conventional symmetric energy-momentum stress tensor may be modified in the following manner:

$$\bar{\theta}_{\mu\nu} = \theta_{\mu\nu} - (1/6)(\partial_\mu \partial_\nu - \eta_{\mu\nu}\partial^2)\chi^2. \tag{6}$$

Here $\bar{\theta}$ is the modified tensor and θ the conventional one. We can likewise modify the dilatation current D_μ to a new current $\bar{D}_\mu$

$$\bar{D}_\mu = D_\mu - (1/6)\partial_\nu(\chi_\nu\partial_\mu - \chi_\mu\partial_\nu)\chi^2 - \partial_\nu F_{\mu\nu} \tag{7}$$

where $F_{\mu\nu}$ is an anti-symmetric quantity which can be specified completely in terms of field operators. It turns out that in this formulation, one can show that quite generally

$$\bar{D}_\mu = \chi_\nu \bar{\theta}_{\mu\nu}.$$

With these definitions, whenever the symmetry-breaking term $A\chi$ is introduced into the Lagrangian, $\partial_\mu \bar{D}_\mu$ is not zero. Instead the PCAC like equation for the dilatation current is given by

$$\partial_\mu \bar{D}_\mu = \bar{\theta}_{\mu\mu} = 3A\sigma, \tag{8}$$

while the usual PCAC equation reads

$$\partial_\mu A_\mu = A\pi. \tag{9}$$

Note how beautifully the two relations seem to form a unit.

Let me turn to one more use of the non-linear realizations. The space-time group one may consider after the conformal group is the general linear group of coordinate transformations GL(4,R) defined by

$$A^\nu_\mu(x) = \frac{\partial x^\nu}{\partial \bar{x}^\mu} \tag{10}$$

If the 16 generators of GL(4,R) are designated as F^α_β, these satisfy the well-known commutation relation given below:

$$-i[F^\alpha_\beta, F^\gamma_\delta] = \delta^\alpha_\delta F^\gamma_\beta - \delta^\gamma_\beta F^\alpha_\delta \tag{11}$$

Consider non-linear realizations of this group, linear with respect to the Lorentz group with its six generators. The number of preferred fields which one would need to introduce to construct non-linear realizations is 16-6=10. These fields will enter the definition of the "co-variant" derivatives. Together with Isham and Strathdee we have studied this aspect of GL(4,R) invariance. We find that the 10 preferred fields can be identified with the set of ten pseudosymmetric vierbein fields well known in general relativity theory, particularly when one wishes to write generally co-variant spinor field equations. The non-linear realization point of view for the general linear group then automatically leads to a vierbein point of view of developing a vierbein universal (gravitational) field interaction. Assuming that quarks and leptons are the fundamental bits of matter and represent non-linear realizations of the group GL(4,R) we find that the generally co-variant fields equations for these spin 1/2 objects can be written thus : Introduce the ten preferred (vierbein) fields $L^{\mu a}$. Introduce a quadratic product of these fields

$$\eta_{ab} L^{\mu a} L^{\nu b} = g^{\mu\nu}, \quad \eta^{ab} = (+, -, -, -) \qquad (12)$$

The expectation value of this quadratic product $g^{\mu\nu}$ for any physical state happens to represent the space-time metric. The co-variant derivative of the spin 1/2 object ψ, representing quarks or leptons can be shown to equal

$$\psi_{;\mu} = \psi_{,\mu} - i(B_{\mu ab} S^{ab}), \qquad (13)$$

where $\quad B_{\nu ab} = L_{\lambda a} \Gamma^{\lambda}_{\nu\mu} L^{\mu}_{b} - L_{\mu a,\nu} L^{\mu}_{b}$ and $\Gamma^{\lambda}_{\mu\nu}$ is the affinity.

The vierbein fields are the massless Goldstone bosons for this non-linear situation. The total Lagrangian (for the ψ-field and $L^{\mu a}$ field) is invariant but the vacuum is not. This is the point of view which a group theorist of our day would take towards general relativity theory. There is no doubt at all that general theory is an amazing theory which can be looked upon from at least a half-dozen different points of view--as a Yang-Mills theory, for example, developed in 1955 by Utiyama or as I said in our day as a non-linear realization where vierbein fields are the primary fields and the metric fields are the secondary entities.

As I shall show later I shall need the vierbein formalism when I consider renormalization and suppression of <u>all</u> infinities of quantum field theories. And I now turn to this subject.

Before I consider vierbein relativity, I wish to show first that the conventional treatment of the infinities problem which unwarrantedly assumed that all Lagrangians in physics are polynomials in field variables needs drastic modifications. The non-linear Lagrangians Professor Nambu and I have been speaking about are of course particularly important examples of non-polynomial Lagrangians. The first modern paper on non-polynomial Lagrangians was written by Okubo in 1954. Between '63 and '68 Fradkin, Efimov and Volkov in the USSR worked on these problems, with contributions from groups at London and Trieste and Lee and Zumino (1969). A recent contribution is that of Steinmann which I hope lends our heuristic efforts the respectability of an axiomatic framework!

To start from the Book of Genesis, consider what Dyson proved for the polynomial Lagrangian of electrodynamics. He showed that all infinities in a perturbation calculation come from only four types of graphs. These are the vacuum graph (Fig. 1a), the self energies (Figs. 1b,c) and vertex graphs (Fig. 1d).

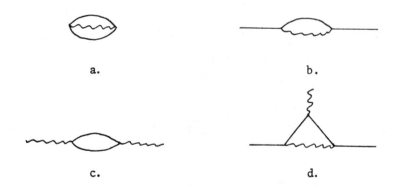

Figure 1. The four basic divergences
a. Vacuum Graph, b. Electron Self Energy,
c. Photon-Self Energy, d. Vertex Modification

He further showed that <u>all</u> infinities can be absorbed into (infinite) renormalizations of mass and charge. It is important to note that there are two distinct parts of this program. First, the number of <u>distinct</u> infinities is not unlimited; secondly, that these can be absorbed in renormalization of constants. It could have transpired that the infinity-absorption was just not possible through a renormalization procedure.

Dyson's work for electrodynamics was soon extended for <u>all polynomial</u> Lagrangians. One could divide theories into three classes-- super-renormalizable, renormalizable, and non-renormalizable. To see the distinction simply, consider the propagator of a spin 0 field $<\phi(x)\phi(0)>=\Delta(x)$. The propagator behaves at $x = 0$ like x^{-2}. If we associate x with a dimension M^{-1} we may associate with each field ϕ a behavior at infinity of the type M, so far as the Feynman propagator for this field is concerned. For the spin 1/2 field we may associate the dimensions $M^{3/2}$ since $<\bar{\psi}\psi> = \frac{1}{x^3}$, and for spin 1 field U_μ a length M^2. ($<UU> \approx \frac{1}{x^4} \approx M^4$.)

In terms of these growths at infinity of the relevant fields, a Lagrangian like ϕ^3 which represents a super-renormalizable theory with only two infinite graphs (Fig. 2) behaves like M^3.

Figure 2. Divergent diagrams of the ϕ^3 theory

Renormalizable theories given by Lagrangians ϕ^4, $(\bar{\psi}\psi)\phi$ and with four finite graphs (⌬ , ─⚬─, ⚬ , ─△─) behave like M^4. Nonrenormalizable theories with Lagrangians like ϕ^5, ϕ^6_3,... or $\bar{\psi}\psi U$, and where <u>every</u> type of graph is infinite behave like M^5, M^6,... or worse at infinity.

SUMMARY OF DYSON CLASSIFICATION OF FIELD THEORIES

$$<\phi(x), \phi(0)> = \Delta(x) \sim \frac{1}{x^2} \sim M^2 \qquad M \to \infty$$

$$\phi \sim M$$

$$\psi \sim M^{3/2}$$

$$V_\mu \sim M^2$$

1) Super-renormalizable $L_{int} = \phi^3 \sim M^3$ only 2 distinct types of infinities.

2) Renormalizable $L_{int} = \phi^4 \sim M^4$ or $\bar\psi\psi \sim M^4$ only 4 distinct types of infinities.

3) Non-Renormalizable $\phi^5 \sim M^5$, $\bar\psi\psi\partial\psi \sim M^5$, $\bar\psi\psi U_\mu \sim M^5$ an unlimited number of distinct types of infinities.

[All Higher Spins]

If a Lagrangian L behaves like M^n at infinity, we shall call n the Dyson index. A conventional renormalizable theory possesses the Dyson index 4. An index higher than 4 means the theory has an infinite number of maladies. An index equal to 4 means the theory is just about curable. A lower index is better; an index of 3 pertains to a super-renormalizable situation. One may conjecture that if the index were 2, there may perhaps be just one infinity--vacuum to vacuum transition; if the index were still lower than 1 or 0, perhaps there may be no infinity at all.

Before I go on to consider this conjecture and show that it is basically correct, let us talk of one other possibility. All Dyson's work was based on the assumption that the perturbation theory is giving us a correct account of the infinities. Could it be that the infinities are being manufactured by the perturbation theory and are not really intrinsic to the field theories? In other words, if we could solve polynomial Lagrangian theories <u>exactly</u>, perhaps there are no infinities. Now this problem has been considered in recent years by Glimm, Hepp, Jaffe, Segal and others. What they have shown is that if one works in 2 dimensions, one in space and one in time, and takes a ϕ^4 theory which is super-renormalizable in two dimensional space-time, then all of the infinities of the perturbation theory are present in the exact solution and some more. It is a very big extrapolation to say that one can go to 4 dimensions from working in two dimensions of space-time. However one may now no longer discount the possibility that infinities are really intrinsic to polynomial Lagrangians. And if the infinities are really intrinsic then things like the π^+ - π^0 mass difference (observable self-mass effects) can never be computed so long as we stick to polynomial Lagrangians. This is another reason to consider Lagrangians which are non-polynomial in field variables.

CONFORMAL SYMMETRY AND NON-POLYNOMIAL THEORIES 15

Let us go back to Dyson indices equal to <u>two</u> or <u>less</u>. The only polynomial theory with index 2 is ϕ^2--a rather trivial mass term. To get an interesting theory we must consider nonpolynomial objects like

$\dfrac{\phi^3}{1 + f\phi}$ or a more physically motivated Lagrangian like that for π-mesons for chiral $SU(2) \times SU(2)$ in Schwinger's representation:

$$L = \frac{(\partial \phi)^2}{(1 + f^2 \underline{\phi}^2)} \qquad (14)$$

Both of these behave like M_n^2.

A theory like $G \dfrac{\phi^n}{1 + f\phi}$ contains two coupling parameters G and f. I shall call these major and minor coupling constants. We shall work to a <u>given</u> order in the major constant but we shall solve the theory exactly to <u>all</u> orders in the minor. There are two steps--:

(1) Given any physical process, specified by the external lines and the momenta P_i associated with them, go to the Symanzik region of these momenta. This is the <u>unphysical</u> region where all momenta are space-like. Carry out all computations of matrix elements in this region and then continue back to <u>time-like momenta</u>. The advantage of working in the Symanzik region is that there are no physical thresholds and all functions, including the Feynman propagator $\Delta(x)$ are <u>real</u>. For a zero mass field ϕ, $\langle(\phi(x)\phi(o))_+\rangle = \Delta(x)$ is $\dfrac{1}{x^2}$ and for simplicity hereafter I shall work with this case. (For massive particles $\Delta(x) = \dfrac{1}{x^2} + m^2 \log(m^2 x^2) + O(x^2)$ and the most singular term is the first one.)

(2) To make computations, expand $L(\phi)$ in powers of f. For example

$$L(\phi) = G \frac{1}{1 + f\phi} = G(1 - f\phi + f^2\phi^2 - \ldots). \quad (15)$$

A fixed order in the major constant G, say G^2- order needs computation of a term like

$$F(\Delta) = <\left\{L(\phi(x_1))L(\phi(x_2))\right\}_+>$$

$$= G^2 <(1 - f\phi(1) + f^2\phi^2(1)\ldots, 1 - f\phi(2) + f^2\phi^2(2)\ldots)_+>$$

This is easily computed and equals:

$$F(\Delta) = G^2 \sum_n n! (f^2 \Delta(x))^n \text{ where } x = x_1 - x_2 \quad (16)$$

and corresponds to a sum of graphs in Fig. 3

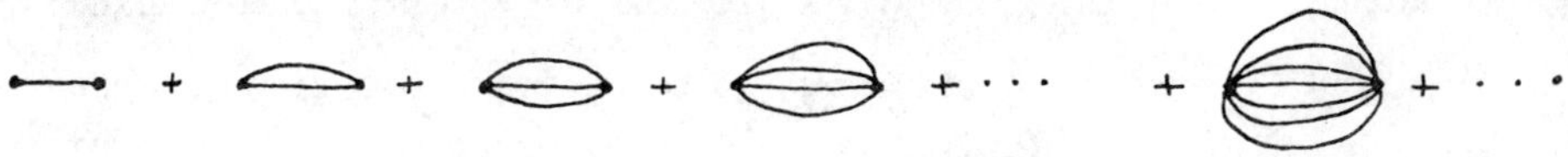

Figure 3. The graphs contributing to the two-point function to second order in the major coupling constant.

Note that each term in the series is more and more infinite because

$$\Delta^n(x) = \left(\frac{1}{x^2}\right)^n,$$

as $x \to 0$ the higher terms get more and more virulent ultra-violet divergent as n increases. Our point of departure is to sum this formal series by the following method:

Write:
$$n! = \int_0^\infty d\rho\, \rho^n e^{-\rho} \qquad (17)$$

Then we can obtain the sum:

$$F(\Delta) = G^2 \int_0^\infty d\rho\, e^{-\rho}\, \frac{x^2}{x^2 - f^2 \rho}. \qquad (18)$$

If we look at this sum, we note that this is no longer infinite at $x = 0$.

There is a <u>smearing</u> of the infinity at $x = 0$ by an amount $f^2 \rho$; roughly speaking there is a cut-off in x^2 - space at f^2 or a cut-off $\frac{1}{f^2}$ in the reciprocal energy space. The problem of ultra-violet infinities has disappeared on summation in the <u>minor</u> coupling constant. In a way this is hardly surprising; a Lagrangian like $\frac{1}{1 + f\phi^2}$, even on the face of it, has a visible damping factor built in its denominator.

The question now arises; is this suspicion innocuous in respect of preserving desirable unitarity and analyticity properties of the theory. Similar looking summations have been tried in the past--for example, the paratization procedure of Paid and Feinberg. They did destroy some of these desirable properties.

One big difference between what they did and our approach is that they take the matrix elements and split them into parts and sum certain parts and not others. This splitting of each matrix element was a dangerous process. We have carried out a complete summation in f. We must however check that, in our own appraoch, the usual unitarity properties for example do hold, i.e. that up to a given order in the major constant G, say the second order, $\text{Im} S_2 = S_1 S_1^+$ in an obvious notation. To do this, take the Fourier transform of Eq. 18 , (interpreting the integral in ζ-space as a principal-value integral. I shall come to this point later). One can easily show that it can be written in a dispersion form:

$$F(s) = \int_0^\infty \frac{e^{-\beta s'}}{s' - s} ds' + e^{\beta s}[E_1(\beta s) \mp i\theta(s)], \quad \text{Im } s \lessgtr 0 \quad (19)$$

$$E_1(z) = \frac{e^{-z}}{z}(1 - 1/z + 2/z^2 - \ldots).$$

The matrix element displays the usual unitarity cut which extends from zero to infinity; there are no other singularities in the complex s plane. Further, unitarity is indeed satisfied. This was the case of massless particles. For massive case there is a general argument given by Lee and Zumino who have pointed out that for a massive particle the existence of thresholds (and unitarity singularities) is connected with the behavior of the propagator for large values of x^2, $\Delta(x) \simeq e^{-mx}/(x^2)^{2/3}$, while the infinities arise from the values of $x^2 \to 0$. Hence the damping manipulations which kill ultra-violet singularities are unlikely to affect the unitarity of the theory. This problem of analyticity and unitarity--and also of ultraviolet infinities--can perhaps best be approached if we do not work in x-space as we have done hitherto but directly in momentum space. If I go back to Eq. 16 we can perform a Sommerfeld-Watson transform (rather than a Borel sum) to convert the sum into an integral, with the contour in the z plane circling the positive real axis as shown in Figure 4.

Figure 4. Sommerfeld-Watson Contour

$$F(\Delta) = \sum_\eta \eta! \Delta^\eta = \frac{i}{2} \int \frac{dz \Gamma(z+1)(-\Delta)^z}{\sin \pi z} \qquad (20)$$

Now the Gelfand-Shilov formula for the Fourier transform of $(\Delta(x))^z$ yields $\Gamma(2-z)/(\Gamma(z)(p^2)^{2-z})$ for $0 < \text{Re} z < 2$. Hence the Fourier transform of $F(\Delta)$ is:

$$\tilde{F}(p^2) = \frac{i}{2} \int \frac{dz}{\sin \pi z} \Gamma(2-z)(-1)^z \frac{1}{(p^2)^{2-z}} \qquad (21)$$

The expression is just like a standard Feynman expression except that the propagator for the entire cocoon--we call it the superpropagator--carries an index z in the factor $(\frac{1}{(p^2)^{2-z}})$ and is multiplied by a few $\Gamma(z)$ factors. One can easily build up a whole Feynman calculus for super-graphs; for example, in the fourth order of the major constant, the scattering of mesons by mesons is given by a super-graph shown in Fig. 5

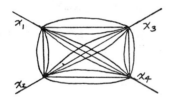

Fig. 5 Fourth-order Super-graph

One obtains this by taking 4 points $x_1 x_2 x_3 x_4$ and connecting them in all possible ways with just one superpropagator between two points. The rule for writing down matrix elements for supergraphs are exactly like the Feynman rules. For each "super-line" we put in one momentum p_i; the propagator is proportional to $\frac{1}{(p_i^2)^{2-z}}$. Energy and momentum is conserved at each vertex. For example, for the vertex-graph, we get:

$$= \int \rho(z_1 z_2 z_3)\, dz_1\, dz_2\, dz_3 \times$$

$$\int d^4k\, \frac{1}{(k^2)^{2-z_1}}\, \frac{1}{[(p-k)^2]^{2-z_2}}\, \frac{1}{[(p'-k)^2]^{2-z_3}} \tag{22}$$

The density functions $\rho(z_1 z_2 z_3)$ consists of (specified) gamma functions. From this expression one can easily convince oneself that the usual Cutkosky rules for the singularity structure of Feynman graphs apply equally well here. The theory is likely to be unitary and analytic in the conventional sense for all orders in the major constant.

Now what about renormalization? It is conjectured that theories with Lagrangians $\mathcal{L} \approx M^{2-\varepsilon}, \varepsilon > 0$, at infinity have no infinities, while those with $\mathcal{L} \approx M^4$ are renormalizable. Naive power counting of these general supergraphs seems to assure this. But one knows from experience that even for conventional polynomial Lagrangians naive power counting can be misleading. We have undertaken a systematic program to check this conjecture for classes of non-polynomial theories; the program is not yet complete but the following statements can perhaps be made with some degree of confidence:

(1) Theories with $\mathcal{L} \propto M^{2-\varepsilon}$ are free of ultra-violet infinities.

(2) Theories with $\mathcal{L} \propto M^2$ and M^3 are renormalizable. Those with $\mathcal{L} \propto M^2$ may possess no infinities if a gauge is operating.

(3) To our surprise we found that a non-polynomial theory with $\mathcal{L} = \phi^5/(1+f\phi)$ behaving like M^4 is not renormalizable.

CONFORMAL SYMMETRY AND NON-POLYNOMIAL THEORIES

Now let us come back to the suppression of the infinities through gravitation. As I said, the conjecture was made by Klein, Pauli, Landau, Deser and others that quantum gravitation may provide a universal cut-off. The argument went roughly like this: The ultra-violet infinities all come from the propagators vanishing on the light cone ($\Delta(x) = 1/x^2 \to \infty$ when $x^2 \to 0$). A gravitational field affects the light-cone; in particular the zero-point fluctuations of gravitation should <u>smear</u> the singularity of $\frac{1}{x^2}$ by an amount K^2 where K is the gravitational constant ($K^{-2} \approx 10^{40} m_c^2$). What I want to do is to show that this smearing <u>does</u> happen and indeed provides for a universal quenching of infinities, provided--and this is important--gravitational theory is treated as a non-polynomial theory as Einstein intended it to be.

What is not commonly realized--and the professional relativists are perhaps guilty of having obscured the matter--is that there are two inequivalent tensors $g^{\mu\nu}$ and $g_{\mu\nu}$, related to each other through $g^{\mu\nu} g_{\nu\rho} = \delta^\nu_\rho$. It is

$$\sqrt{-g}\, d^4 x^\alpha = \sqrt{\det g_{\alpha\beta}}\, d^4 x^\alpha$$

which makes the scalar volume element. If we consider $g^{\mu\nu}$ as the basic tensor field (we write, as usual $g^{\mu\nu} = \eta^{\mu\nu} + K h^{\mu\nu}$ where $\eta^{\mu\nu}$ is the Minkowski tensor) then $g_{\mu\nu}$ is a ratio of two polynomials and $\sqrt{-g} = \sqrt{\det g_{\alpha\beta}}$ has the form

$$\frac{1}{(1 + Kh + K^2 h^2 + K^3 h^3 + K^4 h^4)^{1/2}}$$

If $g^{\mu\nu} \simeq M$, $g_{\mu\nu} \simeq M^{-1}$ and $\sqrt{-g} \simeq M^{-2}$ at infinity.

Consider now the Einstein Lagrangian for gravity. It reads:

$$\mathcal{L} = \frac{1}{K^2} \sqrt{-g}\, g^{\mu\nu} (\Gamma^\lambda_{\mu\rho} \Gamma^\rho_{\nu\lambda} - \Gamma^\lambda_{\mu\nu} \Gamma^\rho_{\lambda\rho}) \tag{23}$$

where

$$\Gamma^\lambda_{\mu\nu} = \frac{1}{2} g^{\lambda\rho} (\partial_\mu g_{\nu\rho} + \quad).$$

Clearly $\mathcal{L} \simeq M$ and using the techniques I have described, graviton-graviton processes should exhibit no infinities.

How about the effect of gravitation on, say, scalar particle interactions? The conventional meson interaction in a gravitational field is given by

$$\mathcal{L}_{meson} = \sqrt{-g}\, \{g^{\mu\nu}(\partial_\mu \phi \partial_\nu \phi) - m^2 \phi^2 + \lambda \phi^4\} \tag{24}$$

Instead of $\mathcal{L}_{meson}$ behaving like M^4, it now behaves as M^3 - a small mercy but not enough to get rid of all infinities. Now consider the vierbein formulation of gravitation needed in any case for fermion-graviton interactions. Since $L^{\mu a} L^{\nu b} = g^{\mu\nu} \eta^{ab}$ where $L^{\mu a}$ is vierbein gravity, the assumption that it's the field $L^{\mu a}$ which is the fundamental field (and order M), implies that $\sqrt{-g} = \det(L_{a\mu})$ is of order M^{-4}. It is easy to check now that the Dirac Lagrangian for an electron is $\simeq M$, while the zero-spin particles have a Lagrangian behaving like M^2. In this formulation of gravitation theory, if we start with a conventional matter field (electron-photon-quark-gluon) Lagrangian which is renormalizable; that is, behaves like M^4 when not in a gravitational field, the effect of quantum vierbein gravity is to bring down

CONFORMAL SYMMETRY AND NON-POLYNOMIAL THEORIES   23

its behavior to M^2 at worst. We expect that there will be no ultra-violet infinities left if the techniques of this paper are used.

To make what I am saying more precise consider the electromagnetic interaction of an electron which now reads

$$\mathcal{L} = e(L/\det L) \, \bar{\psi} \, \gamma \, \psi \, A. \tag{25}$$

Writing $L = 1 + \frac{K}{2} H$ where H is quantized vierbein gravitation:

$$\mathcal{L} = e \left[\frac{(1 + (K/2)H)}{1 + KH + \ldots + \overline{K^4 H^4}} \right] \bar{\psi} \, \gamma \, \psi \, A. \tag{26}$$

The self-energy of the electron when no gravity is present is given in the lowest order by the graph

which in x space gives the contribution

$$\frac{1}{x^2} \cdot \frac{1}{x^3}$$

(Here $\frac{1}{x^3}$ is the factor arising from the electron propagator and $\frac{1}{x^2}$ from the photon). The self mass is proportional to:

$$\delta m \approx \int \frac{1}{x^2} \cdot \frac{1}{x^3} \approx \underset{M \to \infty}{\operatorname{lt}} M \; ,$$

which is a linear infinity. As is well known, a careful calculation shows that the linearly infinite term always drops out and

$$\delta m/m \sim (e^2/4\pi) \log (M/m_e)^2$$

where M is the cut-off.

In our theory one is computing the graph

with millions of gravitons going across, coming from the term

$$\frac{1}{1 + KH + \ldots + K^4 H^4}$$

in the Lagrangian. Thus we expect an additional factor in the matrix element of the form

$$\int \frac{e^{-\zeta}}{1 - (K^2 \Delta)\zeta} d\zeta$$

so that the total contribution is of the form

$$\iint \frac{e^{-\zeta}}{x^5} \cdot \frac{1}{x^2 - K^2 \zeta} d^4 x \, d\zeta.$$

There is in the theory a built in cut-off at $\frac{1}{K^2}$. To get the orders of magnitude right

$$\frac{\delta m}{m} \sim (e^2/4\pi) \log(1/Km_e)^2 \simeq 1/2.$$

The important point is that with the gravitational cut-off acting in electromagnetism, the self-mass is not coming out to be outrageous; it's a reasonable fraction of the electron mass. If we do our algebra right, it's fully possible that we may get the whole electron mass as self-mass. Could one then, conversely, use this as the input principle to compute the gravitational constant? If nature does reveal outrageous coupling parameters like $K^2 \sim \frac{10^{-40}}{m_e^2}$ (outrageous when compared to "normal" constants like $\frac{1}{137}$), surely the "principle" which will yield them from pure thought has also got to be outrageous.

But it's not this numerology I wish to stress. Rather it's the hope that once and for all the sceptre of ultra violet infinities in quantum theory is laid when full quantum theory with all the realistic forces is considered. I do not wish to convey the impression that we have checked in detail that the naive considerations I have made regarding the dimensionality of Lagrangians and its relation to the infinities they may produce. We have much more work yet to do. I also do not wish to convey that every field theoretic problem in the sums over the minor constants has been resolved. There still remain some basic difficulties with the Borel sums: first, there is the well known problem of uniqueness of these sums; second, it appears that though the high frequency behavior of matrix elements for space-like momenta is polynomially bounded, the behavior for time-like momenta is not, unless we take a sum of the series over the major coupling constant as well. These problems do not affect the considerations I have given in respect of ultra violet infinities; so for present purposes I have ignored them.

DISCUSSION OF PROF. SALAM'S TALK

F. Rohrlich, Syracuse University: I have one comment and one question about Prof. Salam's interesting and sweeping paper. My comment is the following:

If a theory needs renormalization, this renormalization must be carried out. However, it may still be divergent, irrespective of whether it needs renormalization or not so that there is no direct connection between divergences and renormalization. We also know that the renormalization process is not a very interesting process from the point of view physics because it is based on a separation between, for example, mass and charge into two parts which are not separately observable. It is possible to formulate theories in a way in which they do not need renormalization, in particular for those interactions which in the usual formulation lead to super-renormalizable or renormalizable theories. One can cast the theory into a form in which renormalization is necessary and one can show that those theories are completely free of divergences.

Now my question is the following. You have succeeded in removing the divergence difficulties at least to some degree by introducing non-linearities which are of such a nature that they do not effect the physical results of the theory because they are small enough and the cut-offs are high enough. In fact, they are not observable at the present time. This builds in an ambiguity as to what type of non-linearities you may use. For instance you can do it with some power of $1 + f\phi$ instead of just $1 + f\phi$ in the denominator. This seems to be a rather bothersome arbitrariness. How do you deal with it?

Salam: I think I agree with your comment. I have tried to stress in the very beginning that the existence of infinities and the possibility of renormalization are unrelated phenomena. With respect to the reformulation of renormalizable theories in a manner in which infinities do not make their appearance, I imagine you have in mind defining products of operators in such a way that these products have a meaning in a distribution theory sense. If this is possible then it is difficult to understand the work of Jaffe and Glimm at its face value. These men, I believe, show that there are infinities which cannot be "talked away" in this manner. I would like Hepp who is here to say something about this.

Regarding the question about non-uniqueness of non-polynomial theories you have raised, I agree this is important. We can in order to get rid of infinitiesm multiply any Lagrangian by a factor like $(1 + f\phi)^{-n}$. This would mean that depending on the constants f and n, the class of Lagrangians in the world is arbitrarily large. Is there some principle to restrict the proliferation of Lagrangians? I believe there is. For example I would like to restrict myself to just those non-polynomial Lagrangians which arise naturally through the non-linear realization method. I believe that, at long last, we have a proceedure for writing strong Lagrangians which we've never had before. That is, write a non-linear chiral $SU(3) \times SU(3)$ invariant Lagrangians. It happens also to be non-polynomial. Use for it the technique I have described and its infinities will also be minimal. Likewise, the massive Yang-Mills theory offers a definite

Lagrangian. Surprising thought it may seem, we can use the summation techniques I have described for it. The gravitational Lagrangian, of course, one can't tinker with; this is perfectly well specified. The conformal non-linearity which I exhibited is again a very definite one. You cannot tinker with it. The only place in physics where a Lagrangian is not yet agreed on is weak interaction theory.

<u>K. Hepp, University of Zurich</u>: I want to make one constructive remark. The constructive remark is the following. There is one well defined form of analyticity which you should check. That is, if you compute in perturbation theory the retarded commutators and the generalized commutators their Fourier transform should have analyticity properties in suitable cones. So if you have those analyticity properties, then you would have a perturbation theory since they are essentially like the assumptions which Steinman made in order to prove the existence of a scattering theory for those nonrenormalizable interactions.

<u>Salam</u>: Has Steinman not done that:

<u>Hepp</u>: Steinman has not proved, as far as I know that these super propagators have the analyticity properties in cones for the generalized retarded commutators.

He has built a theory on axioms and one of the axioms is such an analyticity property, so you have to check it. The last constructive remark is the following. One has to be aware of the possibility that the Hamilitonian of the renormalized theory is not bounded from below. Take $\lambda\phi^3$ theory. Osterwalder has for instance shown that in four dimensional space-time, this theory exists after infinite renormalization and even the infinite renormalizations do not change the intrinsic bad behavior of the theory. The Hamiltonian is unbounded from below after very strong renormalization.

Salam: Is this connected with the oddness of the Lagrangian?

Hepp: Yes and in perturbation theory, one does not see this unboundedness from below in the energy sector so that problem remains.

Salam: That means we have to have a very special class of positive Lagrangians; presumably things like $(1 + \lambda \phi^2)^{-1}$.

Hepp: Finally I wish to make a remark concerning Rohrlich's comments. There exist many formulations where you do not introduce explicitly the infinities. For instance, if you solve recursively the LSZ equations for the generalized retarded functions you can redefine the problem so that the infinities appear as ambiguities in the correct definition of current commutators. Still, in another sense, the infinities are present. They would be present if in such a theory you would compute the vaccuum expectation value and would compute the product of field operators and the behavior of field operators at two space-time points which coincide in such a sense. I think the ultraviolet divergences are as real as in the Hamiltonian scheme, where one starts in Fock Space from $H^o + H^i$.

Rohrlich: May I just answer this very briefly. There are two types of infinities I am referring to. One part that has to do with vaccuum expectation values and that is removed by taking proper account of the distribution nature of the object that enters the theory, or in physicists's language, by putting it in a box and then making the size

of the box go to infinity. That is one type of divergence
and that is always there. I was referring to the ultra-
violet divergences which are not of this nature and that
are usually renormalized. Those can be eliminated. They
are absent from the theory, in the first place, if you cast
the theory into a form which does not require renormalization. If you like, in terms better known, one can write
down super equations for the scattering operator which already contain the subtractions in them, but, they are not
introduced in the process of eliminating divergences. They
come in from the very beginning and the theory is finite.
There are no divergences and no renormalizations. The mass
and the charge are put in essentially by boundary conditions
on the scattering matrix.

R. E. Marshak, Rochester University: I don't think you said
that you have made much progress on the weak interaction,
and yet, let me ask the question this way. It seems to me
that if you just take the weak interaction, by itself, no
matter how you play with it, even if you make it convergent,
the essential cutoff will be around the so-called unitarity
limit. However, from other types of calculations, the mass
difference of neutral kaons and so on, you seem to get a much
lower value when you take into account the strong interactions.
So it seems to me, even from your point of view, one doesn't
see how to solve that problem and therefore, one might have to
introduce something new. One possibility is the strong self
interaction of the intermediate vector boson which when it
combines with the semi-weak, could wind up giving you this
structure that you need in order to get a lower convergence
cutoff. Do you agree with that sort of point of view?

CONFORMAL SYMMETRY AND NON-POLYNOMIAL THEORIES

Salam: In an intermediate boson theory of weak interactions, as is well known, the terrible infinites come from the spin-zero daughter of the vector meson field and its derivative coupling. One can transform this spin-zero part of the coupling to a non-polynomial form and one gets the result that the infinities are greatly suppressed and the effective cut-off is the inverse of the Fermi constant, that is, as you correctly point out, at the unitarity limit. You are speaking of a cut-off arising from a number of other distinct mechanisms which pertain to where the form factor starts falling. I agree with you that this fall at a considerably lower mass may have a different origin altogether.

E. P. Wigner, Princeton University: I don't have a question but I would like to ask you to comment on something. If one could get such theories with axiomatic field theoretic methods, I think that we would introduce an operator let us say $\phi^4/(1 + a\phi^2)$ which avoids one of the problems. If one asks how can such an operator behave, the operator at one space-time point is more or less defined but the integral over space, I think, is not defined. Could you say a few words?

Salam: Again, this is beyond my competence but as I understand it, this type of object is supposed to define localizable field theories in the Jaffe sense.

Hepp: I want to give an answer in terms of perturbation theory. In terms of perturbation theory if you take such objects ϕ would be a free field and the expression would be Wick ordered. The space integral is not defined as an operator in Fock space. The expressions which appear in the perturbation expansion for the Green's function or for the S-matrix are suitably connected graphs with external lines.

For these matrix elements these divergences do not appear even though the space integrals of the Wick ordered quantities themselves do diverge.

SYMMETRY BREAKDOWN AND SMALL MASS BOSONS*

Yoichiro Nambu

THE ENRICO FERMI INSTITUTE AND DEPARTMENT OF PHYSICS
THE UNIVERSITY OF CHICAGO

<u>Abstract</u>: In a talk presented at the symposium "The Past Decade In Particle Theory", Prof. Y. Nambu reviews the meaning and mechanism for spontaneous breakdown of a symmetry. After reviewing the history of the Goldstone boson, he describes the relationship of the Goldstone theories to PCAC, Current Algebra, and soft pion limits. These problems are unified in the approach of utilizing the non-linear relizations of the symmetries. A discussion of Prof. Nambu's talk follows this paper.

1. <u>Introduction</u>

The past decade has been a decade of symmetries. To be more exact, I should say an era of broken symmetries: the discovery and establishment of symmetry patterns in elementary particle phenomena which are recognized right from the beginning only as approximately valid. As was once remarked by Salam,[1] one first tries hard to establish a symmetry, but the next moment he is figuring out how to break it. This is an entirely human attitude. The complexities of high energy phenomena convince everybody that

---

*This work supported in part by the U. S. Atomic Energy Commission.

there cannot be perfect regularities beyond the classical
conservation laws--and even those have been crumbling down
since 1956. (I am, of course, referring to the violations
of P, C and T.) Yet to achieve any meaningful phenomenological description of complex phenomena, one must organize
them into regular patterns, and then allow for exceptions
and deviations. But these exceptions and deviations themselves in turn must follow certain rules, and so on down
the line. In this way we have been accustomed to various
approximate symmetries, some of which are by now well
established, like $SU(3)$ and probably chiral $SU(2) \times SU(2)$ and $SU(3) \times SU(3)$, while others are of lesser validity or significance, as in the case of $SU(6)$ and col-
linear $SU(6)$.

In order to formulate the above problem mathematically,
people have been led to consider a hierarchy of symmetries
which are embodied in an imaginary (hypothetical) Hamiltonian,
which consists of several "pieces" of varying magnitudes.
Each piece is characterized by its symmetry properties, and
is responsible for certain processes and their regularities
(or irregularities depending on the point of view). One
usually hastens to mention apologetically that any explicit
expression for the Hamiltonian only serves as a model, and
should not be taken seriously except for its symmetry properties. (I do not really subscribe to such a philosophy,
but this need not be discussed here.)

Now there are several ways in which the symmetries
(exact or approximate) can manifest themselves.

1) Symmetries (invariances) of H → conservation laws (selection rules)
   → degeneracies
   → relations among S-matrix elements
2) Transformation properties and commutation relations of particular pieces of H
   → Symmetry breaking patterns
   → Current algebra relations
3) Spontaneous breakdown of symmetries of H
   → Incomplete multiplets
   → Massless bosons.

My assigned task is to address myself to the last problem.[1a] Whether this is to be regarded as a manifestation of symmetries (of a Hamiltonian) or a violation of symmetries (of state vectors) depends on one's point of view, but I assume that no explanation is necessary about what is meant by it.

2. Spontaneous breakdown of symmetries

So far as is known, the only example in elementary particle physics where the notion of spontaneous breakdown definitely seems to make sense is, according to my view, the case of chiral symmetry, especially $SU(2) \times SU(2)$. Even in this case there are people who do not see the necessity of invoking such a principle, and I will come back to this point later. What I would like to do first is to give a historical account of the problem as I have seen it from my side.

The nomenclature "spontaneous breakdown", which is generally used now, originated in the paper by Baker and Glashow (1962)[2] bearing the same title. I had been searching

for an appropriate term but had not hit on it, although a
term like the "spontaneous polarization" of a ferromagnet
was in common usage.[4]

The "Goldstone boson" of course owes its origin to
Goldstone's paper (1961)[3] in which he quotes, as his starting
point, the preprint of my talk on the superconductivity
model at the Midwest Theoretical Physics Conference (1960)[4]
at Purdue University. At that time I was kicking around
the same idea, but was kept busy working out the consequences
of the chiral (γ_5) symmetry problem. [Later, Freund and I
(1964)[5] tried the name "zeron".] Curiously, Goldstone's
paper seems to be mainly concerned with the nonperturbative
nature of asymmetric solutions; he works out two examples
of discrete symmetry [the mass reversal $\gamma_5 \rightarrow -\gamma_5$ in a neutral
pseudoscalar meson theory, and the amplitude reversal $\phi \rightarrow -\phi$
(Bronzan and Low's A parity[6]) in a ϕ^4 meson theory], and
only casually discusses the breaking of continuous symmetry
(gauge transformation of the 1st kind) and massless excitations in a charged ϕ^4 theory, although he conjectures the
result to be more general.

The concept of degenerate vacua (or ground states)
seems to have been first stated in a paper by Heisenberg
(1959)[7] and co-workers on the nonlinear spinor theory.
Heisenberg tries to identify the Pauli-Gürsey group[8] inherent in his theory with the isospin group, and in doing
so is led to consider a vacuum state (or a "world") having
very large isotopic spin, because he does not have two
distinct "bare" fields to represent proton and neutron.
This is like saying, in the usual case of chiral symmetry,
in the usual case of chiral symmetry, that a left-handed

nucleon or quark becomes right-handed by picking up a spurion, or that a $1/2^+$ nucleon becomes a $1/2^-$ isobar in a similar fashion. [In the language of nonlinear realizations, $q_r \sim M q_L$, where $M = 1 + if\tau \cdot \pi + ...$]. He also refers to Mach's principle, emphasizing the logical independence of the boundary conditions from the intrinsic symmetries of a Hamiltonian. In later publications, Heisenberg's group has been attempting to interpret photons as Goldstone bosons resulting from broken isospin symmetry. I must confess, however, that I have not understood the mathematical handling of the problem. (Some more discussions on the photon problem later.)

Let me now come back to my own story. The idea of linking massless pion with chiral symmetry came as an accident through the study of the BCS theory of superconductivity (1958)[9]. When I first heard Schrieffer's seminar at Chicago on the theory they were just working out, I was struck by the boldness of taking a trial wave function which did not conserve charge. But how can such a wave function represent a physical solution? This started to bother me more and more as time went on, and eventually led me to an investigation (1960)[10] of the gauge invariance requirements and the role of collective excitations which had previously been taken up by Anderson and others.[11,12] This problem arises in the static London relation

$$J_i(q) + K_{ij}(q) A_j(q) \tag{1}$$

in an infinite medium. The Meissner effect and gauge invariance require the response function K_{ij} to be of the form

$$(\delta_{ij} - \frac{q_i q_j}{q^2}) K(q^2), \quad K(0) \neq 0. \tag{2}$$

There are two possibilities as to the origin of the kinematic pole in the second term: either it is of dynamic origin so that in nonstatic cases q^2 is replaced by $q^2 - \alpha^2 \omega^2$, implying acoustic-type collective excitations, or else the pole remains static. Since the response function is a dynamic quantity, the latter could not happen unless a $1/q^2$ singularity were built into the Hamiltonian from the beginning. Indeed it is the Coulomb interaction that actually produces such an effect. In the absence of the Coulomb interaction, the pole would be dynamic. In this case K_{ij} reduces to a non-gauge invariant form $\sim \delta_{ij}$ if we take the limit $q \to 0$, $\omega \neq 0$, and then $\omega \to 0$. One might say it reveals the symmetry breaking aspect of superconductivity.

In relativistic field theories the above mentioned two alternatives still exist, as was first suggested by Schwinger[13] and Anderson.[14] The fact that second alternative escapes the relativistic Goldstone theorem developed by various people[15] just because of the lack of manifest Lorentz invariance of gauge vector fields was discovered later.[16] In particular, Higgs[17] envisages a realization of massive vector and pseudoscalar mesons from a single dynamical principle. But as far as the actual finite pion mass is concerned, it seems to be due to a non-spontaneous symmetry breaking effect (see below).

The obvious analogy[11] of the Bogoliubov-Valatin quasi-particle in the BCS theory with the Dirac electron naturally led me to consider the latter in light of the former. It is the γ_5 invariance that is broken by the mass (energy gap) in the case of the Dirac electron. If the breaking is spontaneous, we must also have massless pseudoscalar excitations. Very similar ideas seemed to have occurred to Vaks and Larkin,[18] who considered the Heisenberg model and derived massless excitations. At any rate, this reasoning naturally directs one to the nucleon and pion problem rather than to the electron.[19] Taking the matrix element of axial nucleon current $J_{\mu 5}$ for $q_\mu \to 0$, we find a zero-mass pion pole and a Goldberger-Treiman condition[20] just as the two terms in Eq. (2) are uniquely related by gauge invariance. Since the current is an observable, manifest Lorentz covariance requirement closes the escape hatch that is available for gauge fields, and so the finite pion mass has to be attributed to nonspontaneous breaking. More direct physical arguments for this conclusion are: 1) $g_A(0) \neq 0$ for the nucleon β decay, and 2) $\pi^\pm \to \ell^\pm \nu$ is not forbidden.[20a] If the weak axial current were really divergenceless and yet $M_\pi \neq 0$ these could not happen.

The large value of nucleon to pion mass ratio, and the miraculous Goldberger-Treiman relation gave me motivation for building a dynamical theory of strong interactions.[21] The problem was to choose a model which was simple and would not cloud the main theme with side issues. The Heisenberg-type nonlinear model was chosen for this purpose, but without the indefinite metric nor any attempt at reducing the number of independent fermion fields. Heisenberg's suggestion regarding the degenerate vacua was not known to me at first. Rather, the idea was generated by the study of the BCS ground state. This also led me to contemplate other examples such as ferromagnetism, in which obviously the spin waves were the symmetry-restoring collective modes.

I must refer at this point to the series of work developed by Gürsey[22] (1960) independently and from a quite different viewpoint. His elegant mathematical formalism (in terms of the operator M quoted already) was in fact a realization of the spontaneous breakdown. Much later his formalism was revived and developed to a higher degree of sophistication, to which topic I will come back after a while.

3. PCAC, Current Algebra, and Soft Pion

Gell-Mann and Lévy (1960)[23,24] initiated the well-known approach to the Goldberger-Treiman relation. It explicitly exploits the fact that pion mass is nonzero, as one can see from their formula, often called the field theoretical version of the PCAC (partially conserved axial vector current) condition,

$$\partial_\mu J^i_{\mu 5} = C\phi^i, \qquad C \propto m_\pi^2 \qquad (3)$$

In this sense it is orthogonal to the spontaneous breakdown approach. As is often remarked, Eq. (3) actually amounts to a mere definition of the pion field ϕ^i, and becomes meaningful only with the additional "gentleness" assumption about the variation of matrix elements between $q^2 = m_\pi^2$ and $q^2 = q_\mu^2 = 0$, which would be reasonable only if m_π is small compared to other mass scales and no singularities are crossed in the process of extrapolation.

SYMMETRY BREAKDOWN 41

Advantages of Eq. (3) are that one can easily extract various consequences from it, with the help of usual reduction techniques; the mass of external pions may be actually switched off by extrapolation while keeping the internal pion mass fixed, thus leading, for example, to the Adler self-consistency condition.[25] On the other hand, Eq. (3) does not explain why m_π is small, nor predict what would happen if actual pion mass could be turned off.

The spontaneous breakdown approach starts from the situation where $m_\pi = 0$, so that the mathematical handling of the problem becomes more complicated, though the results are the same in most cases. Clearly, however, this is a more restrictive assumption than the formal PCAC condition. What it says is that the nucleon mass should remain finite if the agent responsible for the pion mass were turned off. Most models that have been constructed do possess both features, so there are no practical differences between the two approaches.

The group theoretical structure of chiral symmetry, however, is important in making spontaneous breakdown of chiral symmetry a meaningful statement. For there exists the derivative coupling model[23] for which the symmetry group is an inhomogeneous SU(3) (the inhomogeneous part being realized by the displacement of massless pseudoscalar fields). The fermions do not take part in chiral transformations, so their masses need not be created by a spontaneous breakdown. This may be part of the reasons why some people do not regard spontaneous breakdown as necessary for the success of PCAC. As far as single pion emission is concerned, SU(3) x SU(3) and the inhomogeneous SU(3) are equivalent. But a difference

shows up in the commutator of two axial generators, as
in π-N scattering. It is here that the significance of
current algebra emerges. The success of the Adler-Weisberger
relation[26] (1965) indicates that SU(3) x SU(3) [or at least
SU(2) x SU(2)] is indeed the relevant group for the strong
interaction of hadrons.

The soft pion theorems, including part of the SU(2)
x SU(2) current algebra relations, were created for the
purpose of testing the basic idea of chiral symmetry.[26a]
But these earlier attempts at applications of the soft
pion theorems did not yield much because of a lack of useful data. For example, for the electro-pion production
process, the data has become available only recently.[27]
Applications to semileptonic or nonleptonic processes could
have been done more profitably. In fact this was being
contemplated in the early days. What prevented its success
was the absence of the quark model. Without it the extension
of chiral SU(2) to the baryon octet would have too much
arbitrariness.[28] So one had to wait until Gell-Mann (1964)
introduced the quark model[29] and subsequently laid down the
principle of current algebra,[30] followed by Fubini and Furlan's
discovery[31] of combining current algebra with dispersion
relations, and Adler and Weisberger's remarkable result on
g_A/g_V,[26,32] before Suzuki, Sugawara, Callan and Treiman,
etc., could start with confidence the rush to weak processes.[33]

We have yet to discuss the choice between mere PCAC
and spontaneous (or almost spontaneous) breakdown. Relevant
in this respect is the work of Kim and Von Hippel[34] on
baryon-pseudoscalar meson scattering lengths, which is an
extension of Tomazawa's formulation[35] of pion nucleon

scattering as a low energy theorem. The point can be
explained in terms of the effective Lagrangian method
(see next Section). If the nucleon mass is of spontaneous
origin, the mass is part of the Gürsey operator M. The S
wave pion-nucleon scattering length $a_{1/2} + a_{3/2}$ (the so-
called σ term) computed from this Lagrangian is zero.
On the other hand, if the mass is actually a symmetry
breaking term, it would be a pure constant, behaving
like (2,2*) of SU(2) x SU(2), and would give a large
scattering length. The same argument applies to SU(3)
x SU(3) too. The Kim-Von Hippel analysis indicates that
the actual breaking of SU(3) x SU(3) is essentially in
the bare mass of the strange quark only, and its magnitude
is of the order of Gell-Mann - Okubo mass splitting.
This bulk of the baryon mass appears to have been spontaneously created, and SU(2) x SU(2) symmetry is a better
symmetry than broken SU(3) (because $m_\pi^2 << m_K^2$). More
quantitatively, one can compute the ratios of bare quark
masses from the masses of the octet pseudoscalar mesons.
This scheme was considered by Gell-Mann, Oakes and Renner.[36]
If one assumes the quark masses to be $(m_p, m_n, m_\lambda) = (m_1, m_1, m_2)$, one gets

$$2m_1/(m_1 + m_2) = m_\pi^2/m_K^2 \approx 1/12. \qquad (4)$$

Actually, all quark masses may be different. If the K
mass difference is attributed to this cause, the relation

$$m_n - m_p : m_n + m_p : 2m_\lambda + m_n + m_p = m_{K^0}^2 - m_{K^\pm}^2 : m_{\pi^\pm}^2 : m_{K^0}^2 + m_{K^\pm}^2$$

then yields the result

$$m_p : m_n : m_\lambda \approx 2 : 3 : 60. \tag{5}$$

(There is no pion mass difference in this approximation because it belongs to an I=2 piece.)

The significance of nearly spontaneous breaking of SU(3) x SU(3) as a valid picture of hadrons has been emphasized by Dashen.[37] There are, however, some unsolved puzzles. One is the p wave nonleptonic decays of hyperons for which PCAC is not as successful as for the s wave decays. Another is the ξ parameter in $K_{\ell 3}$ decays, which is $\approx$ -1 experimentally in contrast to the theoretical estimate $\approx$ 0. A more serious problem, it seems to me, is the question why chiral SU(3) is a better symmetry than chiral U(3), as is evidenced by the large mass of η' compared to those of the pseudoscalar octet, with little mixing between them. The symmetry breaking scheme in terms of quark masses alone cannot explain it.

Related to this kind of program is the attempt at seeking causal relations among Gell-Mann - Okubo SU(3) breaking, electromagnetic interaction, and the Cabibbo angle in weak interactions. These create three different preferred axes in the SU(3) space, and might correspond to a self-consistent pattern of symmetry breaking similar to the Jahn-Teller effect. The discovery by Brout[38] and Cabibbo[39] that in SU(3) a symmetry breaking driving force and its response need not be in the same direction made such a mechanism an intriguing possibility. This has been further pursued by Gatto and others[40] from somewhat different angles, but it is still an open question whether the picture is fundamentally correct or not.

4. Unified description of hadron symmetries

The physical consequences of spontaneous breakdown can be most elegantly displayed by the method of nonlinear realizations. It originated in the work of Gürsey (1960)[22] and one of the models considered by Gell-Mann and Lévy.[23] A forerunner to Gürsey's nonlinear Lagrangian is the model of Nishijima.[41] Nishijima's idea was to save chirality invariance in leptonic and hadronic weak currents by considering the phase of a chiral gauge transformation as a neutral pseudoscalar massless field. He also discussed a kind of low energy theorem analogous to the soft pion theorem.

Gürsey considered massless pion fields as the phases of a chiral SU(2) gauge transformation operator,[42] and constructed an invariant Lagrangian out of such gauge operators. As it happens, this model is essentially equivalent to Gell-Mann - Lévy's nonlinear σ model in spite of their apparent differences. They can be transformed into each other by a nonlinear redefinition of the pion field without affecting the S-matrix elements. This point has been clarified after Weinberg[43] (1968) and others[44] started to develop a theory of nonlinear realizations, which reached its most general form in the work of Coleman, et al.[45]

A significant new element which emerged as a result is the marriage of nonlinear realization with the gauge theory of massive vector and axial vector fields, as was initiated by Lee, Weinberg and Zumino[46] (1967). It offered an elegant unified approach to PCAC and vector dominance, enabling one to handle the "hard pion" problem.

Eventually, one must come to an understanding of the role of the chiral group in all hadronic states. The chiral group may be realized either linearly (as parity doublets) or nonlinearly (via small mass mesons). It is possible that some baryon resonances belong to the former. But if that is the case, how does a state decide to choose one or the other? And how can one reconcile the chiral symmetry with the conflicting SU(6)-type symmetries? In spite of some suggestive possibilities offered by Weinberg's attempts[47] and Lovelace's discovery[48] of the connection between chiral symmetry and the Veneziano model, we are still far from having a clear cut picutre.

5. Further search for spontaneous breakdown

Except in nonrelativistic phenomena, we do not seem to have any genuine cases of spontaneous breakdown. The only known massless particles are gravitons, photons and neutrinos. Each has already a perfectly respectful description without invoking any symmetry breaking. Besides, neutrinos could not fit the picture because they are fermions. Some attempts have been made to reformulate electromagnetism, and perhaps also gravity, as a spontaneous breakdown phenomenon.

In either case, the relevant symmetry must be some space-time symmetry. The trouble is that once we break a symmetry even spontaneously, it is no longer a symmetry in the ordinary sense. Thus one loses, at least, manifest spacetime invariance of a theory, and probably the actual invariance as well unless one is lucky. Such a lucky situation, however, seems to prevail in electrodynamics. A preferred direction, characterized by a timelike vector η_μ, generates Goldstone modes $A_\mu(x)$ which are orthogonal to it, and may be identified with the vector potential. A particular choice of η_μ merely corresponds to a particular choice of gauge, and therefore does not destroy the Lorentz invariance of physical predictions. This can be most clearly understood if we work in the Dirac gauge:[49] $A_\mu(x) A^\mu(x) = $ const., and regard this relation as a nonlinear realization of the Lorentz group.[50] So electrodynamics is compatible with spontaneous breakdown, but the latter is not necessary for the understanding of the former.

A more ambitious program in exploiting spontaneous breakdown is the quantum electrodynamics of Baker, Johnson and Willey,[51] who try to create electron mass spontaneously. But the electron mass breaks γ_5 as well as scale invariance, and it is not clear why one can escape the consequences of the Goldstone theorem. Nevertheless, it is perhaps interesting to seek differences, if any, in the predictions of their theory and the conventional quantum electrodynamics with infinite subtractions.

I would like to thank Professors E. C. G. Sudarshan and Jagdish Mehra for giving me an opportunity to stay at the University of Texas and prepare part of this talk.

FOOTNOTES

1. A. Salam, quoted by J. J. Sakurai, Ann. Phys. __11__, 1 (1960)(footnote, p. 5).

1a. My earlier attempts along this line are found in: Group Theoretical Concepts and Methods in Elementary Particle Physics (Ed. F. Gürsey), Gordon and Breach, New York, 1964; Lectures at the Instanbul Summer School, 1966 (Univ. Chicago Preprint EFINS 66-107); Talk at the Amer. Phys. Soc. Meeting, Chicago, 1968 (Univ. Chicago Preprint EFINS 68-11).

2. M. Baker and S. L. Glashow, Phys. Rev. __128__, 2462 (1962).

3. J. Goldstone, Nuovo Cimento __19__, 155 (1961).

4. Proc. Midwest Theoretical Physics Conference at Purdue Univ. (1960).
5. P. G. O. Freund and Y. Nambu, Phys. Rev. Letters $\underline{12}$, 714 (1964).
6. J. B. Bronzan and F. E. Low, Phys. Rev. Letters $\underline{12}$, 522 (1964).
7. H. P. Dürr, W. Heisenberg, H. Mitler, S. Schlieder and K. Yamazaki, Z. Naturf. $\underline{14a}$, 441 (1959).
8. W. Pauli, Nuovo Cimento $\underline{6}$, 204 (1957).
9. J. Bardeen, L. N. Cooper and J. R. Schrieffer, Phys. Rev. $\underline{106}$, 162 (1957).
10. Y. Nambu, Phys. Rev. $\underline{117}$, 648 (1960).
11. P. W. Anderson, Phys. Rev. $\underline{110}$, 827 (1958); $\underline{112}$, 1900 (1958).
    G. Rickayzen, Phys. Rev. $\underline{111}$, 817 (1958); Phys. Rev. Letters $\underline{2}$, 91 (1959).
12. For a historical perspective, see P. W. Anderson in Superconductivity (Ed. by R. D. Parks), Marcell Dekker, Inc., New York, 1969 (p. 1343).
13. J. Schwinger, Phys. Rev. $\underline{125}$, 394 (1962). A massive photon emerges in two-dimensional electrodynamics. Note that the charge has dimensions of mass in this case.
14. P. W. Anderson, Phys. Rev. $\underline{130}$, 439 (1963).
15. J. Goldstone, A. Salam and S. Weinberg, Phys. Rev. $\underline{127}$, 965 (1962). For a detailed review of the mathematical aspects of the Goldstone theorem see articles by T. W. Kibble (p. 277) and D. Kastler (p. 305) in Proc. 1967 International Conference on Particles and Fields (Ed. C. R. Hagen et al.), Interscience, New York, 1967; and by G. S. Guralnik, C. R. Hagen and T. W. Kibble in Advances in Particle Physics, Vol. 2. (Ed. R. L. Cool and R. E. Marshak), Interscience, New York, 1968.
16. P. W. Higgs, Phys. Letters $\underline{12}$, 132 (1964); Phys. Rev. Letters $\underline{13}$, 508 (1964).
    F. Englert and R. Brout, Phys. Rev. Letters $\underline{13}$, (1964).
17. P. W. Higgs, Phys. Rev. $\underline{145}$, 1156 (1966).
18. V. G. Vaks and A. I. Larkin, JETP $\underline{40}$, 282 (1961) [Soviet Phys. $\underline{13}$, 192 (1961)].

19. Y. Nambu, Phys. Rev. Letters $\underline{4}$, 380 (1960).
20. M. L. Goldberger and S. B. Treiman, Phys. Rev. $\underline{110}$, 1178 (1958).
20a. J. C. Taylor, Phys. Rev. $\underline{110}$, 1216 (1958).
21. Y. Nambu and G. Jona-Lasinio, Phys. Rev. $\underline{122}$, 345 (1961); $\underline{124}$, 246 (1961).
22. F. Gürsey, Nuovo Cimento $\underline{16}$, 230 (1960); Ann. Phys. $\underline{12}$, 705 (1960).
23. M. Gell-Mann and M. Lévy, Nuovo Cimento $\underline{16}$, 705 (1960).
    J. Bernstein, M. Gell-Mann and L. Michel, Nuovo Cimento $\underline{16}$, 560 (1960).
    J. Bernstein, S. Fubini, M. Gell-Mann and W. Thirring, Nuovo Cim. $\underline{17}$, 757 (1960).
    Chou Kuang-Chao, JETP $\underline{39}$, 703 (1963) [Soviet Phys. $\underline{12}$, 492 (1961)].
24. For the state of affairs prevailing around 1960, see Proc. 1960 International Conference on High Energy Physics at Rochester (Ed. E. C. G. Sudarshan et al.), Interscience, 1960, especially articles by Gell-Mann (p. 508), Gürsey (p. 572), Heisenberg (851), Goldberger (p. 733), Nambu (p. 858), Okun (p. 743), Vaks and Larkin (p. 873). Landau is also credited with the idea of PCAC (pp 741, 749).
25. S. L. Adler, Phys. Rev. $\underline{137}$, B1022 (1965); $\underline{139}$, B1638 (1965).
26. S. L. Adler, Phys. Rev. Letters $\underline{14}$, 1051 (1965); Phys. Rev. $\underline{140}$, B736 (1965). W. I. Weisberger, Phys. Rev. Letters $\underline{14}$, 1047 (1965); Phys. Rev. $\underline{143}$, 1306 (1966).
26a. Y. Nambu and D. Lurié, Phys. Rev. $\underline{125}$, 1429 (1962).
     Y. Nambu and E. Shrauner, Phys. Rev. $\underline{128}$, 862 (1962).
     E. Shrauner, Phys. Rev. 131, 1847 (1963).
27. Y. Nambu and M. Yoshimura, Phys. Rev. Letters $\underline{24}$, 25 (1970).
28. This is also the reason that the Baker-Glashow program[2] based on the Sakata model, and the complicated octet model of S. Glashow [Phys. Rev. $\underline{130}$, 2132 (1962)] and of N. Byrne, C. Iddings and E. Shrauner [Phys. Rev. $\underline{139}$, B918, B933 (1965)] did not yield interesting results.

29. M. Gell-Mann, Phys. Letters **8**, 214 (1964); G. Zweig, CERN Reports Nos. 8182/TH401, 8419/TH.412, 1964 (unpublished).

30. M. Gell-Mann, Physics **1**, 63 (1964). A detailed exposition of the subject, including many of the topics discussed in this section, is found in S. L. Adler and R. F. Dashen, Current Algebras, W. A. Benjamin, Inc., New York, 1968.

31. S. Fubini and G. Furlan, Physics **1**, 229 (1965).

32. In early days there was a misguided conjecture that $g_A/g_V \to 1$ in the chiral symmetry limit. There is no general proof for it when spontaneous breakdown sets in.

33. M. Suzuki, Phys. Rev. Letters **15**, 986 (1965).
    H. Sugawara, Phys. Rev. Letters **15**, 870; 997 (1965).
    C. G. Callan, Jr., and S. B. Treiman, Phys. Rev. Letters **16**, 153 (1966).

34. F. Von Hippel and J. K. Kim, Phys. Rev. Letters **22**, 740 (1969).
    Yuk-Ming P. Lam and Y. Y. Lee, Phys. Rev. Letters **23**, 734 (1969).

35. Y. Tomozawa, Nuovo Cimento **46A**, 707 (1966).
    S. Weinberg, Phys. Rev. Letters **17**, 616 (1966).

36. M. Gell-Mann, R. J. Oakes and B. Renner, Phys. Rev. **175**, 2195 (1968).

37. R. Dashen, Phys. Rev. **183**, 1245 (1969).
    R. Dashen and M. Weinstein, Phys. Rev. **183**, 1261 (1969).

38. R. Brout, Nuovo Cimento **47A**, 932 (1967).

39. N. Cabibbo, in Hadrons and Their Interactions (Ed. A. Zichichi), Academic Press, New York, 1968.

40. R. Gatto, G. Sartori and M. Tonin, Phys. Letters **28B**, 128 (1968).
    N. Cabibbo and L. Maiani, Phys. Letters **28B**, 131 (1968).

41. K. Nishijima, Nuovo Cimento **11**, 698 (1959).

42. A similar idea is found in G. Kramer, H. Rollnik and B. Stech, Z. Phys. **154**, 564 (1959).

43. S. Weinberg, Phys. Rev. **166**, 1568 (1968).

44. For a survey of nonlinear realizations and related topics see S. Gasiorowicz and D. A. Geffen, Rev. Mod. Phys. **41**, 531 (1969).

45. S. Coleman, J. Wess and B. Zumino, Phys. Rev. 177, 2239 (1969).
    C. G. Callan, Jr., S. Coleman, J. Wess and B. Zumino, Phys. Rev. 177, 2247 (1969).
46. T. C. Lee, S. Weinberg and B. Zumino, Phys. Rev. Letters 18, 1029 (1967).
47. S. Weinberg, Phys. Rev. 177, 2604 (1969).
48. C. Lovelace, Phys. Letters 28B, 265 (1968).
49. P. A. M. Dirac, Proc. Roy. Soc. A209, 291 (1951). For further references, see Ref. 50.
50. Y. Nambu, Progr. Theoret. Phys. Suppl., Extra Number, 190 (1968). The vector η_μ may be interpreted, in the context of Dirac's work, as the velocity of "ether". Compare also F. London, **Superfluids**, John Wiley & Sons, New York, 1950, Vol. 1, p. 62.
51. K. Johnson, M. Baker and R. Willey, Phys. Rev. 136, B1111 (1964).

DISCUSSION OF PROF. NAMBU'S TALK

<u>NAUENBERG(UNIVERSITY OF CALIFORNIA AT SANTA CRUZ & CERN)</u>:
What is the value of the ξ parameter for the $K_{\ell 3}$ decay as predicted from PCAC and what is the experimental value for this parameter?
<u>NAMBU</u>: The weak meson current $j_\mu(x)$ is given by the equation

$$j_\mu(x) = f_+(p_k + p_\pi)_\mu + f_-(p_k - p_\pi)_\mu$$

and the parameter ξ is defined by the expression

$$\xi = \frac{f_-}{f_+}$$

Experimentally there are three different kinds of experiments one can perform to obtain the value of ξ and they are, I think, not completely in accord with each other. Roughly ξ is of the order of -1.

If one believes in PCAC one obtains a theoretical value of ξ which is of the order of 0 to 0.2. The fact that ξ is small is essentially a statement of the Ademallo Gatto Theorem.

By invoking a spontaneous breaking of the SU(3) symmetry there is at least one way to avoid the discrepancy between the experimental and theoretical values for ξ. One says that the current is not conserved because of the difference between the mass of the kaon and pion. If one assumes the current to be conserved in the limit that m is negligible one has

$$(p_k - p_\pi)^\mu <j_\mu(x)> = 0$$
$$m_\pi \to 0$$

In the limit that the pion mass can be neglected in comparison to the kaon mass the above equation yields the result

$$f_+ + f_- = 0$$

that is

$$\xi = \frac{f_-}{f_+} = -1$$

However, I don't think that this is a very good justification for the spontaneous breaking of SU(3) symmetry and mass splitting.

BARUT (UNIVERSITY OF COLORADO): The additional symmetry in the conformal group is used by Nishijima. You put this spacetime symmetry into an inhomogenous SU(2). Is it true that the two methods of characterizing the symmetry are not mutually exclusive? That is, should one put this into a conformal group or into an internal symmetry group?

NAMBU: Which theory are you talking about?

BARUT: If the mass of the pion is zero, one has conformal invariance.

NAMBU: Yes, that is right.

BARUT: You would like to treat it as inhomogenous SU(2).

NAMBU: That is exactly right. As far as the difference in the two groups is concerned, I will have to refer to Professor Salam.

Y. NE'EMAN (TEL-AVIV UNIVERSITY AND THE UNIVERSITY OF TEXAS):
What is your feeling on the origin of the pion mass?

NAMBU: This is a difficult question. In the original model I suggested that the electromagnetic interaction intrinsically breaks the SU(2)xSU(2) symmetry and might induce the non-zero mass pions. In fact, I made the observation that if one assumes the Gell-Mann, Oakes, and Renner Model (not SU(2) but SU(3)) one can compute a mass ratio of the pions and nucleons. One has a bare neutron with zero mass and a bare proton with mass μ. While the non-zero mass pion is automatically created, there is no mass splitting between the pions because of the same group structure. In this naive model, one has

$$\frac{m_\pi^2}{m_n} = \mu = 8 \text{ MeV}.$$

It may not be unreasonable to suppose that the mass of the proton is of electromagnetic origin so I would put effective electromagnetic symmetry breaking into this situation.

<u>NE'EMAN</u>: That would be like including a U(3) term in the Gell-Mann, Oakes, and Renner Model?

<u>NAMBU</u>: Yes, that is right. Also, there may be some independent mechanism which is not of the usual electromagnetic type which is responsible for the pion having non-zero mass. The quarks may have something to do with this. I really don't know. This is speculation.

<u>R. E. MARSHAK (UNIVERSITY OF ROCHESTER)</u>: So you would say that the n, p mass splitting and the π^+, π^0 mass splitting are unrelated.

<u>NAMBU</u>: Yes, I would say that such an idea would be very old fashioned.

ADVANCES IN THE APPLICATION OF SYMMETRY*

Y. Ne'eman

Dept. of Physics & Astronomy,　　Center for Particle Theory,
Tel-Aviv University, Tel-Aviv,　　University of Texas, Austin
　　　　Israel　　　　　　　　　　　　　Texas　78712

Abstract: Prof. Ne'eman reviews the accomplishments of the past decade in the application of symmetry methods to the phenomena of particle physics. After reviewing the impact of the development of algebraic methods from the earlier group methods, he reviews the meaning of invariance and classification in this broader context. He then describes the known symmetries and their interpretation in quark models. From these he goes on to investigate the use of higher symmetries in particle physics.

Introduction

　　　This meeting represents an attempt to summarize the results and lessons of the last decade. The aim is to achieve an extrapolation into the future, based upon this synthesis. Hopefully, we shall gain some insights which will help us pick the more promising tracks in our exploration of physical laws.

---

*Research sponsored by the Air Force Office of Scientific Research, Office of Aerospace Research, United States Air Force, under AFOSR grant number EOOAR-68-0010, through the European Office of Aerospace Research, and supported in part by NSF Grant GU-1598.

In this paper, I shall try to present a summary of achievements in the domain of symmetry at the particle level.

To start with, we shall outline some of the more general lessons. We shall then review the following topics:

Unitary Symmetry (SU(3))
Chiral SU(3)
Quarks, SU(6) and related results
Spectrum Generating Algebras (SGA) and the
  Local (Infinite) Algebra of the Currents
Space-Time Symmetries (the Poincaré group and
  Conformal Invariance).

I shall not dwell on the discrete operations of C, P and T and their combinations. In this context, the sixties have mainly brought phenomenological evidence of a further breakdown, with no new insights.

"Algebraic" Particle Physics, or Symmetries?

The change in nomenclature "group theory" to "algebra" follows a point of view voiced by Gell-Mann,[1] according to which it is through the infinitesimal generators rather than through the system's invariant automorphisms that we may hope to get at the foundations of a dynamical theory. That program is based upon one of the most important ideas of the previous decade, namely a remark first enunciated somewhat crudely by Gershtein and Zeldovitz,[2] then emphasized and perfected by Feynman and Gell-Mann,[3] and best known as the Conserved Vector Current (CVC) hypothesis. It conjectures that the charge-current four-vectors V^μ, representing weak and electromagnetic transitions, coincide with space densities of strong-interaction symmetry generators. As noted by these

authors and later proved explicitly by Okubo and Ioffe[4]
this feature is embodied in the lack of renormalization
of the weak vector-coupling, best seen in the latter's
equal values for neutron and muon decays. Another such
bridge between strong and weak dynamics is provided by
the somewhat more mysterious Goldberger-Treiman[5] relation.
This literally algebraic approach has been fruitful, in
that it has led to applications in which the physical
texture of the system of generators has been beautifully
exploited - such as the Adler-Weisberger calculation[6]
of the renormalization of the axial-vector coupling in
beta-decay. The techniques emulate another algebraic
theory - Heisenberg's Matrix Mechanics. On the other
hand one may still regard classification as the most
important application of the overall method. Remembering
that the corresponding indentification of patterns in
the hadron spectrum depends strongly upon the S-matrix
(or the Hamiltonian) being approximately symmetric
under the action of the group, we are again aware of the
undiminished importance of the more conventional symmetry
approach. This is true both of the "internal" symmetry[7,1]
(SU(3)) and of the more advanced utilization of the
Poincaré group through the methods of harmonic analysis.
Toller's[8] application of this relatively sophisticated
mathematical tool has indeed yielded yet another case in
which group theory provides an understanding of patterns
in the system of hadron excitations and their strong
interactions.

Mathematical Tools

It is worth mentioning at this stage that as in many other cases in theoretical physics, progress has been generally accompanied by the introduction of ever-more varied mathematical tools. Mackey's theory of induced representations, the studies of Gelfand, Naimark, Bargmann and Wigner relating to unitary representations of non-compact groups, and the methods of harmonic analysis, represent cases in which mathematics and mathematical physics have paved the way for the theoretical physicist in the domain of particles and fields. Non-linear "realizations" of Lie algebras (introduced in connection with the reformulation of the ideas of Chiral Invariance), the construction of representations of the infinite-parameter Lie algebra of the local charge-densities - these are cases in which the mathematical and physical frontiers may even coincide. If the present attempt to represent covariantly the entire system of local time- and space-components of the currents, manages to succeed (the prospects are not too good), it will have established yet another "first" for the physicists, since the inclusion of Schwinger[9] terms may well extend beyond the range encompassed by present treatments of infinite Lie algebras.

Classification and Symmetries

Returning to our discussion of the prevalence of the algebraic generators, it is worth noting that the complete parallelism between symmetries and conservation laws, guaranteed by Noether's theorem, still holds. Although it was originally the algebraic thinking which did lead to the set of sum-rules popularly known as "current algebra", Weinberg, Schwinger and others have since been able[10] to restructure the strong Lagrangian so as to provide a symmetry derivation for the same results. Considering our lack of understanding of the actual dynamical role of such a Lagrangian in the absence of perturbation theory, it seems that the algebraic approach is still the more physical of the two. Nevertheless, the attempt to revive invariance considerations, as enacted on Lagrangians, has been vindicated by the insight it has provided in the case of chiral unitary symmetry. It turned out that in this case <u>the tandem "symmetry-conservation law" becomes divorced from the accompanying concept of a classification scheme</u>. This happens through a <u>non-linear action of the chiral group</u> $SU(3) \times SU(3)$ on the Hilbert space of particle states.

The opposite situation had just been attained[11] via a generalization in the treatment of algebraic systems whose conventional linear action on the carrier space provides a classification of the system. These generalized "spectrum-generating algebras" do not generate physically interesting symmetries. They thus represent a <u>classification divorced from meaningful invariance</u> considerations, just as the non-linear implementation of chiral unitary symmetry displays an invariance principle devoid of any classification connotations.

Invariance

The application of invariance considerations suffers from some ambivalence in the choice of the object to which they are applied. Lagrangians are excellent theoretical laboratories because we possess the entire machinery of relativistic quantum field theory and the appropriate version of Noether's theorem. They are as yet useless as actual physical tools, however. Shall we turn to the S-matrix? This is certainly a good physical object, but we have no theory for its fine-structure in terms of some basic dynamical variables of a local nature, and we have only a rough understanding of the way in which it could yield the weak currents, for instance.[12] It is certainly useful for kinematics, and the great advances in dispersion theory have resulted in some understanding of the S-matrix in terms of its analytic structure. Hadron classification efforts have been greatly helped by our growing familiarity with the fine structure represented by the system of singularities. This has of course provided ways of applying unitary symmetry to strong interactions dynamics,[13] quite aside from the exploitation of Poincaré invariance. However, one is then still puzzled by the lack of a clear functional connection with the same unitary symmetry, when its generator densities happen to represent weak currents.

Gell-Mann has initiated a third approach, in which the invariance postulates are applied to the Hamiltonian density,[1] a good local physical observable, since its matrix elements are coupled to gravitation. It remains to be seen whether indeed $\theta^{\mu\nu}(x,t)$ will provide a more useful answer. At the least, this will have renewed the interest in the energy-momentum tensor, leading already to some spin-off results, such as a revival of the Feynmann-Huggins tensor,[14] as an alternative to the "classical" Belinfante-Rosenfeld tensor.[15]

Unitary Symmetry

I shall dwell first upon this subject, since this is probably the main advance of this decade in our subject.

In the best-understood problems of quantum-mechanics, we have often encountered so-called "dynamical" symmetries.[16] The better known examples are SO(4) in the hydrogen atom, and U(3) in the harmonic oscillator. These dynamical symmetries contain a basic (kinematical) kernel, the 3-dimensional rotation group due to the Galilean invariance of non-relativistic problems. Orbital angular momentum L thus makes up the SO(3) subalgebra of both systems, with the additional operators

$$\frac{1}{2}(\underline{p} \wedge \underline{L} - \underline{L} \wedge \underline{p}) - \underline{r}\,|r^2|^{-1} \quad \text{in hydrogen} \quad (1)$$

or

$$\frac{1}{2}(a_i a_j + a_j a_i), \text{ with } \underline{a}=\underline{r}+i\underline{p} \quad \text{for the oscillator} \quad (2)$$

generating the specific "dynamical" symmetry due to the "accidental" structure of the corresponding Hamiltonians.

Unitary symmetry, also a U(3) group, may be considered as "dynamical" in the above sense, with two distinguishing features. First, it is built around a different basic kernel. From the very first, we had noted[7] that it looked like a generalization of the electric charge gauge invariance. The isospin-hypercharge (I,Y) subgroup of SU(3) constituted a first step in that generalization, as embodied by the Gell-Mann Nishijima relation[17]

$$Q = |e|(I_z + \frac{1}{2} Y) = \frac{1}{2}|e|(\lambda_3 + \frac{1}{\sqrt{3}}\lambda_8) \quad (3)$$

With Cabibbo's identification of the weak charges[18]

$$W^\pm = \left(\frac{G}{\sqrt{2}}\right)^{\frac{1}{2}} \frac{1}{\sqrt{2}} \{\cos\theta(\lambda_1+i\lambda_2) + \sin\theta(\lambda_4+i\lambda_5)\} \quad (4)$$

we get additional - or alternative - pieces of that kernel. If we take the Gell-Mann viewpoint, it is the set $(Q,W^\pm)$ which represents the "original" kernel.

Summing up this point we therefore observe that the entire $U(3)$ is orthogonal to $\underline{J}$ (angular momentum). Indeed, we can regard $U(3)$ as commuting with the entire Poincaré group, in the symmetry limit. In a Lagrangian field-theory treatment, in which we ordinarily relegate the $SU(3)$ symmetry-breaking to the mass term of the fundamental triplet field[19] ("quarks"), $SU(3)$ would still commute with $\underline{J}$ and with $\underline{P}$ the space translations. Only P° the time-translation generator and the Lorentz boosts $\underline{A}$ would cease to commute with the 4th to 7th directions in $SU(3)$. On the other hand we do have a kernel with a clear dynamical role, i.e. the $U(2)$ generated by the closure of (W^+, W^-, Q).

The other difference between hydrogen or the oscillator, and the unitary spin of hadrons, is a matter of our ignorance. We do not know the hadron Hamiltonian, and have no direct indication as to the way in which the entire $SU(3)$ algebra arises. The existence of a fundamental triplet[19] would represent the simplest possible explanation, but I think that to be honest we couldn't even take this as a reason. Why a triplet at all? Isn't that just the same thing as saying that the symmetry is $SU(3)$? Various guesses have tried to answer the basic question of "why $SU(3)$?". I have a certain predilection for the attempts to explain

the emergence of SU(3) out of the interplay between the
various interactions.[20] Either a "fifth" interaction,[21]
forcing the system to close on SU(3) instead of just
U(2); or the "primordial" strong interaction itself,[22]
if it lay skew with respect to the weak and electromagnetic U(2). We have however no definite knowledge as
to the origin of the relative angle between these two
subgroups, even though much effort has been spent by the
groups of Cutkosky, Cabibbo, Gatto, etc. in recent years
in search of an answer. In these treatments, SU(3) itself
was assumed, together with the weak U(2), and the direction of strong SU(3) breaking was supposed to arise
spontaneously at the searched-for angle. It seems that
within SU(3) itself[23] such a treatment would yield $\theta = 0°$
or 90° for the Cabibbo angle. Working with SU(3) x SU(3)
and adding some assumptions[24] about cancellation of
divergences leads to closer agreement with experiment.
The entire approach is however highly speculative.

Baryons and mesons form SU(3) multiplets, and the
Rosenfeld wallet cards gradually provide the necessary
states. For each baryon multiplet B* we should also have
such parameters as the f = F/F+D ratio in B*BM, where B
is the lowest $\frac{1^+}{2}$, $\underline{8}$ and M is the meson 0^-, $\underline{8}$. If there
are any other B*' with the same J^P, we may expect various
amounts of mixing. This then adds one or more parameters.
The final couplings are nevertheless still overdetermined,
as there are 12 couplings in $\underline{8} \times \underline{8} \times \underline{8}$ for instance. At
the recent Coral Gables Conference, Samios presented[25]
the data for the following multiplets (the masses correspond
to the Y = 1 state for $\underline{8}$ or $\underline{10}$)

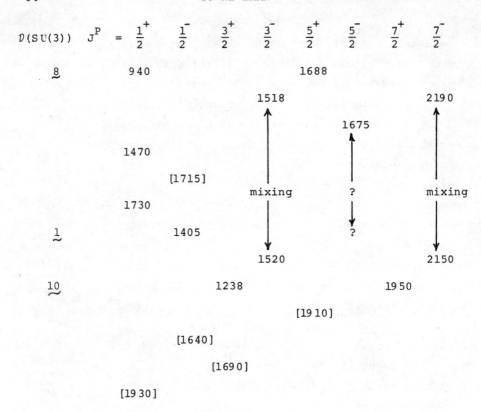

The brackets represent[26] additional states, found by phase shift analysis but as yet unconfirmed by formation experiments.

In the mesons, we know the $J^{PC} = 0^{-+}$, 1^{--}, 1^{++}, 1^{+-}, 2^{++} nonets $\underline{8} + \underline{1}$, with an abundance of additional states to be taken care of in the future when the quantum numbers become better defined.

Strong interaction <u>couplings</u> seem to break SU(3) in a mild fashion; the original impression of violent symmetry breaking in the relative strength of π and K couplings now appears to have been based upon insufficient data, as was shown[27] by Kim, Chan and Meiere or Neih. This has now been confirmed by the smallness of the measured NNη coupling,[28] which corresponds to the same f = 0.4 value. It seems however that some of the bad cases of SU(3) breaking are here to stay, e.g. the opposite behavior of $\frac{d\sigma}{dt}$ in γ+N → N+π versus γ+N → Λ+K in the extreme forward direction. The theory of the breaking should account for these facts.

In Gell-Mann's description of strong interactions using the separation of $\theta^{\circ\circ}$ according to the algebraic behavior of its various pieces,

$$\theta^{\circ\circ} = \bar{\theta}^{\circ\circ} + g^{\circ\circ}(-u_o - cu_8) \tag{5}$$

it now appears[29] that c ~ -1.2. The u_o produces the average 0^{-+} meson mass, u_8 represents SU(3) breaking, and this view registers the observation that they appear to be of the same order. The value c = -$\sqrt{2}$ would make m_π^2 → 0, which is not far from the actual value, on the scale of hadron masses. Baryon masses get strong contributions from $\bar{\theta}^{\circ\circ}$, so that they do display smaller relative mass-splittings. This view differs from the original idea of broken SU(3), where the relative strength of the breaking was small everywhere theoretically. The new approach is due to the addoption of Nambu's interpretation[30] of chiral symmetry, in which meson masses vanish in the symmetry limit. This seems to be preferred by nature over the alternative solution, suggested by lepton physics and favored by many until 1965, in which the fermion masses were made to vanish.[1] We shall return to this point in our discussion of chiral symmetry.

Although c is large, the vacuum turns out to be SU(3) invariant,[29] which explains why the couplings do not break SU(3) badly. Some doubts have recently been aroused with respect to the value of c, from the study of $K_{\ell 3}$ form factors, but I don't think they present a strong case. I had my own doubts based upon difficulties in estimating matrix elements of u_i between 0^- mesons, but I have convinced myself of the consistency of the operation.

In calculating weak or electromagnetic transitions, SU(3) seems to work almost perfectly.[31] This may be again due to the fact that we are mostly dealing with matrix elements between baryons rather than mesons. The present status of the Coleman-Glashow relation,

$$m_{\Xi^-} - m_{\Xi^0} = m_{\Sigma^-} - m_{\Sigma^+} + m_p - m_n \tag{6}$$

which has 6.55 ± 0.88 for the l.h.s. and 6.626 ± 0.41 for the r.h.s., is typical. The agreement for magnetic moments is somewhat less good: in the SU(3) limit, we should have

$$2\mu_\Lambda = \mu_n \tag{7}$$

and we have -1.46 ± 0.32 on the r.h.s. and -1.913 for the r.h.s. The value of μ_Σ is less well established, but it has continuously come closer[26] to the predicted symmetry value $\mu_{\Sigma^+} = \mu_p$ of 2.793. The present averaged value is 2.57 ± 0.52. The situation with respect to semi-leptonic decays is very much the same.

For non-leptonic decays, the picture is roughly the same, i.e. the $\underline{8}$ behavior of the main (96%) piece of the non-leptonic Hamiltonian predicts branching ratios which fit experiment. The prediction for the S-wave amplitudes[33]

$$-(\Lambda \to p\pi^-) + 2(\Xi^- \to \Lambda\pi^-) = \sqrt{3} <\Sigma^+ \to p\pi^o> \qquad (8)$$

is fitted by the values of -2.49 ± 0.064 and -2.0 ± 0.32 or -2.72 ± 0.25 for the two sides. (Depending on the as yet unfixed sign of a certain parameter). However, we do not yet know[35] whether H_W does indeed contain a sizable $\underline{27}$ piece, as befits its $j^{\mu\dagger} j_\mu$ structure. In that case, the preponderance of $\underline{8}$ would be due to an enhancement mechanism only, and some nuclear $|\Delta I| = 2$ transitions should display parity-breaking due to that $\underline{27}$ contribution. This should be roughly 1 : 600 (in the rates) of the observed ($|\Delta I| = 0$) $\underline{8}$ contributions. Moreover, the relative rates of $|\Delta I| = 1$ and $|\Delta I| = 0$ transitions would be 1 : 15 in that case, since this is the value of $\tan^2 \theta$. Alternatively, the supposed $\underline{27}$ part may be absent through cancellation by additional neutral currents, and the non $|\Delta I| = \frac{1}{2}$ effect in charged K decay would be due to radiative corrections. The $|\Delta I| = 1$ nuclear transitions would be of the same order as the $|\Delta I| = 0$ (and there would be no $|\Delta I| = 2$) since they would now be due to some of the five neutral currents (and 5 intermediate bosons) necessary for the suppression of the $\underline{27}$ term. In terms of the next decade, here is a crucial experiment which I hope will be performed before 1980.

Chiral SU(3) x SU(3)

Whenever a conserved quantity is found, there is a point in trying to extend the theory by checking whether in some approximation it can be conserved separately by two systems. Angular momentum J (an SU(2) symmetry) is separated into L (orbital) and S (Spin), with

$$J = L + S \qquad (9)$$

This implies a view in which we approximate the situation by conserving separately L and S. One thus assumes symmetry under an $SU(2)_L \times SU(2)_S$ group representing the separate action of rotations on the coordinates and on internal space respectively. LS coupling then breaks this larger symmetry, and we are left with invariance under the subgroup generated by J only. This is the "diagonal" SU(2) in $SU(2)_L \times SU(2)_S$, represented by the combined added action of the two algebras (the non-conserved generators represent action in opposite directions). Sometimes it even turns out that the two new conservation laws are exact, like their diagonal product (and originator): fermion number conservation $U(1)_F$ has even yielded a triple $U(1)_B \times U(1)_{\ell e} \times U(1)_{\ell \mu}$, where B, ℓe, $\ell \mu$ stand for baryon, electronic lepton and muonic lepton charges.

For SU(3), the chiral $SU(3)_L \times SU(3)_R$ of (L) left- and (R) right-handed currents (suggested by the weak current which coincides with an $SU(3)_R$ density) has proved to be an extremely useful symmetry. We mentioned its truncated role as far as classification applications; it seems we are not faced by sets of parity-doublets, at least as far as the lowest states go. Instead, the almost-zero-mass pion provides composite parity doubles (N+π,N+nσ) for the nucleon[35] etc. The usefulness of the symmetry is in the couplings. Results were first found by saturating matrix elements of the charges (the generator algebra) between one particle states, and accounting for the "leakage" as contributing to the many particle states;[36] using the Goldberger Treiman relation to describe the amount of non-conservation of the axial charges ($L-R$) in terms of pions thus led to the above mentioned "pionized" states in the saturation scheme. Using non-linear realizations of the chiral group on the pion field then reproduced the same results in a symmetry language. We now know that the γ_5 pion-nucleon coupling of the fifties has to be replaced[10] by a combination of several more complicated couplings - to the extent that one likes to use Lagrangians - ,

$$L = -\bar{N}\gamma^\mu D_\mu N - \frac{1}{2} D^\mu \underset{\sim}{\pi} \cdot D_\mu \underset{\sim}{\pi} + iF^{-1}(\frac{g_A}{g_V}) \bar{N}\gamma_5 \gamma^\mu \underset{\sim}{\tau} ND_\mu \underset{\sim}{\pi} - \frac{1}{2} m_\pi^2 [1+F^{-2}\underset{\sim}{\pi}^2]^{-1} \underset{\sim}{\pi}^2 \qquad (10)$$

with

$$D_\mu N \equiv \partial_\mu N + 2i(F_\pi^2 + \underset{\sim}{\pi}^2)^{-1} \underset{\sim}{\tau} \cdot (\underset{\sim}{\pi} \wedge \partial_\mu \underset{\sim}{\pi}) \qquad (11)$$

and

$$D_\mu \underset{\sim}{\pi} \equiv (1+F_\pi^{-2}\underset{\sim}{\pi}^2)^{-1} \partial_\mu \underset{\sim}{\pi} \qquad (12)$$

instead of the "traditional"

$$L = -\bar{N}\gamma^\mu \partial_\mu N - \frac{1}{2}\partial^\mu \underset{\sim}{\pi}\cdot\partial_\mu \underset{\sim}{\pi} - \frac{1}{2}m_\pi^2 \underset{\sim}{\pi}^2 + ig_{NN\pi}\bar{N}\gamma_5 \underset{\sim}{\tau}\cdot\underset{\sim}{\pi}N \quad (13)$$

with

$$g_{NN\pi} = 2m_N F_\pi^{-1}(\frac{g_A}{g_V}) \quad (14)$$

We thus have a trilinear vertex (single pion emission) with coupling F_π^{-1} together with double pion emission from the "free" N Lagrangian with "covariant derivative" $D_\mu N$, with a coupling strength of F_π^{-2}, plus 4-pion emission with F_π^{-4} etc. In addition, the pion field undergoes an interaction $F_\pi^{-2}\cdot\underset{\sim}{\pi}^4$ etc. This Lagrangian is chiral symmetric, with transformations,

$$\delta\vec{\pi} = \frac{F_\pi}{2}(1-F_\pi^{-2}\underset{\sim}{\pi}^2)\underset{\sim}{\varepsilon} + F_\pi^{-1}(\underset{\sim}{\pi}\cdot\underset{\sim}{\varepsilon})\underset{\sim}{\pi} \quad (15)$$

$$\delta N = iF_\pi^{-1}(\underset{\sim}{\tau}\wedge\underset{\sim}{\pi})\cdot\underset{\sim}{\varepsilon}N \quad (16)$$

SU(2) x SU(2) being isomorphic to SO(r), it turns out that the pion's behavior corresponds[37] to that of a 3-vector $\underset{\sim}{\pi}$ in a 3-dimensional flat hyperplane, on which we project the action of the J_{14}, J_{24}, J_{34} "boosts" on a 3-

dimensional hypersphere in the $\underset{\sim}{4}$ space. Plain isospin corresponds to J_{12}, J_{23}, J_{31} in that $\underset{\sim}{4}$-space and thus acts linearly on the $\underset{\sim}{\pi}$ vector. In this manner, the dimensionality of this "realization" of SO(4) stays $\underset{\sim}{3}$, i.e. the same as that of the linear representation of SO(3), the isospin, which is here the diagonal SU(2) subgroup of SO(4) $\sim$ SU(2)$_L$ x SU(2)$_R$ (We have glossed over the global distinctions between SU(2) and SO(3)). In a similar fashion, an octet of 0^{-+} mesons is sufficient to guarantee chiral unitary symmetry (linearly, we would have needed another octet of 0^{++}, for instance).

Chiral invariance supplies the entire dynamics of soft pions. This can be supplemented to a certain extent by the physics of "soft" vector mesons, assumed to dominate the vector currents' matrix elements. The entire system of 1^{--} and 0^{-+} mesons can fit into a chiral-symmetric Lagrangian, and may further be extended to provide a "local" gauge as in the Yang-Mills case.

Quarks, SU(6) and further extensions

Symmetry seems to have provided us here with the most intriguing paradox of this decade. Physics alternates between the concrete or structural ideas, and the abstracted representations of phenomenology. The search for the "internal" symmetry started out with an attempt to account for the interplay of π- versus K-mediated strong nuclear forces; in the Sakata model[38] it merged with an attempt at a concrete structural theory. Then it turned out[7] that the $\frac{1}{2}^+$ baryons we are made of, fit an abstract representation like the octet, rather than a basic triplet. The most straightforward explanation of the latter fact would be given by the existence of an as yet undiscovered basic triplet, with forces saturating in a 3-body configuration. Should we therefore adopt this view?

Explanation by further dissection-decomposition has generally been successful. I already viewed the baryon octet as a product of three basic triplets[19] in the beginning of 1962, and wrote up this idea in an article in II Nuovo Cimento (with H. Goldberg) which went unnoticed. Remember that SU(3) itself, and the octet identification in particular, were still generally considered as highly speculative. Noticing the $B = \frac{1}{3}$ (and related fractional charges) of the basic triplets, I preferred to present them as a field theory model and a group-theoretical mnemonic, and I didn't comment on the possibility of an independent existence as particles besides. When Gell-Mann treated the same question in 1964, he had the same doubts. However SU(3) had been vindicated, and by publication time the Omega Minus experiment[39] had provided final proof of the octet assignment. The issue was thus more compelling, and Gell-Mann did go as far as suggesting a search for free quarks; Zweig was even more structural in his presentation.[40] Noticing that the Omega Minus had established the decimet assignment for $\frac{3^+}{2}$ baryons, Zweig realized that this was in fact another three-quark combination. He therefore went on to suggest an important feature of the present "quark model" - saturation at qqq. Gell-Mann had allowed for $qqqq\bar{q}$ etc., since he had not started out structurally, and still considered the scheme more as a group-theory mnemonic or as a field theory from which he could derive further intuition. As to Zweig, he had revived the "model" structural idea of Fermi and Yang, Goldhaber, Christy and Sakata.

It is this "model" view which has generated our beautiful paradox. It still needed the helping hand of the abstract approach once more - in the assignment[41] of baryons in SU(6) to the $\underline{56}$, contrary to what Fermi statistics would have picked. Zweig himself, and Gürsey and Radicati were careful enough not to follow the structural approach too dogmatically, and were saved from the fate of the Nagoya school in its dealing with SU(3). Ikeda, Ogawa, Ohnuki and their collaborators had done so much beautiful work on SU(3) in 1959-60, without noticing that the $\frac{1}{2}^+$ states fitted an octet - which would have contradicted their basic premise, but did provide the best illustration of the usefulness of the mathematical system they had developed. Yamaguchi told me he in fact had noticed the possibility - he was in Europe in 1960 and was thus perhaps less under the influence of Sakata's ideas. Still, even he stuck to the original model and continued to fence for it until late in 1962.

The additional abstract step consisted in picking the totally symmetric three-quark representation $\underline{56}$, rather than an antisymmetric representation as would have befitted Fermi statistics. It amounted, when returning to the structural level, to the adoption of either Greenberg's parastatistical quarks[42] or (and this seems equivalent to a large extent) the Han-Nambu[43] model of 3 sets of 3 quarks. From this point on the way was clear, and the "quark-model" view has yielded tremendous results.[44] Some have been derived with the help of SU(6) and its extensions; others follow a dynamical reasoning in which quarks play the role of nucleons in nuclei. In almost all cases, it has proved possible to reproduce these results by appropriate algebraic and dynamical postulates. However, none of this is convincing enough to serve as an alternate starting point, and

we have not yet identified the real theory. Is the algebra
a direct result of the existence of quarks? Or are quarks
a fictitious set of objects capable of faking the algebraic
dynamics?

There would have been no paradox if free quarks would
indeed have been found, with a mass of some 500 Mev a piece.
It seems however that even if they do exist, they will be
extremely heavy (> 5 Gev). This fact has led to some guesses
at the nature of the forces binding them in nucleons. Notwithstanding tremendous binding energies, all the characteristics of independent quarks survive in their bound state.
A nucleon appears to consist of 3 noninteracting quarks
whose masses are just $\frac{m_N}{3}$, i.e. their "effective" mass when
bound. Kokkedee and Van Hove have studied[44] the phonon-like
object we are using in the quark model. Does it relate in
any way to free quarks - parastatistical of à la Han-Nambu?
Note also that the Han-Nambu like picture predicts additional
quantum-numbers and a huge extension of the spectrum at higher
energies.

We shall now discuss the following aspects of the
"quark model":

    a)   SU(6) and its extensions

    b)   low energy dynamics

    c)   high energy dynamics

    d)   graphical duality

Classification of the spectrum of hadron excitations
seems to represent one of the best applications of SU(6)
and its extensions. For baryons, the present states up to
about 2 Gev and $J \leq \frac{7}{2}$ do fill up some 50% of the required
assignments for the one- and two-quantum harmonic oscillator
excitations of the 3 quark symmetric system. In fact,

besides the ground state $\underset{\sim}{56}$, L = 0, the odd parity first
excitation $\underset{\sim}{70}$, L = 1 seems complete, and so are the $\underset{\sim}{56}$
L = 0 and $\underset{\sim}{56}$ L = 2 of the double excitation. At the Vienna
Conference in 1968 Harari pointed out that $\underset{\sim}{70}$ L = 0
ought to be in the same region, and the more recent Rosenfeld table already has parts of three multiplets belonging
to this supermultiplet: $\underset{\sim}{8}$, $\frac{1^+}{2}$ with N(1780); $\underset{\sim}{8}$, $\frac{3^+}{2}$ with
N(1860); $\underset{\sim}{10}$, $\frac{1^+}{2}$ with Δ(1910). It is too early however to
regard these assignments as a proof of the oscillator
model, but it seems there is sufficient information to
fix most parameters in a mass formula and predict the
missing states. In a way this is also the strongest point
for the quark-model assumption of saturation at 3 quarks.

The situation is less clear in the meson spectrum,
where the doubling of the A_2 may point to a departure
from the quark-antiquark systematics (this may also happen
for the 3-quark saturation picture if the various Z*
baryons reported in recent years are indeed confirmed by
new experiments).

The usefulness of SU(6) x L as a classification scheme
is not really established, but appears plausible. Before
we go on to the question of the larger classification framework we shall discuss the present theoretical picture with
respect to SU(6) x L. One of the key unresolved problems
in symmetry consists in the physical identification of the
algebra of SU(6) x L and its extensions, including W-spin[45]
etc.

The first attempts identified the SU(6)-"spin" $SU(2)_\sigma$
with the Lorentz group spin. This led to a search for
"relativistic SU(6)", based upon the U(6,6) generated by
the tensor-product of the Dirac matrices algebra (with β

as the invariant) with SU(3). It appeared - erroneously - that the space-components of the vector bilinears $\bar{\psi}\gamma^k \lambda_o \psi$ (antihermitean if β is the invariant) and the $SU(2)_\sigma$ represented here by $\bar{\psi}\gamma_5\gamma^k \lambda_o \psi$, generate an SL(2,c) which was taken to coincide with the Lorentz group. This was wrong[46] because we can't use an antihermitean representation; moreover, the charge conjugation assignments are wrong (the Lorentz boost has $J^{PC} = 1^{-+}$, an assignment which cannot be met with Dirac bilinears).

The next attempt seemed more physical. It consisted in the identification of $SU(2)_\sigma$ with the physical axial vector current.[47] One could still use the entire system of Dirac matrices, but they would now correspond to the remaining weak currents and some other such densities, closing on U(12) rather than U(6,6). It provided Gell-Mann with an elegant definition of a relativistic L assignment:[48] this is the difference between the $SU(2)_\sigma$ assignment and actual spin (rest angular momentum). Out of the entire U(12), some subgroups might represent symmetries in some particular situations. One such example[45] is $SU(6)_W$, generated by λ^i, $\beta\sigma_x\lambda^j$, $\beta\sigma_y\lambda^k$, $\sigma_z\lambda^\ell$. Theoretical analysis allows this to represent an approximate symmetry for trilinear couplings. The experimental situation seems to be favorable. The original enthusiastic application of the symmetry to amplitudes with more than 3 external lines is unwarranted,[49] even within the context of collinear processes. The same is true for the "coplanar" SU(3) x SU(3) group. Considering however that many resonance decays make a 3-external lines situation, these symmetries may still be useful to check the related SU(6) x L rest-assignments.

Nevertheless, the actual connection with the currents is still not established beyond the chiral SU(3) x SU(3) generated by the time components of the axial and vector currents. There are two kinds of difficulties in the construction of sum rules that would check the identifications for other components of SU(6) or SU(6)$_W$. The first arise because of the behavior of the matrix elements of these densities when taken between infinite-momentum states. This kinematic choice seems essential for the application of dispersion relations and several other reasons. The second difficulty involves the probably operator-structured Schwinger terms appearing in the commutators between density-components, about which almost nothing is known.

The latter difficulty has led to an attempt to replace the quark model commutators with contracted[50] ones, abstracted from the commutation relations of the Yang-Mills field components. In a model in which the local gauge is broken by a mass term for the Yang-Mills field itself, field theory makes the currents proportional to these vector meson fields. These commutators have been applied usefully, but imply abandoning the connection between SU(6) and the currents altogether. As shown recently,[51] the SU(6) spectrum cannot even be considered to arise via a de-contraction of the current-commutators. We are thus forced to choose between sticking to the connection or to the simplified "field algebra". What in the latter case would then be the meaning of the system of SU(6) generators?

To return to the applications of SU(6), let us first note that we are still faced with various paradoxes. The beautiful result of $\frac{\mu_n}{\mu_p} = -\frac{2}{3}$ makes us wonder at $\left|\frac{G_A}{G_V}\right| = \frac{5}{3}$.

All of these parameters have been fitted together[52] in SU(3) x SU(3) at the $\vec{p} \to \infty$ limit (the $(1\pm\sigma_z)\lambda$ subgroup of SU(6) which then coincides with the chiral group) by the adoption of a particular mixing scheme between $\underline{56}$ L = 0 and $\underline{70}$ L = 1, but this then doesn't say anything about the static SU(6) results in the pure $\underline{56}$. Are they simple coincidences? Note that F/D ratio for baryon-pseudoscalar coupling seems beautifully confirmed by experiment. Again, Schwinger did construct a chiral Lagrangian yielding at the same time $\left|\frac{G_A}{G_V}\right| = \frac{5}{3\sqrt{2}}$, but this treats the pseudoscalars non-linearly and is thus even more puzzling in its reflection on SU(6), which provides the correct F/D linearly! It is possible that all this can somehow be made to look consistent, but this has yet to be done.

Various calculations based on the quark picture have yielded interesting low-energy results, such as the $\omega^\circ \to \pi^\circ \gamma$ width[53] (experimentally 1.1 ± 0.15 MeV; calculated: 1.18 MeV).

At high energy, the quark model approach (additivity and an impulse approximation, i.e. only one quark interacts each time) was shown[54] to be exactly equivalent to the exchange of bosonic objects behaving like $\underline{8}$ and $\underline{1}$ (for J^{PC} = 1^{--} and 2^{++}) and coupling universally in SU(3) for the 1^{--} set, and to $\beta\lambda^i$ for the 2^{++}. It is the $\beta\lambda^i$ and the universality property which generate effective additivity of quark-quark or quark-antiquark amplitudes. The approach leads to the famous Levin-Frankfurt result[55]

$$\lim_{\nu \to \infty} \frac{\sigma_{\pi N}}{\sigma_{NN}} = \frac{2}{3} \qquad (17)$$

and the numerous equalities and inequalities due[44] to
Lipkin and Scheck, Freund, Kokkedee and Van Hove. In
CHN, some additional results are due to the combination
with Regge dynamics.

The present difficulties with Regge theory - mostly
due to the experimental results regarding $\frac{d\sigma}{dt}$, polarizations and to the Serpukhov total cross sections - make it
difficult to develop CHN as an independent dynamical
theory, and it would be useful if the entire system could
be related to either the currents or phenomenological
Lagrangians. The interplay between chiral $[SU(3) \times SU(3)]_{\gamma_5}$
on the one hand, and the quark-counting[54] $[SU(3) \times SU(3)]_\beta$
should be realized directly upon $\theta^{\mu\nu}$ or a Lagrangian.

The most recent success of the quark model is known[56]
as "graphical duality". In this approach, based upon the
generalized Veneziano amplitudes,[57] the hypothesis of saturation at $\bar{q}q$ and qqq is used to derive via "continuous"
duality the entire set of exchange-degeneracies between
Regge trajectories. For a certain low-to-medium energy
region, the results are impressive. Is there really something fundamental here, and how does this relate to the
SU(6) quark results at low energy? And to the quark-counting ones at high-energy? Can it all be systematized
and understood together?

Spectrum Generating Algebras and the Local (Infinite) Algebra of the Currents

During the fifties, the list of elementary particles
included thirty stable or metastable states. The $J = \frac{3^+}{2}$ $I = \frac{3}{2}$
and higher resonances of the N-π system were treated on a
different footing. Alvarez' opening[58] of the Pandora box of

meson and baryon resonances, together with the conceptual development of hadron physics, brought about the adoption by 1964 of a very different view. SU(3) and SU(6) had not provided the $J = \frac{1^+}{2}$ baryon octet with any special "fundamental" role. The appearance of a metastable state (the Ω^-) in one multiplet with nine fast-decaying resonances emphasized the circumstantial nature of stability as a criterion (just as the existence of both β^- and β^+ decay had put proton and neutron on the same footing in the thirties). The bootstrap ideology helped in the realization that at least all <u>known</u> hadrons are not fundamental, and do form yet one additional spectroscopy, even if the dynamics behind it are still unknown. This was implemented in the Regge formalism. True, the attempts to reproduce the systematics dynamically gave at first qualitatively wrong clues, because they were based on the Yukawa nature of pion interactions - and Yukawa potentials do not produce long sequences of bound states. However, Chew and Frautschi[59] had drawn straight lines for the trajectories as a first guess, and experiment showed that these straight lines (in $J(m^2)$) seemed to go on and on. We have by now several cases[60] of four or five known levels (for Δ_δ, N_γ, p) at least, and the picture is much more reminescent of the harmonic oscillator spectrum than of a Yukawa force. The fact that we are faced here with a continuous superposition of Yukawa exchange forces, and considerations of unitarity in the decay of the highest excited states may be responsible for this apparent dilemma, but even this is highly speculative at this stage.

The algebraic approach, which had been so successful for the "internal" quantum numbers and even for the correlations

with spins (in SU(6) x L) was then extended so as to enable
it to cope with these sequences of excitations. Spectrum
generating algebras[11] have now become a familiar tool.
Some pioneering work in those lines had already been done
by Lipkin and Goshen[61] in the late fifties for nuclei.
The method, using unitary infinite representations (of the
discrete type - "ladders") of non-compact groups, was
reinvented and developed by the particle physicists (and
has in fact recently been reapplied to nuclei in a more
physical context[62]). For the soluble cases in non-relativistic quantum mechanics, spectrum generating algebras provide
an alternative to the more conventional analytical treatment of the Schrödinger equation. This should be especially
useful in cases where the Hamiltonian is too complicated,
but the pattern of bound states is identifiable with a
definite SGA, and an educated guess may be ventured as to
the physical content of the generators. For example, in the
above-mentioned case of nuclei, the sequences of states
were grouped in irreducible representations of SL(3,R), and
it was conjectured that this SL(3,R) was generated by the
time-derivatives of the gravitational quadrupoles.[11,62]
Using the assumption of proportional electric and gravitational quadrupoles, the algebra can be utilized to derive
sum rules etc.

There are two ways in which the spectrum of hadrons
may "explode" at higher energy: we may have several
definite SU(6) representations, with their excitations
appearing as sequences in L; alternatively, the excitations
may occur in the "internal" variables too, so that we face
ever-larger representations of SU(6), probably with L

increasing anyhow (the need for L can be seen[63] at the $p \to \infty$ limit, where the neutron magnetic moment would have vanished otherwise). The first method corresponds to qqq and $q\bar{q}$ saturation in the quark model, whereas in the second method we would have $qqqq\bar{q}$, $qqqq\bar{q}\bar{q}$ etc. Almost all observed states in the hadron spectrum can be accommodated without appealing to the latter, except[26] for the unconfirmed $Z_0(1865)$, $Z_1(1900)$ and the split A_2 meson.

The identification of the SGA in the abstract is of little value. The situation would improve tremendously if the physical content of the generators would become known as well. Indeed, we might then possess an almost-complete dynamical theory, i.e. something that would be equivalent to a full knowledge of the Hamiltonian (and handier than the unusuable perturbative approach). It might[11] involve the true mechanism of the bootstrap, as exemplified in the simplified strong-coupling models leading to $SL(4,R)$, $SL(6,C)$ or $SU(6,6)$.

Gell-Mann and Dashen[63] have conjectured that the SGA is provided by the infinite algebra of current densities. This "local" $SU(3)$ or $SU(3) \times SU(3)$ does indeed have representations which reduce[64] to $[SU(3)]^{\otimes n}$ or some contraction[65] of that group, with infinite sequences of L (embedded in an $SL(2,C)_L$ in the simplest case). To make sure that the states are one-particle states, it was suggested that the $\underset{\sim}{p} \to \infty$ limit be used (with several other expected dividends).

The program has been on since the beginning of 1966, and it seems difficult to conserve one's optimism. Three methods have been used:

a) starting from $\underset{\sim}{p} \to \infty$ and using the "angular condition" to ensure covariance.[63]

APPLICATION OF SYMMETRY   83

    b) using infinite-component wave equations[66]

    c) constructing first a representation[67] of the local current with good states and checking on covariance at $\underset{\sim}{p} \to \infty$.

The first two methods have been shown to generate space-like states connected to the time-like ones by the currents. This was only proved in the simplest case in which the internal quantum-numbers are entirely factorized out (i.e. all states have one single I spin). The alternative is a single mass for all states, or an infinitely degenerate continuous energy spectrum.

Method c has yeilded a complete classification of the representations.[65,68] For the Fourier transforms of the currents, these can be grouped in two large classes. The first group can be written as a sum

$$V^i(\underset{\sim}{k}) = \sum_{a=1}^{p} \sum_{r=1}^{n} k^{r-1} e^{\xi_a k} V^i_{r,a} \qquad (18)$$

with finite p and n, and no k dependence in the $V^i_{r,a}$. The second group contains the cases in which p or n $\to \infty$ and non k-factorizable situations. It looks probable that the above mentioned no-go theorem[69] (orginally due to Grodsky and Streater) can be generalized to the entire first group.

Method c has yielded constructions which do not have unsegregated space-like states[67] The energy spectrum is however still unknown and may depend upon the choice of commutation relations between the space-components of the currents. Hopefully, this will become clear within the coming year.

Working with the space-components raises the question of Schwinger terms and the closure of the system of current components and such terms. It is to be hoped that even though one may have to abandon the simplified contracted commutators[50,51] (this is very probable), some set with simple closure leading to a well-defined Lie algebra will be found.

If the program fails, i.e. if it turns out that the currents cannot be approximately saturated by one-particle states, it will have yielded some insights, but we shall have to look for other physical candidates for the SGA. Otherwise, hadron physics will look very much like low-energy nuclear physics, with different "models" for different regions of the spectrum. This would be sad, considering the progress achieved in the dynamical description, as exemplified by the Veneziano formula.[70] (Note that the Veneziano formula spectrum has been reproduced[71] in an infinite-component equation).

Space-time symmetries

Progress in the treatment and application of the geometrical symmetries was prominent in the following three topics:

a) an algebraic treatment of the complex-J plane, using harmonic analysis methods as applied to the Poincaré group and the Complex Lorentz group

b) applying the conformal group as a broken symmetry

c) a phenomenological determination of the actual breakdown of CP and T. We shall not discuss this topic, since progress was entirely in the observations, not in understanding.

The work of Toller[8] and collaborators produced first an algebraic understanding of the complex-J formalism. Toller showed that this amounted to the study of Poincaré invariance of an amplitude, using a space-like 4-momentum to project out SO(1,2) (or SL(2,R)) as the little group. Harmonic analysis then defines an expansion over the irreducible representations of that group, just as the partial wave expansion could be shown to arise from a time-like four-momentum and SU(2) as a little group. One immediate application was the discovery of an expansion over SL(2,C) representations occuring when $P^\mu = 0$. This is the case which looked "sterile" to Wigner because he was classifying at the time the irreducible representations for one-particle states or for fields. In an amplitude, $P^\mu = 0$ proved very useful since it fitted (for equal masses) the simplest case, i.e. forward scattering. It was conjectured[72] (and later proved by Nakanishi[73]) that these "Lorentz poles" factorize into sequences of Regge poles (i.e. a reduction of SL(2,C) over SL(2,R)).

A somewhat related approach based upon the complex Lorentz group SL(2,C) x SL(2,C) was suggested by Freedman and Wang.[74] The generators are J, Λ and their complexified analogs behaving like "iJ", "iΛ". It turns out that J and "iΛ" close on a compact SO(4) subalgebra. If one continues analytically the vector sum of incoming momenta (i.e. the above-mentioned time-like case) down to $P^\mu = 0$, the entire SL(2,C) x SL(2,C) commutes with P^μ. The space components of the particles' momenta become imaginary; out of the entire complex Lorentz group, only the compact SO(4) will not vary this property, and this is then used for the expansion.

For the conformal group, only scale-invariance was actually applied in most cases. It was first applied[75] directly to the S-matrix, yielding Regge behavior for large s and t, which does not seem to fit experiment. More recently, scale invariance has been used to extend the analysis of the various pieces of $\theta^{\mu\nu}$ in Gell-Mann's approach.[76] One is then just using dilation D as a broken symmetry, which might still be closer to approximate invariance if it acts in conjunction with a 0^+ Goldstone particle (and a 1^- one if full conformal invariance is requested). One by-product of these studies has consisted in the use of a different $\theta^{\mu\nu}$ (first suggested[14] by Huggins) instead of the Belinfante-Rosenfeld[15] prescription. The scale-invariant $\theta^{\mu\nu}$ leads to a renormalizable theory in the coupling to linearized gravitation.[77]

Another improvement on the Belinfante-Rosenfeld formalism has been provided by Wouthuysen and collaborators.[78] The conventional field theory separation of angular momentum into spin and orbital pieces involved expressions which were not covariant tensors in fact. This has now been remedied and will hopefully reduce the confusion arising from using expressions such as

$$\frac{1}{2} \int d\sigma^\rho \; \frac{1}{2} \; \bar{\psi} \, \gamma^\rho \, \sigma^{\mu\nu} \, \psi + \text{h.c.} \tag{19}$$

which do not behave as tensors under the Lorentz group and have led to many misconceptions about Lorentz transformations. The Hilgevoord-de Kerf spin operator[78] for the same case of a spinor field is,

$$\frac{1}{2} \int d\sigma^\rho \, \bar{\psi} \, \gamma^\rho \{ \frac{1}{2} \sigma^{\mu\nu} + \frac{1}{2m} (\gamma^\mu \partial^\nu - \gamma^\nu \partial^\mu) \} \psi + \text{h.c.} \qquad (20)$$

Conclusions

These should be left to the reader.

References

1. M. Gell-Mann, Phys. Rev. $\underline{125}$, 1067 (1962).
2. S. S. Gershtein & J. B. Zeldovich, Zh.Ek.i Te.Fiz. USSR $\underline{29}$, 698 (1955).
3. R. P. Feynman & M. Gell-Mann, Phys. Rev. $\underline{109}$, 193 (1958).
4. S. Okubo, Nuovo Cim. $\underline{13}$, 292 (1959); B. L. Ioffe, Nuovo Cim. $\underline{10}$, 352 (1958).
5. M. L. Goldberger & S. B. Treiman, Phys. Rev. $\underline{110}$, 1178 (1958).
6. S. L. Adler, Phys. Rev. Letters $\underline{14}$, 1051 (1965); W. I. Weisberger, Phys. Rev. Letters $\underline{14}$, 1047 (1965).
7. Y. Ne'eman, Nucl. Phys. $\underline{26}$, 222 (1961); M. Gell-Mann, report CTSL 20 (unpub.), 1961.
8. M. Toller, Nuovo Cim. $\underline{53A}$, 671 (1968).
9. J. Schwinger, Phys. Rev. Letters $\underline{3}$, 296 (1959).
10. S. Weinberg, Phys. Rev. Letters $\underline{8}$, 188 (1967); J. Schwinger, Phys. Letters $\underline{24B}$, 473 (1967).
11. Y. Dothan, M. Gell-Mann & Y. Ne'eman, Phys. Letters $\underline{17}$, 148 (1965); Y. Dothan & Y. Ne'eman, Proc. 1965 Athens (Ohio) Conference on Resonant Particles, p. 17 (1965).
12. See for example R. Omnes, in Proc. of the 1968 Eighth Nobel Symposium (Elementary Particle Theory), Almqvist & Wiksell pub. (Stockholm, 1968), p. 319.
13. See for example V. Barger & M. Olsson, Phys. Rev. Letters $\underline{15}$, 930 (1965).
14. E. R. Huggins, Cal. Tech. Thesis.
15. F. J. Belinfante, Physica $\underline{6}$, 887 (1939); L. Rosenfeld, Mem. Acad. Roy. Belg. $\underline{18}$, 6 (1940).
16. See for example chapters 3 and 10 in Y. Ne'eman, *Algebraic Theory of Particle Physics*, W. A. Benjamin pub. (N.Y. 1967).
17. M. Gell-Mann, Phys. Rev. $\underline{92}$, 833 (1953); T. Nakano & K. Nishijima, Prog. Theoret. Phys. $\underline{10}$, 581 (1953).
18. N. Cabibbo, Phys. Rev. Letters $\underline{10}$, 531 (1963).

19. H. Goldberg & Y. Ne'eman, report IA-725 (1962) and Nuovo Cim. 27, 1 (1963).
20. Y. Ne'eman, Physics 1, 203 (1965).
21. Y. Ne'eman, Phys. Rev. 134, B1355 (1964).
22. See p. 282 in M. Gell-Mann & Y. Ne'eman, The Eightfold Way, W. A. Benjamin pub., N.Y. (1964).
23. Within the framework of SU(3), see for example, R. E. Cutkosky & P. Tarjanne, Phys. Rev. 132, 1354 (1963); N. Cabibbo, in Hadrons and their Interactions, A. Zichichi ed., 1967 Ettore Majorana School, Academic Press (N.Y. 1968); R. Brout, Nuovo Cim. 46A, 932 (1967); etc.
24. For work in chiral SU(3) x SU(3), see for example N. Cabibbo & L. Maiani, Phys. Letters 28B, 131 (1968); R. Gatto, G. Sartori & M. Tonin, Phys. Letters 28B, 128 (1968); etc.
25. N. Samios, Seventh Coral Gables Conference (1970), to be published.
26. A. Barbaro-Galtieri et al., Reviews of Mod. Phys. 42, 87 (1970).
27. J. K. Kim, Phys. Rev. Letters 19, 1079 (1967); C. H. Chan & F. T. Meiere, Phys. Rev. Letters 20, 568 (1968) etc.
28. R. C. Chase et al., Phys. Letters 30B, 659 (1969).
29. M. Gell-Mann, R. J. Oakes & B. Renner, Phys. Rev. 175, 2195 (1968).
30. Y. Nambu & G. Jona-Lasinio, Phys. Rev. 122, 345 (1961); J. Goldstone, Nuovo Cim. 19, 154 (1961).
31. See for example Y. Ne'eman, in Contemporary Physics Vol. II, Proc. of the 1968 Trieste seminar, I.A.E.A. pub. (Vienna, 1969), p. 167.
32. S. Coleman & S. L. Glashow, Phys. Rev. Letters 6, 423 (1961).
33. M. Gell-Mann, Phys. Rev. Letters 12, 155 (1964).
34. R. F. Dashen et al., in The Eightfold Way by M. Gell-Mann & Y. Ne'eman, W. A. Benjamin pub. (N.Y., 1964), p. 254.
35. See for example R. F. Dashen, Phys. Rev. 183, 1245 (1969).

36. S. Fubini & G. Furlan, Physics $\underline{1}$, 229 (1965).
37. K. Hiida, Y. Ohnuki & Y. Yamaguchi, INS Report 126 (1968).
38. S. Sakata, Prog. Theoret. Phys. $\underline{16}$, 686 (1956).
39. V. E. Barnes et al., Phys. Rev. Letters $\underline{12}$, 204 (1964).
40. M. Gell-Mann, Phys. Letters $\underline{8}$, 214 (1964); G. Zweig, CERN report Th 401, 412 (1964).
41. F. Gürsey & L. A. Radicati, Phys. Rev. Letters $\underline{13}$, 173 (1964).
42. O. W. Greenberg, Phys. Rev. Letters $\underline{13}$, 598 (1964).
43. M. Y. Han & Y. Nambu, Phys. Rev. $\underline{139B}$, 1006 (1965).
44. See for example J.J.J. Kokkedee, The Quark Model, W. A. Benjamin pub. (N.Y., 1969).
45. H. J. Lipkin & S. Meshkov, Phys. Rev. Letters $\underline{14}$, 670 (1965).
46. D. Horn & Y. Ne'eman, Phys. Rev. (in press); see also in ref. 16 p. 203.
47. R. P. Feynman, M. Gell-Mann & G. Zweig, Phys. Rev. Letters $\underline{13}$, 678 (1964).
48. M. Gell-Mann, Phys. Rev. Letters $\underline{14}$, 77 (1965).
49. S. Coleman, Phys. Letters $\underline{19}$, 144 (1965).
50. T. D. Lee, S. Weinberg & B. Zumino, Phys. Rev. Letters $\underline{18}$, 1029 (1967).
51. L. Michel & Y. Ne'eman, Phys. Rev. (in press).
52. See for example H. Harari, Phys. Rev. Letters $\underline{16}$, 964 (1966) and $\underline{17}$, 56 (1966).
53. W. Thirring, Phys. Letters $\underline{16}$, 335 (1965); C. Becchi & G. Morpurgo, Phys. Rev. $\underline{140}$, B687 (1965); etc.
54. N. Cabibbo, L. Horwitz & Y. Ne'eman, Phys. Letters $\underline{22}$, 336 (1966).
55. E. M. Levin & L. L. Frankfurt, Zhur.Eksp.i Teoret.Fiz. Pisma v.Redak. $\underline{2}$, 105 (1965).
56. H. Harari, Phys. Rev. Letters $\underline{22}$, 562 (1969); J. Rosner, Phys. Rev. Letters $\underline{22}$, 689 (1969).
57. G. Veneziano, Nuovo Cim. $\underline{57A}$, 190 (1968).

58. See for example papers cited in L. W. Alvarez report UCRL 18696.
59. G. F. Chew & S. C. Frautschi, Phys. Rev. Letters $\underline{7}$, 394 (1961).
60. See for example V. Barger & D. Cline, Phys. Rev. $\underline{155}$, 1792 (1967).
61. S. Goshen & H. J. Lipkin, Ann. of Physics $\underline{6}$, 301 (1959).
62. Y. Dothan, M. Gell-Mann & Y. Ne'eman - unpublished report.(1965), communicated to L. Biedenharn; L. Biedenharn, preprint.
63. R. E. Dashen & M. Gell-Mann, Third Coral Gables Conf. (1966), p. 168; also Phys. Rev. Letters $\underline{17}$, 340 (1966).
64. See for example S. Chang, R. Dashen & L. O'Raifeartaigh, Phys. Rev. $\underline{182}$, 1805 and 1819 (1969) and papers mentioned therein.
65. A. Joseph, submitted to Com. Math. Phys. (1970).
66. See for example E. Abers, I. T. Grodsky & R. E. Norton, Phys. Rev. $\underline{159}$, 1222 (1967); Y. Nambu, Phys. Rev. $\underline{160}$, 1171 (1967); A. O. Barut & H. Kleinert, Phys. Rev. $\underline{156}$, 1541 (1967) and $\underline{161}$, 1464 (1967).
67. R. Mendes & Y. Ne'eman, Report CPT 20, to be pub. in J. Math. Phys.
68. R. Mendes & Y. Ne'eman, to be pub. (See Univ. of Texas report CPT 49).
69. I. T. Grodsky & R. Streater, Phys. Rev. Letters $\underline{20}$, 695 (1968).
70. G. Veneziano, Nuovo Cim. $\underline{57A}$, 190 (1968).
71. H. Leutwyler, Phys. Letters $\underline{31B}$, 214 (1970).
72. A. Sciarrino & M. Toller, J. Math. Phys. $\underline{8}$, 1252 (1967).
73. N. Nakanishi, Kyoto report RIMS-33 (1968); see also B. Wolf, Tel-Aviv Univ. Thesis.
74. D. Z. Freedman & J. M. Wang, Phys. Rev. $\underline{160}$, 1560 (1967).
75. H. A. Kastrup, Phys. Letters $\underline{3}$, 78 (1962) and $\underline{4}$, 56 (1963); G. Mack. Nucl. Phys. $\underline{B5}$, 499 (1968).
76. K. G. Wilson, Phys. Rev. $\underline{179}$, 1499 (1969); see also G. Mack & A. Salam, Ann. of Phys. $\underline{53}$, 174 (1969); D. J. Gross & J. Wess, CERN report Th 1076 (1969); L. S. Brown

& M. Gell-Mann, to be pub.; S. Ciccariello et al., Phys. Letters 30B, 546 (1969).

77. C. G. Callan, S. Coleman & R. Jackiw, MIT report CTP 113 submitted to Ann. of Phys.

78. J. Hilgevoord & S. A. Wouthuysen, Nucl. Phys. 40, 1 (1963); J. Hilgevoord & E. A. de Kerf, Physica 31, 1002 (1965).

PARTICLE SPECTROSCOPY*

N. P. Samios

Brookhaven National Laboratory, Upton, New York 11973

Abstract: In a talk presented at the symposium "The Past Decade in Particle Theory," Dr. N. P. Samios reviews the data concerning the meson and baryon resonances. Special attention is paid to the newer resonances and the many outstanding questions associated with the more recently observed resonances. The relationship of the observed resonance spectrum and the multiplets structure required by the theoretical symmetry schemes is reviewed. The baryon resonances are also placed into possible regge families.

The subject of my talk will be boson and baryon spectra. I note that a two-day conference on each of these subjects is to take place in the near future at Duke University and the University of Pennsylvania. It is clear that 45 minutes is a small fraction of four days and as such I recommend the proceedings of these conferences for detailed discussion of these two subjects. Today I wish to discuss a) the general gross features in the boson and baryon categories, i.e. the systematics of the observed spectra and b) some recent interesting experimental results as well as some of the outstanding difficulties in understanding these spectra. I begin with the boson spectrum.

---

*Presented at the Symp. on Particle Theory, Austin, Texas, April 14-17, 1970.

Bosons

Although there has been an enormous increase in the available experimental data, there are still only three certified SU(3) families, namely $J^P = 0^-$, 1^-, and 2^+ nonets. The only outstanding problem with the 0^- ground state is the identity of the ninth member, is it the $X^o(960)$ or the E(1420). The situation is summarized in the first table. The Dalitz plot distribution for the

---

π(140)	K(495)	η(550)
$X^o(960) \rightarrow \pi\pi\eta$ Dalitz Plot 0^-		
$\pi\pi\gamma$ $\sin^2\theta$ distribution 0^-2^-		
$\gamma\gamma$ rules out 1^+		

Is E(1420) also 0^-?

$\bar{p}p \rightarrow E\pi\pi$
$\qquad\quad\hookrightarrow KK\pi$

Decay Distribution	0^- 2% invoke (KK) enhanc.	K(890)K 50%
	1^+ .2% $(\pi\pi) = 2$	(KK)π 50%

Production: 0^- 20%
$\qquad\qquad\quad\, 1^+$ 30%
   Can be either 0^- or 1^+

Table I
Summary of the Status of 0^- Mesons

$\pi\pi\eta$ and $\pi\pi\gamma$ decays of the X^o fits the $J^P = 0^-$ and 2^- hypothesis. Since the decay distributions are isotropic, the fit to 2^- is slightly fortuitous in that a judicious choice of an arbitrary parameter transforms two anisotropic distributions into a flat distribution. The status of the E(1420) is unaltered. I only remind you that the analysis of the production and decay distributions of the E meson favors J^P 0^- or 1^+ equally well. In both cases the simplest matrix elements without interference have been employed; therefore, even these determinations are not on the foremost footings. The choice is then to have a second 0^- multiplet or to place the E(1420) into a possible 1^+ family. The latter alternative is certainly less upsetting.

A recurring question is the status of the splitting of the A_2. The three pertinent decay modes of the A_2 are the $\pi\rho$, $\pi\eta$, and $K\bar{K}$. The original evidence for this splitting, as presented by the CERN Group, is shown in Fig. 1 as well as the evidence presented by the 6 GeV/c BNL π^-p bubble chamber experiment. The evidence for the narrow $A_2 \to K\bar{K}$ is also shown. The pertinent subsequent experiments are shown in Fig. 2. This involves a) the CERN $\bar{p}p$ experiment at 0.7 and 1.2 GeV/c in which an excess of $K\bar{K}$ events are observed for both halves of the A_2. b) The BNL study of K^-n production of the $\pi\rho$ decay of the A_2 which is narrow and in which a spin parity of 1^- is favored. Note, however, the large background. c) The recent study of the $K\bar{K}$ decay mode of the A_2 by the CERN Group in which both halves of the A_2 are seen to decay via the $K\bar{K}$ mode.

The experiments are contradictory, and the situation is surveyed in Table II. There is agreement in the J^{PC} of the upper half of the A_2, this being 2^{++}, however the differences in the $K\bar{K}$ mode in the lower half lead to spin parity assignments 2^{++} or 1^{-+}. The latter value is of extreme interest since it is not allowed by the simple quark-antiquark

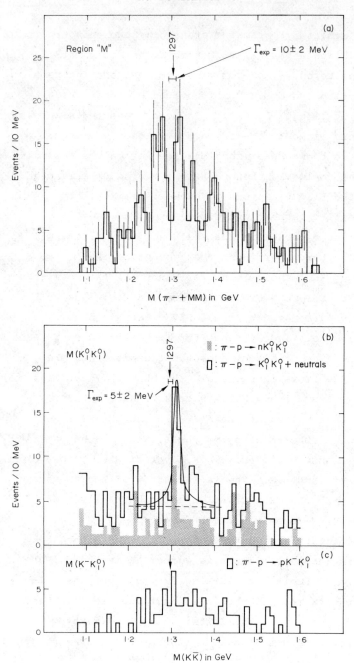

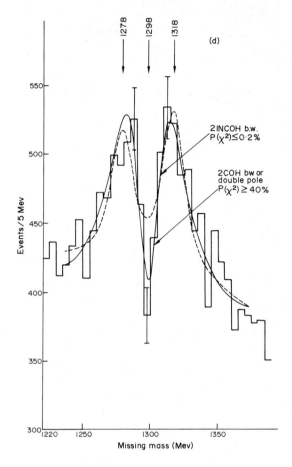

Figure 1. The original evidence for A_2 splitting. (a), (b), and (c) are the Brookhaven National Laboratory bubble chamber experiment. (a) Effective mass distribution of X^- from the reaction $\pi^- p \to pX^-$ at 6 GeV/c. (b) Effective mass distribution of $K_1^0 K_1^0$ from the reaction $\pi^- p \to n\ K_1^0 K_1^0$ and $K_1^0 K_1^0$+ neutrals. (c) Effective mass distribution of $K_1^0 K^-$ from the reaction $\pi^- p\ pK_1^0 K^-$. (d) is the combined CERN missing mass spectrometer and boson spectrometer data for the reaction $\pi^- p \to p(mm)^-$. From B. Maglic <u>Proceedings of the Lund International Conference on Elementary Particles.</u> G. Von Darbel, Editor. (1969).

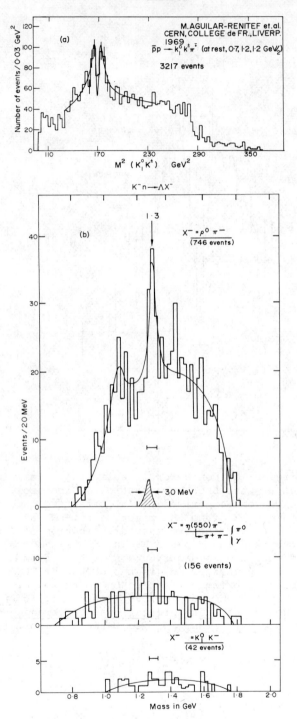

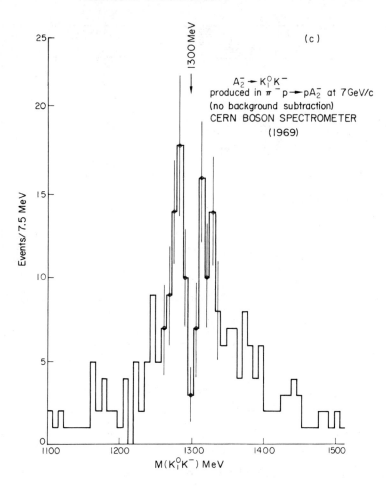

Figure 2. Subsequent evidence for A_2 splitting. (a) CERN experiment at 0.7 and 1.2 GeV/c. From B. Maglic <u>Proceedings of the Lund International Conference on Elementary Particles</u>. G. Von Darbel, Editor. (1969). (b) BNL study of K^-n production. (ibid.) (c) Recent CERN analysis of K<u>K</u> decay mode of A_2 meson.

	A_2^H	A_2^L					
$\pi\rho$	✓	✓	1^-	2^+	1^+	2^-	etc.
$\pi\eta$	✓	✓	0^+	1^-	2^+	3^-	etc.
$K\bar{K}$	✓	✓/x	0^+		2^+	4^+	etc.
	2^+	$2^+/1^-$					
LRL	$\pi^+ p \to K^+\bar{K}^0 p$						
BNL	$\pi^- p \to K^- K^0 p$						
K(1420) → Kπ split							
LRL	$K^+ p \to K^+ \pi^- \pi^+ p$						

Table II

Summary of A_2 Meson Measurements

($q\bar{q}$) model of the boson states. Also noted are two experiments underway, $\pi^+ p$ at 7 GeV/c and $\pi^- p$ at 4 GeV/c, both exploring the $A_2 \to K\bar{K}$ with good resolution, $\simeq \pm 5$ MeV and moderate statistics, $\simeq 200$ events. One could summarize the A_2 situation as critical but unclear.

As already noted, there are numerous examples of 0^- and 2^+ states; the question naturally arises as to the existence of 1^+ mesons. In essence, these are more difficult to detect experimentally. This is illustrated in Table III where the appropriate reactions for such study are listed. The resonant search has been focused on the three-body decays, $\pi\pi\pi$ and $K\pi\pi$ (the decay of a 1^+ meson into two pseudoscalar particles is forbidden). In all the noted

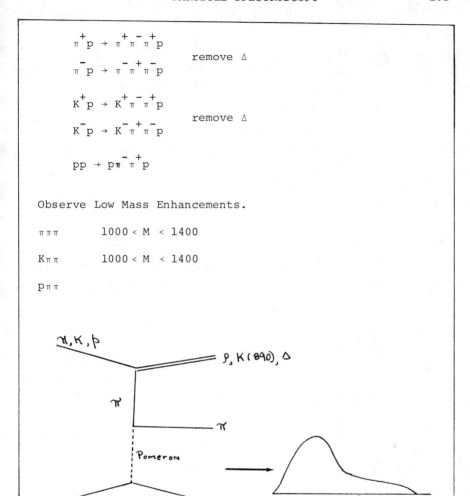

Table III
Summary of Observed 1^+ Enhancements

cases, broad, low mass enhancements are observed so that boson resonances with masses between 1000 and 1400 MeV will occur on very massive backgrounds. This is illustrated

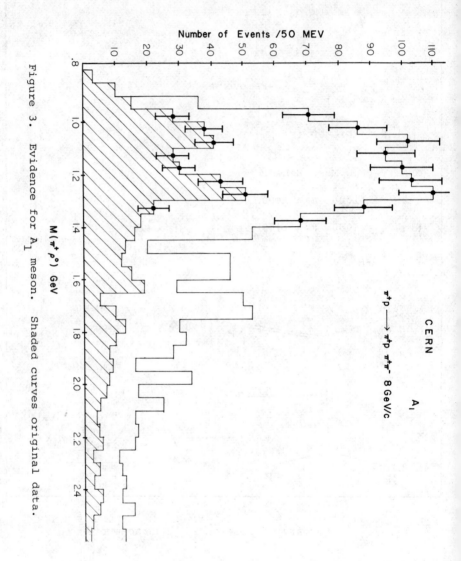

Figure 3. Evidence for A_1 meson. Shaded curves original data.

in Fig. 3 which shows the original evidence for the A_1 by the 8 GeV/c CERN experiment as well as the results from the same Group when the data was substantially increased. In Fig. 4 the $K\pi\pi$ mass spectrum for several experiments as compiled by G. Goldhaber are shown. Again, one observes a peaking at low values of $K\pi\pi$ effective mass. There have been numerous possible explanations for such an effect, among these being the Reggeized Deck model as diagramed in Table III. In spite of these difficulties, there have been two recent experiments showing evidence for the A_1: a) 16 GeV/c SLAC π^-p experiment and b) BNL 6 GeV/c π^-p exposure, both results being displayed in Fig. 5. The new results are such as to resurrect the A_1, however the same cannot be said for any structure in the Q ($K\pi\pi$) low mass region.

There are numerous resonances with I = 1/2, 1 and masses in this intermediate range 1000-1500 MeV which I consider well accepted. These are listed in Table IV. In contrast to the previous discussion, these resonances are mainly initiated by $\bar{p}p$ interaction or produced recoiling against a neutron n or Λ^o, so that the previously discussed low enhancement difficulties are not present. One notes there are three I = 1 states, the $\delta(960)$, $A_1(1100)$, B(1220); one I = 0, the D(1285); and two K states, the K(1240) and K(1320) listed. The indicated spin parities are the most likely possibilities, the actual measurements however are not sufficiently conclusive to exclude other values. In addition, there are several other K members reported in the literature, the K(1080), K(1180), and K(1260) which are more dubious.

Among the higher mass bosons there are numerous reported and verified states, and these are enumerated in Table V.

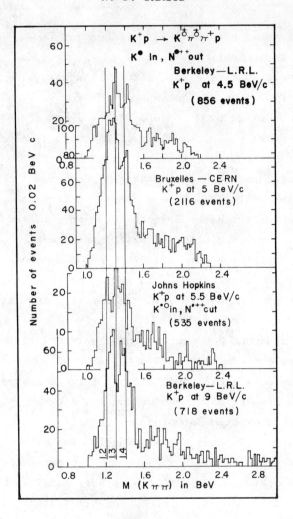

Figure 4

Compilation of Kππ Mass Spectrum Data

Figure 4. Compilation of Kππ mass spectrum data, by G. Goldhaber.

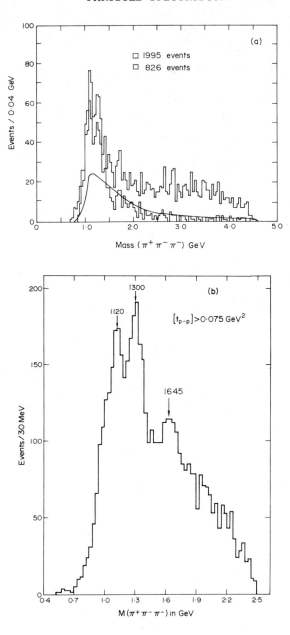

Figure 5. Evidence for the meson (a) 16 GeV/c SLAC $\pi^- p$ experiment (b) GeV/c BNL exposure.

> Boson Resonances: I = 1/2, 1. (1000 < M < 1500)

$A_1(1100) \to 3\pi$ I=1 G=−1 ($J^P = 1^+$)

    Previous: $\pi^- p \to p\bar{A}_1$ backward

    New Evidence: 16 GeV/c $\pi^- p$ SLAC

                      8 GeV/c $\pi^- p$ BNL.

$B(1220) \to \pi\omega$ I=1 G = +1 ($J^P = 1^+$)

    Previous: $\bar{p}p \to \pi B$
                          ↳ $\pi\omega$

    New: $\pi^- p \to pB$
                     ↳ $\omega\pi^-$
      Illinois
                  $B \not\to \pi\pi$

$\delta(960) \to \pi\eta$ I=1 G=−1 ($J^P = 0^+$)

    Several Expts: Widths vary?

$D(1285)$ I = 0 G = + [$J^P(1^+)$]

$\to K\bar{K}\pi$ $\bar{p}p \to D\pi$
                   ↳ $K\bar{K}\pi$

       $\pi^- p \to n(K\bar{K}\pi)^0$

$K_c(1240) \to K\pi\pi$ I = 1/2 ($J^{PC} = 1^+$)
                $\bar{p}p \to \bar{K}K_c$
                      ↳ $K\rho$
                      ↳ $K(890)\pi$

$K(1320) \to K\pi\pi$ I = 1/2 ($J^{PC} = 1^{+-}$)

          $\pi\bar{p} \to \Lambda(K\pi\pi)$

$K(1080) \to K$? $K(1180)$? $K(1260)$?

Table IV

Boson Resonances with I=$\frac{1}{2}$, 1 and Mass Between 1000 and 1500.

g(1650) → 2π → K^+K^0 ↛ K_1K_1	$I = 1$ $G = +1$ ($J^P = 3^-$)		$\pi p \to \pi\pi p$
A_3(1645) → fπ ↛ ρπ	$I = 1$ $G = -1$		$\pi p \to \pi\pi\pi p$
F_1(1560) → K(890)K	$I = 1$		$\bar{p}p \to KK\pi\pi$
φ(1670) → ρπ ↛ $\rho^0\pi^0$ ↛ fπ	$I = 0$ $G = -1$ ($J^P = 1^+, 2^0$)		$\pi^+ d \to 3\pi pp$
ρ(1700) → 4π	$I = 1$ $G = +1$		
ωππ(1695) ↛ fπ ↛ 3π	$I = 1$ $G = -1$		

<u>Missing Mass Spectrometer</u> $I = 1$

R region
S (1930)
T (2195)
U (2380)
X (2620)
X (2510)
X (3000)

Table V

High Mass Bosons

J^{PG}	Quark L S L·S			I = 1	I = 1/2	I = 0		$\|\theta\|$
0^{-+}	0	0	0	π(140)	K(495)	η(550)	η(960)	10°
1^{--}	0	1	0	ρ(750)	K(895)	ω(785)	φ(1020)	40°
1^{+-}	1	0	0	B(1220)	K(1320)			
0^{++}	1	1	-2	δ(960)	K(1080)?	S(1030)	S(1170)?	40?
1^{++}	1	1	-1	A_1(1100)	K_C(1240)	D(1285)	E(1420)	~5°
2^{++}	1	1	1	A_2(1320)	K(1420)	f(1250)+	F(1500)	30°
2^{-+}	2	0	0	A_3(1645)		φ(1670)		
1^{--}	2	1	-3	ρ(1700)				
2^{--}	2	1	-1	F_1(1560)				
3^{--}	2	1	2	g(1650)	L(1750)			
3^{+-}	3	0	0					
2^{++}	3	1	-4	A_2?				
3^{++}	3	1	-1					
4^{++}	3	1	3					

Table VI

Quark Model Assignments of the Mesons

These are all in the so-called R region. The g(1650) and A_3(1645) have both been extensively studied, and their widths and decay modes are well known, but the same cannot be said concerning their spin parities. Three new I = 1 states have been uncovered, the F_1(1560), ρ(1700), and ωππ(1695) as well as one I = 0 boson, the φ(1670), all with the indicated decay modes. The CERN MM Group has also reported additional I = 1 resonances, namely X(2620), X(2810), and X(3000) to be added to the previously published S(1930), T(2200), and U(2375) mesons. Spins, parities, and decay modes are not known.

All the better known established states are categorized in Table VI into SU(3) multiplets. There is some arbitrariness in the procedure since many of the spin parities are unknown, in which case the favored value has been adopted. Also indicated is the simple quark model listing of possible J^{PC} families as well as the L·S splitting expected within a given L grouping. Several comments are now in order: The possible SU(3) nonets have increased from the previously established 0^{-+}, 1^{--}, and 2^{++} families to include 1^{++} and segments of 0^{++} and 1^{+-} multiplets. The latter requires three additional I = 0 resonances in the mass range 1100-1400 MeV. Only the triplet states of the d-shell (L = 2) have been uncovered; the doublets and singlets are expected to occur in the 1500-1800 MeV mass range. This simple model does not accommodate a 1^{-+} state. If the lower half of the A_2 is experimentally shown to possess such quantum numbers, then this simple picture would be untenable. If on the other hand, both halves of the A_2 were 2^{++}, then one would be assigned to the p-shell (L = 1) and the other, the f-shell (L = 3). The remaining question would be to explain the small observed mass difference between these two states if they belong to such different shells. In this latter instance the spin orbit term with L = 3 is such that it would depress the mass of the 2^{++} state; however the problem is the magnitude, it must lower it beyond the L = 2 states.

The Chew-Frautschi plots for this assignment of the triplet and doublet states are shown in Figs. 6 and 7. The more reliable and complete data is available in the J^P = 0^+, 1^-, 2^+, 3^- series where it is experimentally observed that the trajectory slopes are ≈ 1, and furthermore that the trajectories are degenerate. In the case of the ρ trajectory, the intercept is also in agreement with the value expected from the analysis of π^-p charge exchange reactions. The information on the 0^-, 1^+, 2^- series is more fragmentary; however, the π, B(1220), A_3(1645) sequence seem to form another degenerate trajectory, and the A_1(1200) and F_1(1560) states may be the beginning of another such trajectory.

It is of some interest to investigate how well this spectra of particles satisfy the SU(3) classification. This can be done with two levels of sophistication. The first is the placement of the particles into the various representations of SU(3) (1, 8, 10, etc.) and to correspondingly check their agreement with the Gell-Mann--Okubo mass formula. This simple prescription is modified by the possibilities of the mixing of the I = 0, Y = 0 numbers of the 1 and 8 representation. These details are summarized in Table VII. In the case of mixing, another parameter is introduced, the mixing angle θ. With this added parameter, the mass verification is lost. The next level of sophistication then lies in the investigation of the decay rates. The formalism adopted is shown in Table VIII which includes the expression used for the analysis of baryon states (to be discussed later) as well as bosons. In

PARTICLE SPECTROSCOPY   111

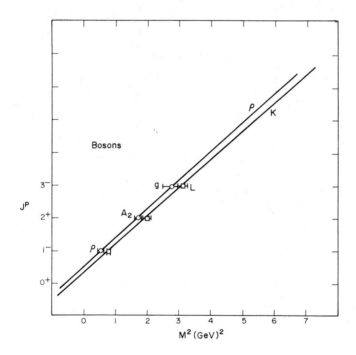

Figure 6. Chew Frautschi Plot of the natural spin parity meson resonances.

Figure 7. (On facing page) Chew Frautschi Plot of the unnatural spin parity meson resonances.

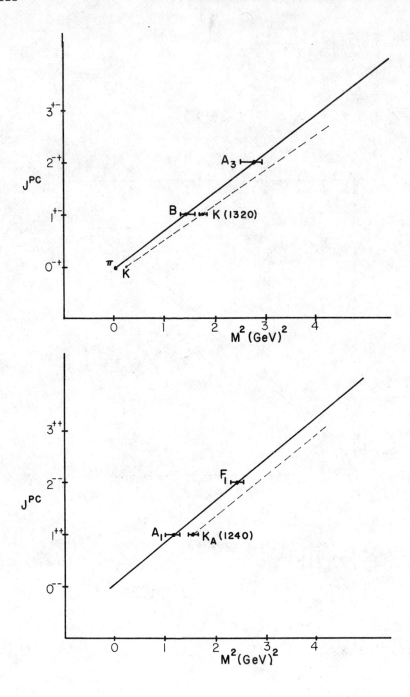

Octet

Bosons
$$\frac{M_K^2 + M_{\bar{K}}^2}{2} = \frac{3M_\eta^2 + M_\pi^2}{4}$$

Baryons
$$\frac{M_N + M_\Xi}{2} = \frac{3M_\lambda + M_\Sigma}{4}$$

Decimet

Equal Spacing
$$M_\Delta - M_\Sigma = M_\Xi - M_\lambda$$

Octet - Singlet mixing

$$I = 0 \quad Y = 0$$

$$\sin^2 \theta = \frac{M_8^P - M_8^O}{M_8^P - M_1^P} \quad \text{baryons}$$

$$\sin^2 \theta = \frac{(M_8^P)^2 - (M_8^O)^2}{(M_8^P)^2 - (M_1^P)^2} \quad \text{bosons}$$

Table VII
SU(3) Mass Formula

this instance one is not only checking SU(3) but also the dynamical assumptions implied by the written rate formula. For a detailed discussion of this subject, see a review article submitted to Reviews of Modern Physics by E. Flaminio, W. Metzger, M. Goldberg, and myself. Suffice it to say that the approach is that of unbroken SU(3) symmetry, where any symmetry breaking is taken into account in the use of the

$$\Gamma = |g^x_{yz}|^2 \times p^{2\ell} \times (p/m)$$

SU(3) sector   barrier sector   phase space sector

Baryons: $x \to y\ z$
$8\ \ \ 8\ 8$ $8 \otimes 8 \to 1 \oplus 8^a \oplus 8^s \oplus 10 \oplus \overline{10} \oplus 27$

$$g^{x=m_8p}_{yz} = \{-C_1 A_1 \sin\theta + (C_s A_s + C_a A_s) \cos\theta\}$$

$$g^{x=m_1p}_{yz} = \{C_1 A_1 \cos\theta + (C_s A_s + C_a A_s) \sin\theta\}$$

$$g^{\text{other}}_{yz} = (C_s A_s + C_a A_s)$$

Bosons:

      Symmetric Couplings   Anti-symmetric Couplings

$g^{x=m_8p}_{yz} = -C_1 A_1 \sin\theta + C_s A_s \cos\theta$ $g^{x=m_8p}_{yz} = C_a A_s \cos\theta$

$g^{x=m_1p}_{yz} = C_1 A_1 \cos\theta + C_s A_s \sin\theta$ $g^{x=m_1p}_{yz} = C_a A_s \sin\theta$

$g^{\text{other}}_{yz} = C_s A_s$ $g^{\text{other}}_{yz} = C_a A_s$

$$\chi^2 = \sum \left(\frac{\Gamma_{SU(3)} - \Gamma_{\text{exptl}}}{\Delta\Gamma_{\text{exptl}}} \right)^2$$

Table VIII

SU(3) Decay Rates

actual physical particle masses in the phase space factor. The particular cases considered are the decay of a member of an

PARTICLE SPECTROSCOPY 115

octet → octet + octet and decimet → octet + octet. This latter
sequence only occurs in the baryon spectra. In the former case,
for bosons there are two unknowns, A_1 and A_s for symmetric
couplings and only one, A_a, for antisymmetric couplings. The
procedure then is to examine the partial widths for the decay
of all members of a given spin-parity multiplet and then form
the χ^2 between the experimentally observed and SU(3) predicted
rates. The system is overly constrained in that there is of
the order of eight decay rates to be fit by one or two parameters.
The results of such an examination are shown in Table IX for
$J^P = 0^-$

(a)	(b)			
m_K^2	$\dfrac{3m_\eta^2 + m_\pi^2}{4}$	$\dfrac{a - b}{(a + b)/2}$	$\lvert\theta_o\rvert$	$m_K^2 - m_\pi^2$
0.24	0.23	0.04	$10°$	0.22

	$\Gamma_{exp'tl}$	$\Gamma_{SU(3)}$
$\pi^\pm(140)$		
$K^\pm(495)$		
$\eta(550)$		
$\eta(960)$		
$\pi^o(135) \to \gamma\gamma$	7.2 ± 1.2 eV	---
$\eta(550) \to \gamma\gamma$	1.0 ± 0.24 KeV	---
$\eta(960) \to \gamma\gamma$	---	60 ± 15 KeV
		340 ± 50 KeV

Table IX
SU(3) Predicted and Experimental Decay Rates for 0^- Mesons

$J^P = 0^-$ multiplet, Table X for $J^P = 1^-$ multiplet, Table XI for $J^P = 2^+$ multiplet, and Table XII for a conjectured $J^P = 0^+$ nonet. The first three of these multiplets have been extensively

$$J^P = 1^-$$

(a)	(b)			
m_K^2	$\dfrac{3m_\omega^2 + m_\rho^2}{4}$	$\dfrac{a-b}{(a+b)/2}$	$\lvert \theta \rvert$	$m_K^2 - m_\rho^2$
0.80	0.60	0.29	40	0.23

$1^- \to 0^- 0^-$		$\Gamma_{\text{exp'tl}}$ (MeV)	$\Gamma_{SU(3)}$ (MeV)
$\rho(750) \to \pi\pi$	$\dfrac{\sqrt{6}}{3} A_a^8$	130 ± 30	149
$K(890) \to K\pi$	$\dfrac{1}{\sqrt{2}} A_a^8$	50 ± 3	50
$\varphi(1020) \to K\bar{K}$	$1 \cos\theta_1 A_a^8$	3 ± 1	2.8

$$\chi^2 = 0.5$$
$$NC = 2$$

$1^- \to 1^- 0^-$		$\Gamma_{\text{exp'tl}}$ (MeV)	$\Gamma_{SU(3)}$ (MeV)
$\varphi(1020) \to \rho\pi$	$\sqrt{3}\left(\sqrt{\dfrac{1}{8}} A_1 \sin\theta_1 - \sqrt{\dfrac{1}{5}} A_s^8 \cos\theta_1\right)$	0.6 ± 0.3	---

Table X

SU(3) Predicted and Experimental Decay Rates for 1^- Mesons

PARTICLE SPECTROSCOPY

$$J^P = 2^+$$

(a)	(b)			
m_K^2	$\dfrac{3m_{f^o}^2 + m_{A_2}^2}{4}$	$\dfrac{a-b}{(a+b)/2}$	$\lvert\theta_2\rvert$	$m_K^2 - m_{A_2}^2$
2.02	1.61	0.22	30°	0.27

$2^+ \to 1^- 0^-$		$\Gamma_{\text{exp'tl}}$ (MeV)	$\Gamma_{SU(3)}$ (MeV)	$\Gamma_{\text{exp'tl}}$ (MeV)	$\Gamma_{SU(3)}$ (MeV)
$A_2(1320) \to \rho\pi$	$\tfrac{\sqrt{6}}{3} A_a^8$	56 ± 19	53	30 ± 10	38
$K(1420) \to K(890)\pi$	$\tfrac{1}{2} A_a^8$	34 ± 12	16	34 ± 10	12
$K(1420) \to K\rho$	$\tfrac{1}{2} A_a^8$	10 ± 9	7	10 ± 9	3
$K(1420) \to K\omega$	$\tfrac{1}{2} A_a^8$	4 ± 3	7	4 ± 3	4
$f(1500) \to K(890)K$	$1\, A_a^8 \cos\theta_2$	9 ± 9	10	9 ± 9	4
		$\chi^2 = 3.5$		$\chi^2 = 6.5$	
		NC = 4		NC = 4	

$2^+ \to 0^- 0^-$		$\Gamma_{\text{exp'tl}}$	$\Gamma_{SU(3)}$	$\Gamma_{\text{exp'tl}}$	$\Gamma_{SU(3)}$
$A_2(1320) \to K\bar{K}$	$\tfrac{\sqrt{15}}{5} A_s^8$	---	6	1.5 ± 0.8	3.4
$A_2(1320) \to \pi\eta$	$\tfrac{\sqrt{10}}{5} A_s^8$	---	4	6 ± 2.5	2.4
$K(1420) \to K\pi$	$\tfrac{3}{\sqrt{10}} A_s^8$	49.5 ± 10	50	49.5 ± 10	27
$K(1420) \to K\eta$	$\tfrac{1}{\sqrt{10}} A_s^8$	0.1 ± 9	1.6	0.1 ± 9	0.8
$f(1250) \to \pi\pi$	$\sqrt{3}\left\{\sqrt{\tfrac{1}{8}} A_1 \cos\theta_2 - \sqrt{\tfrac{1}{5}} A_s^8 \sin\theta_2\right\}$	137.8 ± 24	143	137.8 ± 24	137
$f(1250) \to K\bar{K}$	$1\left\{\tfrac{1}{\sqrt{2}} A_1 \cos\theta_2 + \sqrt{\tfrac{1}{5}} A_s^8 \sin\theta_2\right\}$	7.3 ± 15	7	7.3 ± 15	9
$f(1250) \to \eta\eta$	$-1\left\{\sqrt{\tfrac{1}{8}} A_1 \cos\theta_2 + \sqrt{\tfrac{1}{5}} A_s^8 \sin\theta_2\right\}$	---	0.3	---	0.5
$f(1500) \to \pi\pi$	$-\sqrt{3}\left\{\sqrt{\tfrac{1}{8}} A_1 \sin\theta_2 + \sqrt{\tfrac{1}{5}} A_s^8 \cos\theta_2\right\}$	0.1 ± 0.01	---	0.1 ± 7	6
$f(1500) \to K\bar{K}$	$1\left\{-\tfrac{1}{\sqrt{2}} A_1 \sin\theta_2 + \sqrt{\tfrac{1}{5}} A_s^8 \cos\theta_2\right\}$	69 ± 25	45	69 ± 25	40
$f(1500) \to \eta\eta$	$1\left\{\sqrt{\tfrac{1}{8}} A_1 \sin\theta_2 - \sqrt{\tfrac{1}{5}} A_s^8 \cos\theta_2\right\}$	0.1 ± 7	0.01	0.1 ± 7	0.01
		$\chi^2 = 1$		$\chi^2 = 14$	
		NC = 5		NC = 6	

Table XI

SU(3) Predicted and Experimental Decay Rates for 2^+ Mesons

$$J^P = 0^+$$

(a)	(b)			
m_K^2	$\dfrac{3m_s^2 + m_8^2}{4}$	$\dfrac{a-b}{(a+b)/2}$	$|\theta_o^+|$	$m_K^2 - m_\delta^2$
1.17	1.03	0.06	$40°$	0.24

$0^+ \to 0^- 0^-$		$\Gamma_{\text{exp'tl}}$ (MeV)	$\Gamma_{SU(3)}$ (MeV)
$\delta(960) \to \pi\eta$	$\dfrac{\sqrt{10}}{5} A_s^8 \cos\theta_o$	50±25	64
$K(1080) \to K\pi$	$\dfrac{3}{\sqrt{10}} A_s^8$	---	173
$K(1080) \to K\eta$	$-\dfrac{\sqrt{5}}{10} A_s^8 \cos\theta_o$	---	6
$S(1030) \to \pi\pi$	$3\{\sqrt{\dfrac{1}{8}} A_1 \cos\theta_o^+ - \sqrt{\dfrac{1}{5}} A_s^8 \sin\theta_o^+\}$	42±21	32
$S(1030) \to KK$	$1\{\dfrac{1}{\sqrt{2}} A_1 \cos\theta_o^+ + \sqrt{\dfrac{1}{5}} A_s^8 \sin\theta_o^+\}$	28±16	13
$S(1170) \to \pi\pi$	$-\sqrt{3}\{\sqrt{\dfrac{1}{8}} A_1 \sin\theta_o^+ + \sqrt{\dfrac{1}{5}} A_s^8 \cos\theta_o^+\}$	---	123
$S(1170) \to KK$	$1\{-\dfrac{1}{\sqrt{2}} A_1 \sin\theta_o^+ + \sqrt{\dfrac{1}{5}} A_s^8 \cos\theta_o^+\}$	---	6

$$\chi^2 = 1.4$$
$$NC = 1$$

Table XII

SU(3) Predicted and Experimental Decay Rates for 0^+ Mesons discussed in the literature; the main points I wish to emphasize are 1) SU(3) requires a nonet structure for the 0^- multiplet since the $\eta \to \gamma\gamma/\pi^o = \gamma\gamma$ ratio is observed to be ~ 150 and U spin consideration for an octet set an upper limit of 20 for this ratio. Inclusion of the $\eta(960)$ adds another parameter which can accommodate the larger ratio and predict the $\eta(960) \to \gamma\gamma$ rate as indicated, 2) the 1^- nonet is unchanged, some explanation is still needed to explain the low $\phi \to \rho\pi$ rate, which in SU(3) occurs with $A_1 \simeq A_s^8$, and 3) the SU(3) fit to the A_2 nonet is rather good for either split or unsplit A_2. This is due to the rather poor knowledge of the K(1420) branching ratios, the critical number being the $K(1420) \to K(890)\pi$ partial width.

The possible $J^P = 0^+$ nonet is pure speculation. It includes the $\delta(960)$, $\delta(1030)$, and a possible K(1080), thereby predicting an S(1170) (lying underneath the $f^o(1250)$) with a mixing angle of 40°, all within the realm of possibility and of some benefit in describing the $f^o(1250)$ decay distribution which always requires an s-wave background.

We can summarize the situation among the boson spectra as follows:

Only 1, 8 representations are observed. Searches for members of $\overline{10}$ or 27 representations such as $(\pi^+\pi^+)$, $(K^+\pi^+)$, and (K^+K^+) resonances have been conducted with negative results. Upper limits of ~20 μb have been set for the production of such states.

The detected SU(3) families have J^{PC} values in accordance with the simple quark model, i.e. definite examples of 0^{--}, 0^{+-}, 1^{-+}, etc., not allowed by such a model, have not been found.

The A_2 situation is critical and unclear. The most popular picture is that it is split and both halves are J^{PC} 2^{++}.

The status of the $A_1(1100)$ is firmer now than a year ago with $m(A_1) \simeq \sqrt{2}\, m(\rho)$.

The ρ, A_2, and g trajectories are degenerate as is the conjectured π, β, and A_3 trajectories.

The masses and decay rates for the 0^-, 1^-, and 2^+ nonets are fit rather well by $SU(3)$.

The cross section for producing 0^- states is greater than producing 1^- states is greater than producing 2^+ states

$$\sigma(0^-) > \sigma(1^-) > \sigma(2^+) \quad .$$

There are numerous additional states expected, namely those with

$$I = 0 \qquad M = 1600\text{-}1700 \text{ MeV}$$
$$I = 1/2 \qquad M = 1700\text{-}1800 \text{ MeV}$$

Baryons

The search for baryon resonances has been conducted by three different types of techniques, namely 1) formation experiments, 2) production experiments, and 3) amplitude analysis techniques. Each method has proved quite fruitful and in general they tend to confirm and complement each other. In formation experiments the incoming energy of the bombarding particle is varied and the total cross section measured. Such an example is shown in Fig. 8 for the case of the π-nucleon interaction where the $I = 1/2$ and $I = 3/2$ separation has already been made. Numerous resonances are clearly evident, decreasing in intensity as their mass is increased. This method is limited to the search for N, Δ, Λ, and Σ resonances since only π's and K's are available as projectiles and p, n as targets. Production experiments are those in which multiple particles occur in the final state and one investigates the effective mass M, of a subset of such particles $M^2 = \Sigma E_i^2 - \Sigma \vec{p}_i^2$. This technique is especially useful for finding Ξ's and Ω's and also for unveiling Λ and Σ states that are weakly coupled to the $\bar{K}N$ system and as such are not prominently seen in formation experiments. Such an example of two Σ states is shown in Fig. 9 and possible Ξ states in Fig. 15 (to be discussed later). The amplitude analysis technique relies on angular distribution and polarization information in evaluating the variation of amplitudes in the particular partial waves contributing to the processes in question. For instance, the most extensive work has been done in the study of π-nucleon scattering, examining $\pi^+ p$, $\pi^- p$, and $\pi^- p$ charge exchange angular distribution as well as the recoil proton polarization in the former two reactions. The complex amplitudes are then derived first at low energies, where only one or two waves predominate, and energy independent values are then also derived at all energies where data are available. The number of solutions as well as amplitudes are quite numerous, and they are all joined by some smoothness criterion. The subsequent complex amplitudes are then displayed on an Argand diagram, as shown in Fig. 10 for the πN and $\bar{K}N$ reactions. The question is then, what is a resonance? The general requirement is that the amplitude traverse the complex plane counterclockwise in a circular fashion, with the mass of the resonance being that point where the amplitude reaches its maximum values or where its rate of change with respect to energy variation is the fastest. These are indicated on the same figure. It should

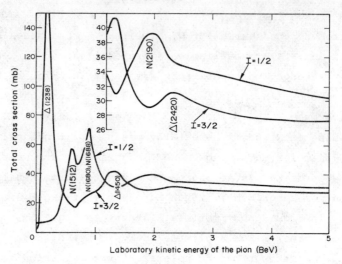

Figure 8. Resonance structure in the low energy π-N cross-sections. A. N. Diddens et al., Phys. Rev. Letters **10**, 262 (1963).

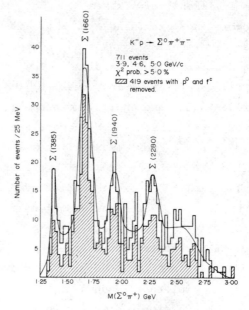

Figure 9. Resonance structure in the low energy π Σ missing mass. V. E. Barnes et al., Phys. Rev. Letters **22**, 479 (1969)

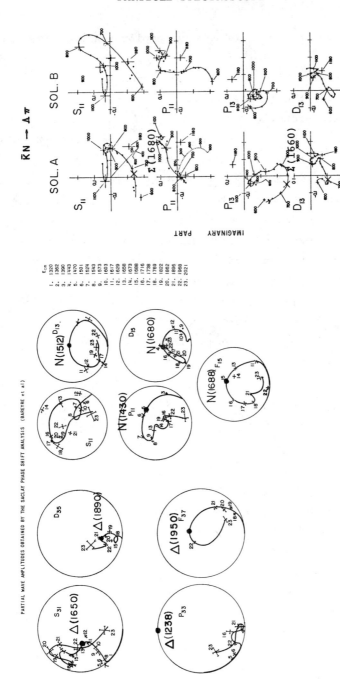

Figure 10. Argand diagrams for πN (Bareyre et al. - Saclay) and $\bar{K}N$ (CHS collaboration) reactions.

be emphasized that this is a very complicated procedure involving hundreds of solutions, and complex minimizing computer programs; however four groups obtain essentially similar solutions with the quoted masses and widths varying somewhat from group to group.

A list of N, Δ, Λ, Σ, and Ξ resonances found by these methods is shown in Table XIII. Those which are underlined are resonances which have yet to be confirmed. They either are small effects in the amplitude analysis, not yet observed by either of the other two techniques, or lack firm statistical validity by themselves and should be observed in another experiment before being accepted. The spin parities of many states are well known, those which are in doubt or have yet to be measured are noted in parenthesis. The particular values quoted are either most likely values or those expected on an SU(3) basis (to be discussed shortly). From this tabulation, one can construct numerous multiplets, namely

$$J^P = 1/2^+ \ (8), \quad 3/2^- \ (9), \quad 5/2^+ \ (8), \quad 7/2^- \ (9)$$
$$1/2^- \ (9)?, \quad 3/2^+ \ (10), \quad 5/2^- \ (8), \quad 7/2^+ \ (10)$$

where the number in parenthesis indicates the representation. One notes that this catalogue utilizes nearly all the accepted resonant states listed in Table XIII.

It seems worthwhile to investigate how well the baryon spectra agree with SU(3), to the same degree of sensitivity as was done for the bosons, mainly with respect to the Gell-Mann--Okubo mass formula as well as with respect to the decay rates. This has been done for the above multiplets, and the results are displayed in Tables XIV through XXI. A more detailed discussion of this comparison can be found in the

PARTICLE SPECTROSCOPY 125

N	Δ	Λ	Σ	Ξ
N(939) $1/2^+$		Λ(1115) $1/2^+$		
N(1470) $1/2^+$	Δ(1238) $3/2^+$	Λ(1405) $1/2^-$	Σ(1190) $1/2^+$	Ξ(1320) $1/2^+$
N(1518) $3/2^-$				
N(1530) $1/2^-$		Λ(1520) $3/2^-$		
N(1680) $5/2^-$	Δ(1650) $1/2^-$	Λ(1670) $1/2^-$	Σ_1(1660) $3/2^-$	Ξ(1530) $3/2^+$
N(1688) $5/2^+$	Δ(1670) $3/2^-$	Λ(1690) $3/2^-$	Σ_2(1660) $3/2^-$	Ξ(1640)
N(1700) $1/2^-$			Σ(1750) $1/2^-$	
N(1780) $1/2^+$			Σ(1765) $5/2^-$	
N(1860) $3/2^+$	Δ(1890) $5/2^+$	Λ(1815) $5/2^+$	Σ(1940) $(5/2^+)$	Ξ(1820) $(3/2^-)$
N(1990) $7/2^+$	Δ(1910) $1/2^+$	Λ(1830) $5/2^-$		Ξ(1930) $(5/2^-)$
N(2040) $3/2^-$	Δ(1950) $7/2^+$		Σ(2020) $7/2^+$	Ξ(2030) $(5/2^+)$
N(2190) $7/2^-$		Λ(2100) $(7/2^-)$	Σ(2250) $(7/2^-)$	
		Λ(2350) $(7/2^-)$		
	Δ(2420) $11/2^+$			Ξ(2430) $(7/2^-)$

Table XIII

The Observed Mass Spin Parity Spectrum of the Baryon

$$J^P = 1/2^+$$

$\dfrac{m_N + m_\Xi}{2}$	$\dfrac{3m_\Lambda + m_\Sigma}{4}$	$\dfrac{a-b}{(a+b)/2}$
1127.2	1134.8	0.007

```
N    939.6
Λ°   1115.6
Σ°   1192.5
Ξ°   1314.7
```

	Kim	Martin/Sakitt
$\dfrac{g^2_{\Lambda K^- p}}{4\pi}$	13.5 ± 2.1	5.2 ± 1.9
$\dfrac{g^2_{\Sigma K^- p}}{4\pi}$	0.2 ± 1	2.1 ± 3
α	0.39 ± 0.05	$0.23 \pm 0.08 \rightarrow 0.3 < \alpha < 0.7$

$$\dfrac{g^2_{NN\eta}}{4\pi} < .01 \quad \rightarrow \quad = .25$$

Table XIV

SU(3) Predicted and Experimental Decay Rates for $1/2^+$ Baryons

PARTICLE SPECTROSCOPY   127

$$J^P = 3/2^-$$

(a) $\dfrac{m_N + m_\Xi}{2}$	(b) $\dfrac{3m_\Lambda + m_\Sigma}{4}$	$\dfrac{a-b}{(a+b)/2}$	$|\theta|$	α
1667.5	1682.5	0.009	$23°$	0.78

		$\Gamma_{exp'tl}$ (MeV)	$\Gamma_{SU(3)}$ (MeV)
N(1515)	→ Nπ	57.5 ± 15	46
	→ Nη	Small	0.06
Λ_1(1520)	→ N$\bar{K}$	7 ± 1.5	6
	→ $\Sigma\pi$	6 ± 1.5	7
Λ_8(1690)	→ N$\bar{K}$	10 ± 3	11
	→ $\Sigma\pi$	14 ± 5	20
	→ $\Lambda\eta$	---	0.01
Σ(1660)	→ $\Sigma\pi$	24.5 ± 9	30
	→ $\Lambda\pi$	Small	1
	→ N$\bar{K}$	4.5 ± 1.5	5
Ξ(1820)	→ $\Lambda\bar{K}$	Detected	10
	→ $\Xi\pi$	"	7
	→ $\Sigma\bar{K}$	"	7

$$\chi^2 = 3.2$$

$$NC = 4$$

Table XV

SU(3) Predicted and Experimental Decay Rates for $3/2^-$ Baryons

$$J^P = 5/2^+$$

(a) $\frac{m_N + m_\Xi}{2}$	(b) $\frac{3m_\Lambda + m_\Sigma}{4}$	$\frac{a - b}{(a + b)/2}$	$|\theta|$	α
1859	1846	0.007	---	0.60

		$\Gamma_{exp'tl}$ (MeV)	$\Gamma_{SU(3)}$ (MeV)
N(1688)	$\to N\pi$	75 ± 17	73
	$\to N\eta$	Small	1
$\Lambda(1815)$	$\to N\bar{K}$	47 ± 5	48
	$\to \Sigma\pi$	8 ± 3.6	6
	$\to \Lambda\eta$	Small	0.1
$\Sigma(1940)$	$\to \Sigma\pi$	60 ± 35	79
	$\to \Lambda\pi$	Small	10
	$\to N\bar{K}$	10 ± 5	3.5
	$\to \Sigma\eta$	---	0.4
$\Xi(2030)$	$\to \Lambda\bar{K}$	17.5 ± 11.5	16
	$\to \Xi\pi$	Small	2.5
	$\to \Sigma\bar{K}$	52.5 ± 20	36
	$\to \Xi\eta$	---	1.4

$$\chi^2 = 3$$
$$NC = 5$$

Table XVI

SU(3) Predicted and Experimental Decay rates for $5/2^+$ Baryons

$J^P = 7/2^-$

(a) $\frac{m_N + m_\Xi}{2}$	(b) $\frac{3m_\Lambda + m_\Sigma}{4}$	$\frac{a-b}{(a+b)/2}$	$|\theta|$	α
2310	2333	0.010	$23°$	0.85

	$\Gamma_{exp'tl}$ (MeV)	$\Gamma_{SU(3)}$ (MeV)
$N(2190) \to N\pi$	60 ± 28	57
$\to N\eta$	---	13
$\to \Lambda K$	---	6
$\to \Sigma K$	---	2
$\Lambda_1(2100) \to N\bar{K}$	42 ± 10	45
$\to \Sigma\pi$	---	19
$\to \Lambda\eta$	---	2
$\Lambda_8(2350) \to N\bar{K}$	22.5 ± 9	25
$\to \Sigma\pi$	---	32
$\to \Lambda\eta$	---	2
$\to \Xi K$	---	16
$\Sigma(2280) \to N\bar{K}$	18 ± 13	15
$\to \Sigma\pi$	Large	53
$\to \Lambda\pi$	---	0.4
$\to \Sigma\eta$	---	0.08
$\to \Xi K$	---	1.3
$\Xi(2430) \to \Lambda\bar{K}$	75 ± 39	25
$\to \Xi\pi$	---	14
$\to \Sigma\bar{K}$	75 ± 39	25
$\to \Xi\eta$	---	6

$\chi^2 = 3.5$
NC = 3

Table XVII

SU(3) Predicted and Experimental Decay rates for $7/2^-$ Baryons

$$J^P = 1/2^-$$

(a) $\dfrac{m_N + m_\Xi}{2}$	(b) $\dfrac{3m_\Lambda + m_\Sigma}{4}$	$\dfrac{a - b}{(a+b)/2}$	$|\theta|$	α
1675	1697.5	0.013	$20°$	-0.7

	$\Gamma_{exp'tl}$ (MeV)	$\Gamma_{SU(3)}$ (MeV)
$N(1525) \to N\pi$	27 ± 19	9
$\to N\eta$	52 ± 38	5
$\Lambda_1(1405) \to \Sigma\pi$	38 ± 6	35
$\Lambda_8(1670) \to N\bar{K}$	2.6 ± 1	3
$\to \Sigma\pi$	12 ± 4	5
$\to \Lambda\eta$	9 ± 4	3
$\Sigma(1780) \to N\bar{K}$	18 ± 10	32
$\to \Lambda\pi$	18 ± 10	11
$\to \Sigma\eta$	12 ± 21	4
$\to \Sigma\pi$	---	10
$\Xi(1825) \to \Xi\pi$	---	37
$\to \Lambda\bar{K}$	---	10
$\to \Sigma\bar{K}$	---	5

$$\chi^2 = 10$$
$$NC = 6$$

Table XVIII

SU(3) Predicted and Experimental Decay Rates for $1/2^-$ Baryons

$J^P = 3/2^+$

$\Omega^- - \Xi^-$	$\Xi^- - \Sigma^-$	$\Sigma^- - \Delta^-$
140 ± 2.5	145.8 ± 4	146.7 ± 6

	$\Gamma_{exp'tl}$ (MeV)	$\Gamma_{SU(3)}$ (MeV)
$\Delta(1238) \rightarrow N\pi$	120 ± 5	116
$\Sigma(1385) \rightarrow \Lambda\pi$	32.5 ± 5.5	37
$\rightarrow \Sigma\pi$	3.6 ± 1.5	3.8
$\Xi(1530) \rightarrow \Xi\pi$	7.5 ± 3	13
$\Omega(1673) \rightarrow \Xi\pi$	---	---
weak ΛK	---	---

$\chi^2 = 9$
NC = 3

Table XIX
SU(3) Predicted and Experimental Decay Rates for $3/2^+$ baryons

$$J^P = 5/2^-$$

(a) $\frac{m_N + m_\Xi}{2}$	(b) $\frac{3m_\Lambda + m_\Sigma}{4}$	$\frac{a-b}{(a+b)/2}$	$|\theta|$	α
1803	1814	0.006	--- (16°)	-0.15 (-0.30)

	$\Gamma_{exp'tl}$ (MeV)	$\Gamma_{SU(3)}$ (MeV)	$\Gamma_{SU(3)}$ (MeV)
$N(1675) \to N\pi$	65 ± 20	45	45
$\to N\eta$	Small	1.6	3.1
$(\Lambda_1[1600 - 1800]) \to N\bar{K}$	---	---	11
$\to \Sigma\pi$	---	---	32
$\Lambda_8(1830) \to N\bar{K}$	8 ± 2	4	8
$\to \Sigma\pi$	32 ± 9	46.	29
$\to \Lambda\eta$	Small	2.6	5
$\Sigma(1765) \to N\bar{K}$	45 ± 7	26	40
$\to \Sigma\pi$	Small	1	4
$\to \Lambda\pi$	15 ± 3	16	21
$\Xi(1930) \to \Lambda\bar{K}$		6	12
$\to \Xi\pi$	80 ± 40	36	54
$\to \Sigma\bar{K}$		11	11

$\chi^2 = 14$ $\chi^2 = 5.7$
NC = 3   NC = 3

Table XX

SU(3) Predicted and Experimental Decay Rates for $5/2^-$ Baryons

$J^P = 7/2^+$

$\Sigma - \Delta$

120 ± 20 MeV

	$\Gamma_{exp'tl}$(MeV) (8 ⊗ 8)	$\Gamma_{exp'tl}$(MeV) (8 ⊗ 10)	$\Gamma_{SU(3)}$(MeV) (8 ⊗ 8)	$\Gamma_{SU(3)}$(MeV) (8 ⊗ 10)
$\Delta(1900) \to N\pi$	80 ± 36		80	
$\to \Sigma K$	Small		2	
$\to \Delta(1238)\pi$		80 ± 36		80
$\to \Sigma(1385)K$		3 ± 2		0.12
$\to \Delta(1238)\eta$		---		1.2
$\Sigma(2020) \to \bar{N}K$	15.5 ± 4.5		18	
$\to \Lambda\pi$	45.5 ± 15		27	
$\to \Sigma\pi$	---		11	
$\to \Sigma\eta$	---		2.5	
$\to \Sigma(1385)\pi$		---		26
$\to \Delta(1238)K$		---		17
$\Xi(2140) \to \Lambda K$	---		17	
$\to \Xi\pi$	---		14	
$\to \Sigma K$	---		10	
$\to \Xi\eta$	---		1.5	
$\to \Xi(1530)\pi$		---		5.7
$\to \Sigma(1385)K$		---		2.9
$\Omega(2260) \to \Xi K$	---		29	
$\to \Xi(1530)K$		---		5

$\chi^2 = 2$ $\chi^2 = 1.7$
NC = 2   NC = 1

Table XXI

SU(3) Predicted and Experimental Decay Rates for $7/2^+$ Baryons

forthcoming RMP article by Flaminio et al. As before, singlet-octet mixing is taken into account but octet-octet mixing, which can occur in principle, is ignored since the data does not warrant its inclusion. The formalism adopted has been shown in Tables VII and VIII. In the most complex case, nonets, there are three unknowns A_1, A_s, and A_a to be determined with 5 to 10 decay rates. With the indicated spin-parity assignments, the agreement with SU(3) is uniformly good. In the few cases where the spin parity of all members of a multiplet are reasonably well known, the mass formula works well, and in the more numerous instances where the spin parity is assigned on the basis of the mass formula, the experimental SU(3) decay rate fits agree. The exceptions are the $1/2^-$ and $5/2^-$ multiplets. In the former it is difficult to accommodate a predominent $N \to N\eta$ decay mode and in the $5/2^-$ case the difficulty occurs because of the large $\Lambda(1850) \to \Sigma\pi$ and $\Sigma(1765) \to N\bar{K}$ modes. This can be corrected by invoking another Λ state with mass 1600-1800 MeV which mainly decays via the $\Sigma\pi$ mode. One final word is in order on the well established $3/2^+$ decimet. The slightly poor fit results from the discrepancy in the $\Xi(1530)$ width which is experimentally 7.5 ± 3.0 MeV and SU(3) demands $\simeq 13$ MeV. There are however some more recent measurements, unpublished, which indicate values larger than the 7.5 MeV. As such, I don't consider this to be a major discrepancy.

A partial summary of these baryon multiplets is shown in Table XXII where the J^P, SU(3) representation, mixing angle θ, and α [the $F/(D + F)$ ratio] are displayed. In addition, the Chew-Frautschi plots for these families are presented in Figs. 11 and 12. One notes certain regularities for the $1/2^+$, $3/2^-$, etc. series, namely $\alpha \simeq 0.6$ with the $5/2^+$ being a Regge recurrence of the $1/2^+$ and the $7/2^-$ a Regge recurrence of the $3/2^-$.

PARTICLE SPECTROSCOPY   135

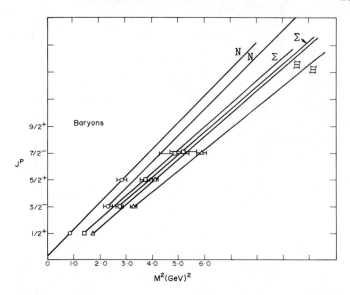

Figure 11. Chew Frautschi Plot for Baryon Regge recurrances.

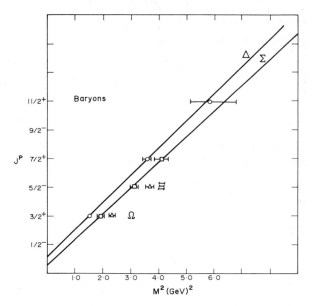

Figure 12. Chew Frautschi Plot for Δ and Σ trajectories and other states.

The slopes of the N, Σ, and Ξ trajectories for these multiplets are ≃ 1 but they are not exchange degenerate. As noted earlier, the $1/2^-$, $3/2^+$, etc. series are not as clear cut, the only reputable family being the $3/2^+$ decimet. The preferred α's are negative, either -.15 or -.7 with there being some evidence for Δ Regge recurrences ($3/2^+$, $7/2^+$, $11/2^+$), and the Δ and Σ trajectory slopes being ≃ 1.

New Developments

In the study of baryon resonances there have been several interesting new findings. These have occurred in S = 0, -1, -2, and +1 states, and I will discuss them in that order. Among the nucleon states, two low spin and relatively high mass states were uncovered by the amplitude analysis technique. The first of these was the so-called Roper resonance, N(1470), $J^P = 1/2^+$ and later an N(1700) with $J^P = 1/2^+$ or $1/2^-$. The existence of two states with the above masses has now been observed in production experiments. This is exhibited in Fig. 13 where prominent peaks are observed in (pπ) and (pππ) effective mass combinations at these two mass values. Due to the large background under these peaks, it is difficult to make a spin determination, however all the distributions are flat, consistent with the spin values, 1/2, derived in the amplitude analysis technique. One now has a situation where there is evidence for possibly more than one multiplet with the same spin parity, $1/2^+$, which somehow must be accommodated in any symmetry scheme.

Since its discovery, the Σ(1660) has been an enigma. Its decay branching ratio into Σππ and Σπ modes has varied from experiment to experiment as well as its spin determination from either modes. There is now general agreement on the spin of

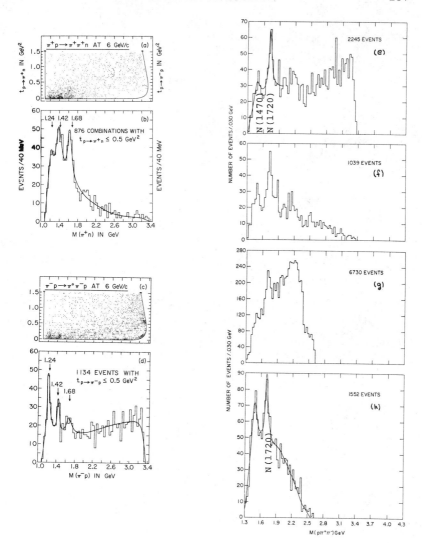

Figure 13. Evidence for Roper, N(1470), and N(1720) in production experiments; (a) - (d) show the resonance in a (Nπ) final state and (e) - (h) in (Nππ) final state. (e) and (f) are at 7.3 GeV/c and (g) and (h) at 3.9-4.6 GeV/c. (f) and (h) have angle cuts. From R. B. Bell et al. Phys. Rev. Letters 20, 164(1968) and V. E. Barnes et al. Phys. Rev. Letters 22, 200(1969).

$3/2^-$ from both formulation experiments examining the $\Sigma\pi$ mode and production experiments on $\Sigma\pi\pi$ modes. The branching ratios were still troublesome until the demonstration of the existence of two $\Sigma(1660)$'s by the LRL group (Eberhard et al.). This is shown in Fig. 14 where the $\Sigma\pi$ and $\Sigma\pi\pi$ production angular distributions are seen to be radically different. This has since been confirmed by a BNL experiment where again the $\Sigma(1660) \to \Sigma\pi\pi$ is produced more peripherally than the $\Sigma(1660) \to \Sigma\pi$. So, there is good evidence for two $\Sigma(1660)$'s with $J^P = 3/2^-$.

Cascade states are necessary members of all 8 and 10 representations. As indicated earlier, quite a few such states have been uncovered: $\Xi(1320)$, $\Xi(1530)$, $\Xi(1820)$, $\Xi(1930)$, $\Xi(2030)$, $\Xi(2430)$. Recently a Brandeis, Maryland, Syracuse, Tufts collaboration has presented evidence for another such resonance with a mass $\simeq$ 1640 MeV. This is shown in Fig. 15. The main observed decay mode is ($\Xi\pi$), and its other properties are unknown.

J^P		α	θ
$1/2^+$	8	0.3-0.4	--
$3/2^-$	9	0.78	23°
$5/2^+$	8	0.60	--
$7/2^-$	9	0.85	23°
$1/2^-$	(9)	(-0.7)	(20°)
$3/2^+$	10	--	--
$5/2^-$	8	-0.15	--
$7/2^+$	10?	--	--

Table XXII

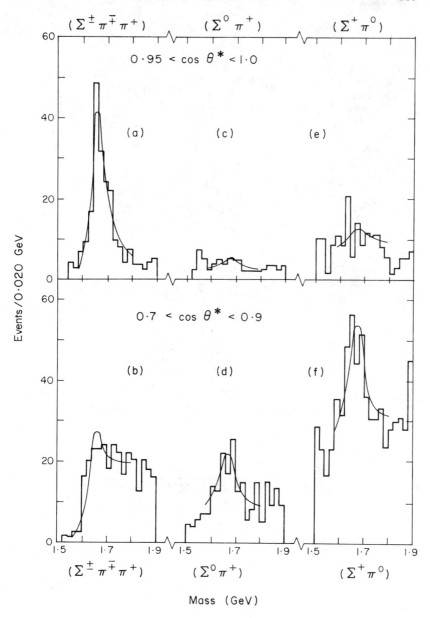

Figure 14. Angular distributions showing evidence for two Σ(1660)'s. P. Eberhard et al., Phys. Rev. Letters 22, 200 (1969).

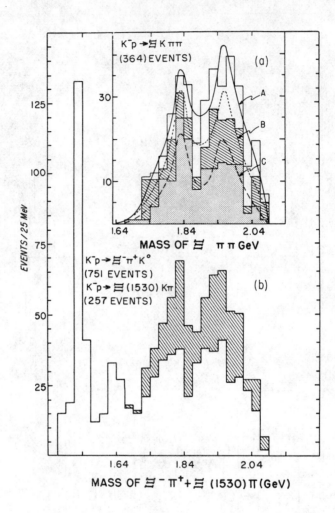

Figure 15. Evidence for a Ξ resonance of mass ≃ 1640 GeV. S. Apsell et al., Phys. Rev. Letters **24** 777(1970).

PARTICLE SPECTROSCOPY   141

If this resonance does indeed exist, it is interesting to
associate it with one of the Σ(1660)'s and therefore with a
possible Δ(1670) noted earlier in Table XIII, all with spin
$3/2^-$. Such a decimet would then require an Ω(1670), in other
words, two Ω's would then be expected to exist with roughly
the same mass, both decaying weakly - quite a speculation!

The final topic that I wish to discuss is the possible
existence of Z*'s, baryon resonances with Y = 2, I = 0, 1
which cannot be accommodated in either an 8 or 10 representation but require either $\overline{10}$ or 27 representation. The total
cross section measurements for K^+d and K^+p remain unchanged.
These are shown in Fig. 16 where the evidence for an I = 1
state with mass $\simeq$ 1900 MeV is seen in the K^+p data, and two
possible resonances in the I = 0 state are also visible.
The latter two resonances have the following mass and width,
M_1 = 1780 MeV and Γ_1 = 500 MeV and M_2 = 1870 MeV and Γ_2 = 160
MeV. The I = 1 bump is associated with the rapid rise of
single pion production via the KΔ final state, while the second of the I = 0 bumps is associated with the rapid rise of
the inelastic channel but this time K*(890)p. The new information involves measurements of the proton polarization in the
K^+p elastic scattering channel as well as detailed measurements
of the K^+p angular distributions. The polarization work has
been conducted by groups at Argonne-Northwestern, Yale, and
CERN. These groups have each utilized the available angular
distribution data as well as each others polarization data in
performing amplitude analysis. Using an energy independent
approach, they each found numerous solutions, however, only
one set joins smoothly onto values at lower energies. The
results of the Yale and Argonne groups are shown in the last
figure. The main feature is a common $P_{3/2}$ solution which

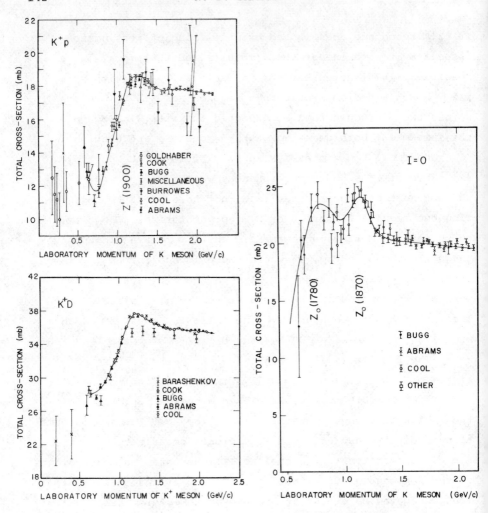

Figure 16. Total cross-section measurements of Kp and KD reactions showing indications of Z*'s. R. J. Abrams et al. Physics Letters 30B, 564. (1969).

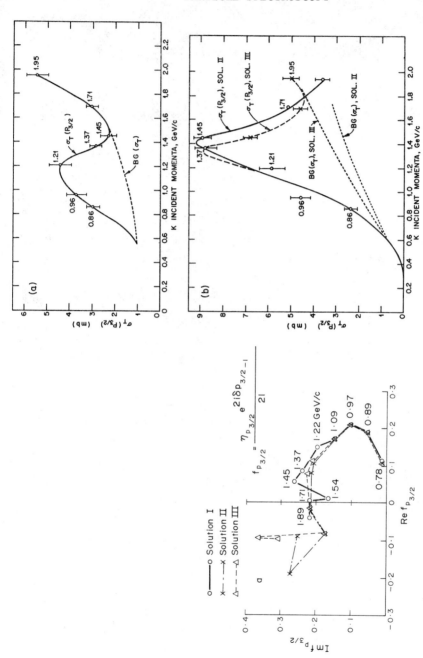

Figure 17. Evidence for Z* resonance. S. Kato et al. Phys. Rev. Letters 24, 615 (1970).

appears to traverse the Argand diagram in a counterclockwise direction. The problem is the behavior of the $P_{3/2}$ amplitude at the higher K^+ momenta > 1.7 GeV/c. Does it continue smoothly across the imaginary axis or does it execute violent excursion? Several experiments are underway at higher momenta to determine this behavior. Of similar importance is the detailed study of the lower mass I = 0 bump which is mainly elastic. There is certainly much to be done experimentally in this interesting area.

One can summarize the baryon spectra as follows: only 1, 8, and 10 representations are observed with the more recent uncovering of the Z*, indicating possible $\overline{10}$ and 27. This latter possibility has yet to be demonstrated.

The observed spectra are fit quite well by SU(3), both with respect to observed masses and decay rates. In this respect the $1/2^+$, $3/2^-$, $5/2^+$, $7/2^-$ series seem reasonably complete with the $1/2^-$ and $5/2^-$ multiplet containing some difficulties.

There exists some evidence for Regge recurrences but not exchange degeneracy among the N, Δ, Σ, and Ξ states examined.

Finally, there is good evidence for high mass, low spin states, such as the N(1470) which have to fit into any proposed symmetry schemes. It is clear that much has been accomplished in the fields of boson and baryon spectra in the last few years, however it is just as apparent that much still needs to be uncovered in determining the number and properties of boson states with masses $\leq$ 2 GeV and baryons with masses $\leq$ 3 GeV.

DISCUSSION OF DR. SAMIOS' TALK

G. Yodh, University of Maryland: I wanted to make a remark on the difficulties of spin-parity determinations--problems in the Ξ^* systems that apply to other channels also. We have tried to analyze all the data that the Brandeis, Maryland, Syracuse, Tufts collaboration has in three body and four body states to try to determine spins and parities of a number of bumps that we have. We can't even determine the spin and parity of the $(3/2)^+$ decuplet member, i.e., Ξ (1530) for which we have a signal which has about 300 events over a background of about 20 or 30. Then we looked at the previous determination of the $(3/2)^+$ as done by the UCLA group, and I think I can only say that the UCLA group was lucky in getting a $3/2^+$ answer at one low momentum. It seems that the power of the determining spin and parity of these resonances disappears rapidly above threshold.

The second remark I want to make is that the only information we have is from sequential decays, that is, when a higher resonance decays to a lower resonance with a spin greater than 1/2. In that case, I think the chances are much better, and I would say in general that the spin is greater than or equal to 3/2 in the 1820, 1960 region. But that is all one can say with a million pictures analyzed over a period of two and a half years.

One more comment. There is a new experiment in the Z^* game by the Arizona Group, Jenkins and co-workers, whose results were shown to us the other day in a colloquium where they extended K^+p and K^+d cross section measurements below 800 MeV/c down to about 300 MeV/c at the Bevatron. They find that their results join smoothly onto the data above that energy and there is just no indication of any other structure below that energy. It is quite smooth.

SAMIOS: I agree. The spin-parity of Ξ states are very difficult things to measure. For example the parity of the Ξ (1320) has not yet been measured. In fact one can turn it around. One might say the only way you'll find the spin parity of Ξ states is by measuring enough decay rates and using the SU(3) formulation.

Yodh: Yes and regarding decay rates the situation now is much more complicated. As you pointed out, with our data, masses come out to be displaced from the masses previously determined and different channels give different masses. So one may have many more resonances than comes to the eye and maybe none of the decay rates can be quantitatively estimated as of now.

W. Thirring, CERN: How does the situation look now with the classification in the 3 quark model with parastatistics and ℓ excitation if you put in the new cascade resonances which have been missing say at the time of the Vienna Conference?

Samios: I refer you to Greenberg's talk at the Lund Conference.

Yodh: Can you account for the Z* resonances?

Samios: For the Z* you always need the 10 or 27 representations. However, SU(3) accounts for all the other cascades, lambdas and sigmas.

E. P. Wigner, Princeton University: I have several questions which may reflect my lack of expertise, but I will ask them just the same. What is the definition of a resonance? By this I mean, supposing you would know exactly the collision matrix, every single element of it, and its energy dependence. When would you call it a resonance?

Samios: I was going to ask you that question. However, I can tell you what people do. They use various different definitions. For instance, if one does formation experiments, the resonant mass is taken as the peak of the cross section and the width is derived by a Breit-Wigner curve fitted to the observed enhancement, similarly for the bumps found in production experiments. In the case of amplitude analysis, people plot the complex amplitude and then the mass of the resonance is either the position where the amplitude varies most rapidly or where it has the maximum elasticity. As a result, the masses and widths disagree from group to group.

<u>Wigner</u>: I don't think that an ordinary theoretician knows. Because with your definition, let us say we have a single scattering. Even if the phase shift decreases rather than increases, you would get a maximum of the cross section. Now my friends tell me that it would be very wrong to consider that to be a resonance. But of course I don't know that there is a moral consideration in this connection. Let me ask the next question if you don't mind. You explain very beautifully the case of the Z* baryon which has no quark explanation. You mentioned in passing, the meson which you said also has no quark explanation. Could you repeat that over?

<u>Samios</u>: The lower half of the A_2 with $J^{PC} = 1^{-+}$

Wigner: Thank you very much. Then you answered also my third question, the A_2. I have one more question. You explained that the Gell-Mann Okubo mass formula fits so beautifully. Now this seems to indicate that the symmetry breaking interaction of the breaking operator is much smaller between multiplets than within a multiplet. Is that true, or what is true?

Samios: My comment is that it's amazing that it works so well when we know that the mass splitting is of the order of 20%.

Wigner: Yes, but why don't different multiplets influence each other and destroy this good agreement.

R. H. Dalitz, Oxford University: They have different quantum numbers.

J. Rosenman, University of Texas: Could you comment on the sign of the mass difference between the ρ^0 and the ρ^-.

Samios: I don't believe that is known.

Yodh: You made a statement about the exchange degeneracy of baryon trajectories. You said that exchange degeneracy was not too good. Does that apply to the lambda trajectory, also?

Samios: I didn't plot the lambda but it is certainly true of the nucleons, sigmas and cascades. They are parallel to each other but not as degenerate, for instance, as the boson trajectories.

Exact Results in the Analytic S-Matrix Theory
of Strong Interactions for High Energies*

Virendra Singh

Tata Institute of Fundamental Research, Bombay, India

Abstract: After reviewing the history of the development of upper bounds for the scattering amplitude, Professor Singh reviews recent work on new bounds derived from the usual constraints of unitarity and analyticity. This material is an expanded version of a talk presented at the Symposium "The Past Decade in Particle Theory."

1. Introduction

The analytic S-matrix theory has provided a convenient framework for a discussion of strongly interacting particles and their reactions.[1] We shall be here interested in the high energy bounds on scattering amplitudes which can be obtained within this framework. The first important result in this direction was obtained by Froissart,[2] at the beginning of the past decade (in 1961), who showed that the total cross-section

$$\sigma_{tot}(s) \underset{s\to\infty}{\leq} C \left[\ln\left(\frac{s}{s_0}\right)\right]^2 \qquad \ldots (1.1)$$

---

*Talk presented at the symposium "The Post Decade in Particle Theory" at the University of Texas, Austin, April 14-17, 1970 where $\sqrt{s}$ is the c.m. energy, i.e. there exist constants C, s_0, s_1 such that the upper bound (1.1) is true for all $s > s_1$. Since that time a considerable body of similar exact results have accumulated. Further the necessary assumptions underlying such results have been considerably sharpened. The present report shall review this progress. We shall further restrict ourselves to reactions of the type a+b→c+d where a,...,d are hadrons since the main progress has been made in the study of such reactions.

2. Basic Assumptions

Let us for simplicity first consider the spinless case. Let F(s,t) be the invariant scattering amplitude for an elastic reaction a+b→a+b where s and t are respectively the squares of c.m. energy and momentum transfer. The partial wave expansion of F(s,t) is given by

$$F(s,t) = \frac{\sqrt{s}}{k} \sum_{\ell=0}^{\infty} (2\ell+1) \, a_\ell(s) \, P_\ell(\cos\theta) \quad \ldots (2.1)$$

where $a_\ell(s)$ is the partial wave amplitude with angular momentum ℓ, k the magnitude of c.m. three-momentum and θ the angle of scattering in c.m. frame. For an elastic reaction $t = -2k^2(1-\cos\theta)$

The basic assumptions, necessary for deriving the asymptotic upper and lower bounds, can be formulated as follows:

1) <u>Unitarity</u>

This is the nonlinear constraint which fixes the scale of F(s,t). We note that it implies, in particular, the restrictions

$$\text{Im } F(s,0) = \frac{k\sqrt{s}}{4\pi} \sigma_{tot}(s) \qquad \ldots(2.2)$$

i.e. $\sigma_{tot}(s) = \frac{4\pi}{k} \sum_{\ell=0}^{\infty} (2\ell+1) \text{ Im } a_\ell(s)$

and $1 \geq \text{Im } a_\ell(s) \geq |a_\ell(s)|^2 \geq 0$ ...(2.3)

2) **Complex cos θ plane analyticity**

The amplitude $F(s,t)$ is analytic in the complex cos θ plane inside an ellipse with foci cos θ = +1, -1 and passing through the point given by $t = t_o$ where $\sqrt{t_o}$ is the lowest mass which can be exchanged in the t-channel i.e. $a\bar{a} \to b\bar{b}$. For $\pi\pi$, πN or KN scattering $\sqrt{t_o} = 2m_\pi$. We shall refer to this ellipse as Lehmann-Martin ellipse.

3) **Polynomial boundedness**

We assume
$$|F(s,t)| \underset{s\to\infty}{<} s^n$$
where N is some positive integer for t in the range $t_o > t \geq 0$.

4) **Complex s-plane analyticity**

The amplitude $F(s,t)$ for t fixed in some neighbourhood of the line segment $t_o > t \geq 0$ is analytic in complex s-plane except for cuts required by s-channel and u-channel unitarity.

5) **Crossing**

For the case of scattering of particles with spin one has more than one invariant amplitude but there is no essential complication and one can give a corresponding discussion. One can similarly discuss the case a+b→c+d with the unitarity relations (2.2), (2.3) appropriately modified.

3. High Energy Bounds on Total Cross-sections and for Fixed Positive $t<t_o$

Consider

$$A(s,t) \equiv \operatorname{Im} F(s,t) \qquad \ldots (3.1)$$

$$= \frac{\sqrt{s}}{k} \sum_{\ell=0}^{\infty} (2\ell+1) \operatorname{Im} a_\ell(s) P_\ell\left(1+\frac{t}{2k^2}\right) \qquad \ldots (3.2)$$

It was shown by Martin[3] that

$$A(s,t) \geq \frac{\sqrt{s}}{k}\left[\sum_{\ell=0}^{L-1} (2\ell+1) P_\ell\left(1+\frac{t}{2k^2}\right) + (2L+1)\eta P_L\left(1+\frac{t}{2k^2}\right)\right] \qquad \ldots (3.3)$$

for $t_o > t \geq 0$

where L, a positive integer, and $1 > \eta \geq 0$ are determined by

$$L^2 + (2L+1)\eta = \frac{k^2 \sigma_{tot}(s)}{4\pi} \qquad \ldots (3.4)$$

The lower bound (3.3) is true for all energies. For high energies and if $\sigma_{tot}(s) \to \infty$ as $s \to \infty$ the lower bound (3.3) reads as

$$A(s,t) \underset{s\to\infty}{>} \text{const. } s\left[\sigma_{tot}(s)\right]^{1/4} \exp\left[\sqrt{\frac{t\sigma_{tot}(s)}{4\pi}}\right] \qquad \ldots (3.5)$$

EXACT RESULTS FOR HIGH ENERGIES

for $t_o > t \geq 0$

This can be combined with Jin-Martin upper bound[4]

$$|F(s,t)| \underset{s\to\infty}{\lesssim} \text{const. } s^2 \qquad \ldots (3.6)$$

for $t_o > |t|$

to yield the following upper bound, due to Lukaszuk and Martin[5], on total cross-section

$$\sigma_{tot}(s) \underset{s\to\infty}{\lesssim} \frac{4\pi}{\tau_o} (\ln s)^2 \qquad \ldots (3.7)$$

where

$$\tau_o = t_o - \varepsilon \qquad \ldots (3.8)$$

and ε is an arbitrarily small positive quantity. It was shown by Mahoux and Martin that the upper bound (3.7) is also true for the general spin case[6].

In fact if we combine the lower bound (3.5) with the following result due to Jin and Martin[4]

$$\int^\infty \frac{A(s',t)\, ds}{s'^3} \quad \text{is convergent for } t_o > t \geq 0 \quad \ldots (3.9)$$

We obtain the stronger result[7]

$$\sigma_{tot}(s) \underset{s\to\infty}{\lesssim} \frac{4\pi}{\tau_o} \left[\ln\left(\frac{s}{(\ln s)^{3/2}}\right)\right]^2 \qquad \ldots (3.10)$$

Further improvements on the upper bound (3.7) are given by the chain of inequalities (4.4) given in the next section, where a lower bound on elastic cross-section $\sigma_{el}(s)$, and consequently on $\sigma_{tot}(s)$ as well, is also given.

Recently upper bounds, under the extra assumption that the d wave $\pi\pi$ scattering length α_2 is finite, on the quantity

$$\frac{(n+1)}{(s-4m_\pi^2)^{n+2}} \int_{4m_\pi^2}^\infty ds' \left\{s'-4m_\pi^2\right\}^n \sigma_{tot}^{\pi\pi}(s')ds' \quad (n=0,1,2\ldots)$$

which can be regarded as an average total $\pi\pi$ cross-section, have also been derived.[8] These involve a knowledge of α_2 but do not involve any arbitrary constants.

4. High Energy Bounds on the Forward Amplitude

We have the following asymptotic upper bound

$$|F(s,o)| \underset{s\to\infty}{\leq} \text{const. } s \, (\ln s)^2 \qquad \ldots(4.1)$$

due to Froissart[2,3,6,9]. This result has been considerably improved recently by Roy and Singh[10,11] who obtain

$$|F(s,o)| \underset{s\to\infty}{\leq} \sqrt{\frac{\sigma_{el}(s)}{16\pi\tau_o}} \; s \ln s \qquad \ldots(4.2)$$

The analogous result for the general spin case is given by[10,11]

$$\left(\frac{d\sigma_{el}}{dt}\right)_{t=0} \underset{s\to\infty}{\leq} \frac{\sigma_{el}(s)}{4\tau_o} (\ln s) \qquad \ldots(4.3)$$

where $\left\{d\sigma_{el}/dt\right\}_{t=0}$ is the elastic unpolarised differential cross-section. The following chain of inequalities follows from the basic result (4.2),

$$\sigma_{tot} \leq \frac{[\sigma_{tot}]^2}{\sigma_{el}} \leq \frac{[\sigma_{tot}]^2}{\sigma_{el}} \left[1+\left(\frac{\text{Re}F(s,o)}{\text{Im}F(s,o)}\right)^2\right] \underset{s\to\infty}{\leq} \frac{4\pi}{\tau_o}(\ln s)^2 . (4.4)$$

We also note that the bound

$$|F(s,0)| \underset{s\to\infty}{\leq} \frac{s}{2\tau_0} (\ln s)^2 \qquad \ldots(4.5)$$

also follows from (4.2). This removes the major arbitrariness in the bound (4.1).

A similar result can be established for the general reaction a+b→c+d also [11,12]. We have

$$\left(\frac{d\sigma_{ab\to cd}}{dt}\right)_{t=0} \underset{s\to\infty}{\leq} \frac{\sigma_{ab\to cd}(s)}{4\tau_0} (\ln s)^2 \qquad \ldots(4.6)$$

where

$$\sigma_{ab\to cd}(s) = \int \left(\frac{d\sigma_{ab\to cd}}{dt}\right) dt \qquad \ldots(4.7)$$

and the integral (4.7) involving $d\sigma_{ab\to cd}/dt$, the unpolarised differential cross-section for a+b→c+d, is over all those t-values which are physically allowed. The upper bound (4.6) again removes the major arbitrariness in the bound due to Logunov et al.[13]

$$\left(\frac{d\sigma_{ab\to cd}}{dt}\right)_{t=0} \underset{s\to\infty}{\leq} \text{const.} \left[\sigma_{ab\to cd}(s)\right] (\ln s)^2 \qquad \ldots(4.8)$$

It is also possible to establish a lower bound on $|F(s,0)|$. For complex s (Ims ≠ 0) we have [14]

$$|F(s,0)| \underset{s\to\infty}{\geq} \frac{\text{const.}}{|s|^2} \qquad \ldots(4.9)$$

This bound is also true, in an average sense, for real s. On combining this result with the upper bound (4.2) we obtain[7]

$$\sigma_{el}(s) \underset{s\to\infty}{\geq} \frac{\text{const}}{s(\ln s)^2} \qquad \ldots(4.10)$$

5. High Energy Particle-Antiparticle total Cross-section Differences and the Pomeranchuk Theorem

We will now discuss what one can say about particle-antiparticle total cross-section differences at high energies using only the general principles[12]. It was conjectured by Pomeranchuk[15] that at high energies the particle-antiparticle total cross-section difference on the same target would approach zero i.e. $\left[\sigma_{tot}^{AB}(s) - \sigma_{tot}^{\overline{A}B}(s)\right] \to 0$ as $s \to \infty$ (Pomeranchuk Theorem). In view of the Serpukhov measurements of total cross-sections[16], which seem to suggest that the Pomeranchuk theorem may not be obeyed, such an analysis is of considerable interest.

It has unfortunately not been possible to derive the Pomeranchuk theorem from the set of basic assumptions given in Section 2.[17] One normally has to make, for this purpose, the following extra assumption, called X,

$$(X) \qquad \frac{\text{Re}T(s,0)}{s \ln s} \xrightarrow[s\to\infty]{} 0$$

where $T(s,0) = F_{\text{Meson-Baryon}}(s,0) - F_{\text{Antimeson, Baryon}}(s,0)$.[18] This assumption X is questionable. If one really finds a violation of the Pomeranchuk theorem then this is the assumption which one would most easily give up. If we drop this assumption then we must get whatever handle we can get on the Re T(s,o) from the assumptions (1)-(5) listed before. The forward bounds (4.3) and (4.6) are very useful in this connection. We have

$$\left(\frac{d\sigma_{el}}{dt}\right)_{t=0} \leq \frac{\sigma_{el}}{4\tau_o} (\ln s)^2 \qquad \ldots (5.1)$$

and

$$\left(\frac{d\sigma_{ch.ex.}}{dt}\right)_{t=0} \leq \frac{\sigma_{ch.ex.} (\ln s)^2}{4\tau_o} \qquad \ldots (5.1')$$

The subscript ch.ex. denotes charge exchange. Let me mention that these upper bounds remain valid even if only the contribution of the helicity nonflip part is kept in evaluating σ_{el} or $\sigma_{ch.ex.}$. Now,

$$\left(\frac{d\sigma}{dt}\right) = \frac{\pi}{k^2 s} \left[(\text{ReF})^2 + (\text{ImF})^2\right] \cdot \qquad \ldots (5.2)$$

$$\left(\frac{d\sigma}{dt}\right)_{t=0} \geq \frac{\pi s}{k^2} \left(\frac{\text{ReF}(s,o)}{s}\right)^2 \cdot \qquad \ldots (5.3)$$

Combining the inequality (5.3) with (5.1) and (5.2) we obtain the required results

$$\left(\frac{\text{ReF}_{el}(s,o)}{s \ln s}\right)^2 \underset{s \to \infty}{\leq} \frac{\sigma_{el}(s)}{16 \pi \tau_o} \qquad \ldots (5.4)$$

and

$$\left(\frac{\text{Re}F_{\text{ch. ex.}}(s,o)}{s\ln s}\right)^2 \underset{s\to\infty}{\leq} \frac{\sigma_{\text{ch. ex.}}(s)}{16\pi\tau_o} \qquad \ldots(5.5)$$

In order to obtain results about asymptotic total cross-section differences we have to use the forward dispersion relations. The basic assumptions (1)-(5) allow us to write the following subtracted forward dispersion relation

$$\left(\frac{\text{Re}T(s,o)}{E}\right) = \left(\frac{\text{Re}T(s,o)}{E}\right)_{E^2=E_1^2} + \frac{E^2-E_1^2}{\pi} \times$$

$$\int_{M_P}^{\infty} dE' \left(\frac{M\sqrt{E'^2-M_P^2}}{2\pi}\right) \frac{\left[\sigma_{\text{tot}}^{\bar{P}T}(E') - \sigma_{\text{tot}}^{PT}(E')\right]}{(E'^2-E_1^2)(E'^2-E^2)} \qquad \ldots(5.6)$$

where $E=(s-M_P^2-M_T^2)/2M_T$ is the lab. energy of the projectile P(mass M_P) on target T(mass M_T). We can use the forward dispersion relation (5.6) to evaluate $\text{Re}T(s,o)$ at high energies in terms of the asymptotic total cross-section difference $\left[\sigma_{\text{tot}}^{\bar{P}T}(E)-\sigma_{\text{tot}}^{PT}(E)\right]$. On combining this information with the upper bound (5.4) or (5.5) we get asymptotic upper bounds on total cross-section differences[12]. These are given below for the cases of πN and KN scattering.

Pion-Nucleon Scattering

Let $\Delta\sigma^{\pi p}(E) \equiv \sigma_{tot}^{\pi^- p}(E) - \sigma_{tot}^{\pi^+ p}(E)$

We then have the upper bound[12]

$$\left| \lim_{E\to\infty} \Delta\sigma^{\pi p}(E) \right| \leq \frac{\pi^{3/2}}{m_\pi \sqrt{2}} \left[\lim_{E\to\infty} \sigma^{\pi^- p \to \pi^0 n}(E) \right]^{1/2} \quad \ldots (5.7)$$

provided the required limits on (5.7) exist and asymptotic isospin invariance is good. Experimentally $\sigma_{ch.ex.}$ is very small at high energies. The experimental result[19] $\sigma_{ch.ex.} \leq 0.03$ mb at a lab. kinetic energy of 17.96 GeV makes the right-hand side of (5.7) less than 3.1 mb. Another improvement of about 30% will result if $\sigma_{ch.ex.}$ is replaced by $\sigma_{ch.ex.n.f.}$, the non-flip contribution. Further, $\sigma_{ch.ex.}$ may be even smaller at the Serpukhov energies. From the Serpukhov data, with the use of charge symmetry, $|\Delta\sigma^{\pi p}|$ 1.3 ± 9.7 mb. This upper bound is therefore quite restrictive.

A similar result not involving the assumption about isospin invariance is given by[12]

$$\left| \lim_{E\to\infty} \Delta\sigma^{\pi p} \right| \leq \frac{\pi^{3/2}}{2m_\pi} \left[\left\{ \lim_{E\to\infty} \sigma_{el}^{\pi^- p}(E) \right\}^{1/2} + \left\{ \lim_{E\to\infty} \sigma_{el}^{\pi^+ p}(E) \right\}^{1/2} \right] \quad \ldots (5.8)$$

provided the required limits in (5.8) exist.

These bounds make precise the feeling that if elastic or charge-exchange cross-sections are asymptotically zero then the Pomeranchuk theorem must be true. We also note that the assumption $\sigma_{ch.\ ex.} \to 0$ is weaker than $\frac{d\sigma}{dt}_{ch.ex.} \to 0$ in the sense that, consistent with the upper bound (5.1') one can have, for example, $\sigma_{ch.\ ex.} \sim \frac{1}{\ln s}$ and $\frac{d\sigma}{dt}_{t=0,\ ch.\ ex.} \sim \ln s$.

We now give a result for the case where $\lim_{E\to\infty} \Delta\sigma^{\pi p}$ does not exist. We have the upper bound[12]

$$\left| \lim_{E\to\infty} \frac{\Delta\sigma^{\pi p}}{\ln E} \right| \leq \frac{\pi^{3/2}}{m_\pi} \lim_{E\to\infty} \left[\sqrt{\frac{\sigma_{el}^{\pi^- p}(E)}{(\ln E)^2}} + \sqrt{\frac{\sigma_{el}^{\pi^+ p}(E)}{(\ln E)^2}} \right] \leq \frac{2\pi^2}{m_\pi^2} \quad \ldots (5.9)$$

provided $\Delta\sigma^{\pi p}$ is a monotonic function of E for sufficiently large E. This result is quite interesting from a theoretical point of view. It shows that while both $\sigma_{tot}^{\pi^- p}$ and $\sigma_{tot}^{\pi^+ p}$ can be increased asymptotically like $(\ln E)^2$ i.e. the Froissart bound, their difference can at most increase like $(\ln E)$.

Kaon-Nucleon Scattering

We now give similar results for KN scattering.[12] We have

$$\left| \lim_{E\to\infty} \left[\sigma_{tot}^{K^+ p}(E) - \sigma_{tot}^{K^- p}(E) - \sigma_{tot}^{K^० p}(E) + \sigma_{tot}^{\bar{K}^० p}(E) \right] \right| \leq$$

$$\frac{\pi^{3/2}}{2m_\pi} \lim_{E\to\infty} \left[\sqrt{\sigma_{K^० p\ K^+ n}(E)} + \sqrt{\sigma_{K^- p\ \bar{K}^० n}(E)} \right] \quad \ldots (5.10)$$

provided the required limits exist and asymptotic isospin invariance is good. Presently available data seems to be consistent with the value zero for the left hand side of the inequality (5.10).

We also have the analogue of the result (5.8) for πN scattering given by[12]

$$\left| \lim_{E \to \infty} [\sigma_{tot}^{K^-p} - \sigma_{tot}^{K^+p}] \right| \leq \frac{\pi^{3/2}}{2m_\pi} \lim_{E \to \infty} \left[\sqrt{\sigma_{el}^{K^-p}(E)} + \sqrt{\sigma_{el}^{K^+p}(E)} \right]. \quad (5.11)$$

provided the required limits exist.

Again the asymptotic total cross-section difference $\sigma_{tot}^{K^-p} - \sigma_{tot}^{K^+p}$ cannot increase faster than $\ln E$ as $E \to \infty$.

I have mentioned only few of the results obtained but these are sufficient to show that one can learn quite a great deal about asymptotic cross-section differences from the general principles alone.

6. Fixed Negative t Bounds

We have so far discussed only the bounds for non-negative t values. We now come to fixed negative t results. These results are relevant for a discussion of the diffraction peak region.

We first note the upper bound on $A(s,t)$ due to Martin[3],

$$\frac{4}{3} \cdot \frac{[1 + \lambda]^{3/4} - 1}{\lambda} \underset{s \to \infty}{\geq} \frac{A(s,t)}{A(s,0)} \quad (t \leq 0) \quad \ldots (6.1)$$

where

$$\lambda = \frac{(-t)\sigma_{tot}(s)}{4\pi} \qquad \ldots (6.2)$$

In obtaining this bound Martin had used the bound

$$|P_\ell(\cos\theta)| \leq [1 + \ell(\ell+1)(\sin\theta)^2]^{-1/4} \qquad (\pi \geq \theta \geq 0)$$

in order to deal with the oscillations of $P_\ell(\cos\theta)$ with ℓ for a given θ. If one takes better care of these oscillations of $P_\ell(\cos\theta)$ then one obtains the improved result[20]

$$\frac{2J_1(\sqrt{\lambda})}{\sqrt{\lambda}} \underset{s\to\infty}{\geq} \frac{A(s,t)}{A(s,o)} \qquad \text{for} \quad 3.46 \geq \lambda \geq 0 \qquad \ldots (6.3)$$

Here $J_1(x)$ is the Bessel function of the first order. Similar results can be given for $\lambda > 3.46$. From the practical point of view it is more useful to give an upper bound on $A(s,t)$ involving both $\sigma_{tot}(s)$ and $\sigma_{el}(s)$. We have[21]

$$\frac{A(s,t)}{A(s,o)} \underset{s\to\infty}{\leq} \left[1 - \frac{\rho}{9} + \frac{3}{8}\left(\frac{\rho}{9}\right)^2 - \frac{21}{320}\left(\frac{\rho}{9}\right)^3 \cdots \right] \qquad \ldots (6.4)$$

$$\text{for} \quad 2.5 \geq \rho \equiv \frac{(-t)[\sigma_{tot}(s)]^2}{4\pi\,\sigma_{el}(s)} \qquad \ldots (6.5)$$

A similar result is valid for $\rho > 2.5$. A comparison of the bound (6.4) with high energy πp, Kp, pp and $\overline{pp}$ data is extremely satisfactory[21,22].

From the upper bound (6.4) we can also deduce

$$\left[\frac{d}{dt} \ln A(s,t)\right]_{t=0} \underset{s\to\infty}{\geq} \frac{[\sigma_{tot}(s)]^2}{36\pi\sigma_{el}(s)} \qquad \ldots (6.6)$$

which is the MacDowell-Martin "Diffraction peak width" bound.[23] We also have, for higher derivatives, the result[20]

$$\left[\frac{d^n A(s,t)}{dt^n}\right]_{t=0} \underset{s\to\infty}{\geq} \frac{s\sigma_{tot}}{8\pi} \frac{1}{n!\,(2n+1)} \left[\frac{2n+2}{2n+1} \cdot \frac{(\sigma_{tot})^2}{16\pi\sigma_{el}}\right]^n \qquad \ldots (6.7)$$

All the above bound (6.1), (6.3), (6.4), (6.6), and (6.7) can be generalised to finite energies[3,20]. We further note that these bounds follow from the unitarity considerations alone.

We now come to bounds on $|F(s,t)|$ for $t<0$. It is well-known that the basic assumptions imply for $\frac{d\sigma_{el}}{dt}$, which is proportional to $|F(s,t)|^2$, a bound of the form[2,9]

$$\frac{d\sigma_{el}}{dt} \underset{s\to\infty}{\leq} \frac{c'}{\sqrt{-t}} (\ln s)^3 \qquad (t<0) \qquad \ldots (6.8)$$

It has recently been shown that one can choose[10]

$$c' = \frac{2}{(\tau_0)3/2} \qquad \ldots (6.9)$$

In fact one has the stronger result[10,11]

$$\frac{d\sigma_{el}}{dt} \underset{s\to\infty}{\leq} \frac{1}{2\pi\sqrt{\tau_o}} \frac{\sigma_{el}(s)}{\sqrt{-t}} \text{(lns)} \qquad \ldots (6.10)$$
$$(t<0)$$

This bound (6.10) is also valid for the case of particles with spin and with obvious modifications also for the inelastic reactions $a+b \to c+d$.

7. Fixed Angle Bounds

We first note that upper bounds analogous to (6.1), (6.3) and (6.4) can also be written down for any energy and fixed angle θ[3,20,21]. For unpolarised differential cross-section we have[10],

$$\frac{d\sigma_{el}}{d\Omega} \underset{s\to\infty}{\leq} \frac{\sigma_{el}(s)}{4\pi} \sum_{\ell=0}^{m} (2\ell+1) [P_\ell(\cos\theta)]^2 \qquad \ldots (7.1)$$

where[11]

$$m = \sqrt{\frac{s}{4\tau_o}} \text{ lns} \qquad \ldots (7.2)$$

Generalisation for the inelastic case $a + b \to c + d$ is easy to write down as before. Using

$$\sum_{\ell=0}^{m} (2\ell+1)[P_\ell(z)]^2 = (m+1)^2 [P_m(z)]^2 + (1-z^2)[P'_m(z)]^2$$

and considering $\theta \neq 0, \pi$ we obtain[11]

$$\frac{d\sigma}{d\Omega} \underset{s\to\infty}{\leq} \frac{1}{4\pi^2\sqrt{\tau_o}} \cdot \frac{\sqrt{s}\,\sigma_{el}}{\sin\theta} \cdot \ln s \qquad \ldots (7.3)$$

Further one can obtain from (7.3) the weaker bound

$$\frac{d\sigma}{d\Omega} \underset{s\to\infty}{\leq} \frac{1}{\pi(\tau_o)^{3/2}} \frac{\sqrt{s}\,(\ln s)^3}{\sin\theta} \qquad \ldots (7.4)$$

We may note in passing that if one is willing to make special assumptions in addition to assumptions (1-5) in Section 2 then one can derive stronger results than (7.4).[24] One can also then obtain a lower bound on $|F(s,t)|$ for negative t.[25]

Acknowledgements

It is a pleasure to thank Dr. S. M. Roy for many discussions.

Footnotes and References

1. We refer the reader, for a description of this approach, to

    G. F. Chew, "S-matrix Tehory of Strong Interactions" (W. A. Benjamin, New York, 1961).

    R. J. Eden, P. V. Landshoff, D. Olive, J. C. Polkinghorne, "Analytic S-matrix" (Cambridge Univ. Press, 1966).

2. M. Froissart, Phys. Rev., $\underline{123}$, 1053 (1961).
   See also O. W. Greenberg and F. E. Low, Phys. Rev., $\underline{124}$, 2047 (1961).

3. A. Martin, Phys. Rev., $\underline{129}$, 1432 (1963).

4. Y. S. Jin and A. Martin, Phys. Rev., $\underline{135}$, B 1375 (1964).

5. L. Lukaszuk and A. Martin, Nuovo Cimento, $\underline{52A}$, 122 (1967).

6. G. Mahoux and A. Martin, Phys. Rev., $\underline{174}$, 2140 (1968)
   See also   J. S. Bell, Nuovo Cimento, $\underline{61A}$, 541 (1969).
                 N. R. Bodine, Nuovo Cimento, $\underline{57A}$, 48 (1968).
                 H. Cornille, Nuovo Cimento, $\underline{31}$, 1101 (1964).
                 Y. Hara, Phys. Rev., $\underline{136}$, B507, (1964);
                 K. Yamamoto, Nuovo Cimento, $\underline{27}$, 1277 (1963).

7. S. M. Roy and V. Singh (to be published).

8. F. J. Yndurian, Phys. Letters, $\underline{31B}$, 368, 620 (1970),
   A. K. Common, Th-1145-CERN (April 1970).

9. A. Martin, in "Strong Interactions and High Energy Physics," Edited by R. G. Moorhouse (Oliver and Boyd, Edinburgh, 1963).

10. V. Singh and S. M. Roy, Annals of Physics, $\underline{57}$, 461 (1970).

11. The factor ln s is to be replaced by $\ln\left(\dfrac{s}{\sigma_{el}}\right)$ if σ_{el} goes to zero asymptotically like a power of s.

12. S. M. Roy and V. Singh, "Upper Bounds on Particle - Antiparticle Total Cross-section Differences at High Energy," Phys. Letters, 32B, (1970).

13. A. A. Logunov, Mestvirishvili, Nguen Van Hien and Nguen ngoc Thuan, Proceedings of CERN Conference on High Energy Collision of Hadrons (1968).

14. Y. S. Jin and A. Martin, Phys. Rev., 135, B 1369 (1964).

15. I. Ya. Pomeranchuk, J. Exptl. Theoret. Phys. (U.S.S.R.) 30, 423 (1956) and 34, 725 (1958) (Translation: Soviet Phys. - JETP, 3, 306 (1956) and 7, 499 (1958).

16. J. V. Allaby et al, Physics Letters, 30B, 500 (1969).

17. We refer the reader to some of the refined "derivations" of the Pomeranchuk theorem:

    D. Amati, M. Fierz and V. Glaser, Phys. Rev. Letters, 4, 89 (1960);
    S. Weinberg, Phys. Rev., 124, 2049 (1961);
    N. N. Meiman, J. Exptl. Theoret. Phys. (U.S.S.R.), 43, 2277 (1962) -- (Translation: Soviet Phys. - JETP, 16, 704 (1965).
    A. Martin, Nuovo Cimento, 39, 704 (1965);
    R. J. Eden, Phys. Rev. Letters, 16, 39 (1966).
    T. Kinoshita, in "Perspectives in Modern Physics," Editor R. E. Marshak (J. Wiley, 1966).

18. For a discussion of the asymptotic behavior of ReF(s,o) see for e.g.

    N. N. Khuri and T. Kinoshita, Phys. Rev., 137, B720 (1965) and 140, B706 (1965);
    Y. S. Jin and S. W. MacDowell, Phys. Rev., 138, B1279 (1965);
    H. Cornille, DPh-T/69/51 (Aug. 1969) CEN-Saclay preprint.

19. G. Bellini, E. Fiorini and A. Orkin-Lecourtois, Phys. Letters, 4, 164 (1963).

20. V. Singh and S. M. Roy, "Unitarity upper and lower bounds on the absorptive parts of elastic scattering amplitudes," Phys. Rev. (1970).

21. V. Singh and S. M. Roy, Phys. Rev. Letters, 24, 28 (1970).

22. M. A. Jacobs, A. Joseph, S. Nussinov and A. A. Rangwala, TAUP-128-70 (Tel. Aviv, Univ. preprint, 1970).

23. S. W. MacDowell and A. Martin, Phys. Rev., 135, B960 (1964).

24. T. Kinoshita, J. J. Loeffel and A. Martin, Phys. Rev. Letters, 10, 460 (1963); Phys. Rev., 135, B1464 (1964).

25. F. Cerulus and A. Martin, Physics Letters, 8, 80 (1963).

PHENOMENOLOGY OF HIGH-ENERGY SCATTERING*

Loyal Durand, III

University of Wisconsin

Department of Physics

Madison, Wisconsin 53706

Abstract: Presenting a review of the phenomenology of high energy scattering, Prof. Durand describes the strengths and weaknesses of the many techniques that have been used to analyze these data. Besides the standard Regge model, he discusses the models based on the use of Regge cuts and complex Regge poles. In addition he reviews the recent Serpukhov data and its implications on Regge based models. He concludes with brief discussion of the diffraction models and the newer hybrid models.

My topic is high energy phenomenology. Actually I will be concerned mainly with Regge type models for high energy scattering. A number of reviews on high energy phenomenology have been written recently, and I have listed a sampling of these in the references.[1-7] For a more detailed discussion

---

*Talk presented at the Symposium "The Past Decade in Particle Theory" at the University of Texas at Austin on April, 14-17, 1970.

of many topics, and especially for a more complete list of references, one should consult one or the other of these reviews. In this talk I shall not even attempt to cite all the names of the important contributors to this field over the past decade.

I shall begin with a bit of historical background. Several types of models for high energy scattering have been popular during the past ten or twelve years. You are all familiar with single particle exchange models and with the more classically oriented diffraction models. I shall discuss these briefly at the end of the talk. A third type of model, which has been the focus of most of the effort in phemomenology in the past decade, is the Regge model. Originally, one considered simple Regge pole models. More recently there have appeared modified versions of the Regge pole model in which one considers cuts and other singularities.

The original work on the Regge model began of course with Regge[8], but in the context of potential scattering theory. The initial observation that the model could be used for high energy scattering was due to Chew and Frautschi[9] and Gribov.[10]

Many other people were concerned with the application of the Regge pole model to theoretical problems, and later, to phenomenology, and the basic ideas of the model had been fairly well explored by 1963. During this early period, the people interested in phenomenology generally considered only very simple models on the expectation that at energies of perhaps 5 GeV one would reach an asymptotic region and could describe all high energy scattering in terms of the exchange of one or two Regge poles. This simple idea created a great deal of excitement, especially since the relevant experiments could be done at the then-new Brookhaven and CERN accelerators. If the initial predictions had been correct, high energy physics would not exist as we know it now. As it happened, things did not work out too well when the high energy cross sections were measured. For example, particle and antiparticle cross sections were found to differ sufficiently up to 30 GeV that they could not be regarded as asymptotic in the sense of the Pomeranchuk theorem at the highest energies then attainable. Moreover, one needed then-undiscovered Regge poles (those connected with the f^0 and A_2 mesons) as well as the ω and ρ poles to explain these cross section differences--many poles were clearly necessary to explain all the measured cross sections simultaneously. Finally, the pp differential scattering cross sections showed the narrowing of the diffraction peak with increasing energy expected in a Regge pole exchange model, but the $\pi^{\pm}p$ cross sections did not, and the pp diffraction peak became broader at higher energies. These difficulties killed the earliest versions of the Regge pole model. As far as its use in phenomenology was concerend, the model was largely ignored for a period of two or three years. This was the time during which the absorption model was most popular. Much of the data on inelastic reactions at 2-4 GeV, the energy range then most thoroughly explored, could be explained quite well using single particle exchange models with absorptive corrections. A number of theorists continued, of course, to work on the theoretical aspects of the Regge model, and there were some applications to high energy phenomenology, but the major successes came later.

Around 1965 Roger Phillips and William Rarita (and, shortly thereafter, Barger and Cline, Reeder, and a large number of other people) began fitting great quantities of high energy data using more complicated versions of the Regge pole model. These people came into the phenomenology business, fortunately, without the biases of those of us who had worked with the theoretical aspects of the Regge model. They were not bothered in the least by exponentially falling residue functions and other apparent peculiarities of the model. They simply proceed to fit data, and they fitted the data extremely well. We will see some of the typical results in a few moments.

The new Regge pole models seemed to work quite well. The results obtained seemed generally satisfactory until about two years ago[1,7]. However, as better data became available on various processes, it became possible to look in more detail at the fine points of the model. Once one could ask whether the simple pole models really fitted the high energy reaction data well, or merely provided general qualitative descriptions of the data, it became clear very quickly that the pole models did not really work. There were some classes of reactions for which they were simply not adequate.

At this point, one had to go back and take into account something which the phenomenologists had studiously avoided during the previous 8 years, namely, the possibility there were moving cuts or fixed cuts in the complex angular momentum plane. Now, as soon as one introduces Regge cuts, the situation becomes quite complicated as far as phenomenology is concerned. The models which are receiving the most attention now, and about which I'll have the most to say, are models which involve Regge cuts as well as Regge poles. Such models are not at all unreasonable from a theoretical point of view. In fact, the existence of cuts was pointed out very early by the theorists. Unfortunately, a cut is a nasty kind of singularity, and we don't know very much about how we should describe the discontinuity. It is consequently easy (and pleasant) to ignore cuts until forced to use them, and that is indeed what the phenomenologists did. But no longer. Cuts and complex Regge poles are now very much in style.[6] In the rest of this talk, I will concentrate on these very interesting recent developments, and will come eventually to some speculations about the future.

Theoretical Developments

Several recent theoretical developments are of particular interest. I shall mention some of these developments briefly at this point, and then discuss a few of them in more detail later.

First, a number of theorists have obtained what appear to be rather important formal results by summing graphs in the infinite energy limit in perturbation theory. These calculations can be performed by considering low order graphs, as in the original calculations of Cheng and Wu[11], by summing large classes of graphs using eikonal methods and simple model theories[12], or by using various infinite momentum techniques as in the work of Chang and Ma and Chang and Fishbane[13]. Many other people have also been involved in work of this type. Now the results which one gets by eikonal methods or summation of graphs depend very much on the graphs one uses. Not surprisingly, one can generate Regge behavior by a judicious choice of the set of graphs which is summed. One can also generate high energy behavior which looks much like that given by the classical diffraction models for scattering amplitudes. I have nothing very definite to say about these results at this point. They are not at the stage at which one can use the results in phenomenology. They are simply very interesting observations which, I think have a great deal to do with the connection between Regge models and the other models which we will discuss later.

A great deal of work has also been done recently on the theory of Regge cuts. The most popular and successful of the current models for high energy scattering all involve cuts in some fashion. I shall return to this topic and discuss it in more detail below.

Another topic which has recently begun to receive attention is connected with oscillating total cross sections, and complex Regge poles. The fact that Regge trajectory functions can be complex below threshold has been known for many years, and is familiar, for example, in potential scattering. In particular, when two Regge poles collide in the angular momentum plane they bounce off each other (the typical "level repulsion" phenomenon), each initially at an angle of 90 degrees to its former path. If the poles were originally on the real axis they bounce off into the complex plane after the collision as complex conjugate poles. This is a well known phenomenon, but one which again was studiously ignored by the phenomenologists until recently, when Chew and Snider[14] pointed out that one could generate oscillating behavior in high energy cross sections if one took complex Regge poles seriously. With the advent of the Serpukhov data, the cross sections seem to have departed from what one expected from simple pole models, and this had become a very interesting possibility, and one which is under quite active study at the moment.

I have just alluded to another development which has had

considerable impact on Regge phenomenology, the measurement of total cross sections at Serpukhov at energies considerably higher than those previously available[15]. The predictions for the high energy cross sections which were made using simple Regge pole models did not work out very well. There is, of course, a possibility that the Serpukhov data are incorrect. At the present time, however, we don't know that, and it is reasonable to try to see what one can do about fitting the data within the context of the models that one has available. The behavior of the Serpukhov data also raised the question of unconventional asymptotics, the question of whether the Pomeranchuk theorem is violated in one way or another. I won't say anything about the theoretical background of these considerations, since Professor Singh will cover that material in his talk, and will confine my remarks to phenomenology.

Finally, recent work provides some insight into the connections between Regge-type models and diffraction scattering models. One particular dynamical model which has been pushed very strongly by Chew, Pignotti and various collaborators is the multiperipheral bootstrap[16]. A number of calculations in the last year or so by Ball and others show that

the multiperipheral bootstrap provides a fairly reasonable
model for the high energy behavior of elastic scattering
amplitudes[17]. These calculations produce the Regge exchange
contributions to elastic scattering amplitudes by starting
with inelastic processes in which many particles are produced. These are described by a Reggeized multiperipheral
model. Such multiperipheral diagrams contribute to the
full unitarity relation as shown schematically in Figure 1.

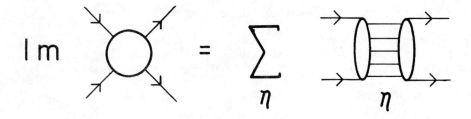

Figure 1. Schematic representation of the multiperipheral
bootstrap model. The absorptive part of the elastic scattering amplitude is approximated through
the unitarity relations by a sum over squares of
n-particle production amplitudes. The production
amplitudes in turn are approximated by multi-Regge
exchange amplitudes. The essential requirement of
the bootstrap calculation is that the output Regge
trajectories identified in the elastic scattering
amplitude correspond to the input trajectories in
the multiperipheral production amplitude.

One can use unitarity to generate integral equations for the Regge trajectory and residue functions which appear in the model, and proceed to calculate some aspects of the Regge behavior. Because the sum over intermediate states includes a sum over many inelastic processes, one would expect the output of the bootstrap calculation to be related in some way to shadow scattering, the diffraction scattering in the elastic channel which results from the absorption into the inelastic channels. Diffraction scattering is described in the simple Regge pole models by the exchange of the Pomeranchuk trajectory. The bootstrap models yield, in fact, a Pomeranchuk type term along with associated cuts, but with an intercept slightly below one.

Characteristics of Regge Models

At this point I will go on and discuss some details of high energy phenomenology. To begin with, let me remind you of the characteristics of Regge type models. The first is the correlation of Regge pole exchanges with resonances. Figure 2 shows an old, but typical, example of this correlation. The curve in Figure 2a shows the differential cross section for the process $\pi^+ p \to K^+ \Sigma^+$. This reaction can proceed

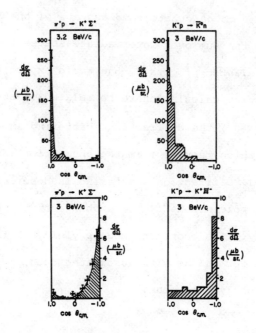

Figure 2. Association of forward and backward peaks with t- and u-channel resonances. The reactions $\pi^+p \to K^+\Sigma^+$ and $K^-p \to \bar{K}^0 n$ can proceed by the exchange of meson resonances in the t-channel (K^* and ρ-A_2 respectively), and the cross sections have strong forward peaks. There are no low-mass resonances in the t-channels for the reactions $\pi^-p \to K^+\Sigma^-$ and $K^-p \to K^+\Xi^-$. These reactions show no forward peaks, and proceed most strongly by the exchange of the Λ and Σ states in the u-channel. The relatively small, backward-peaked cross sections are characteristic of baryon exchange processes in meson-baryon scattering.

by the exchange of a neutral K* meson. The incident π^+ and a low momentum virtual K* can combine to give a K^+ meson which carries off almost all the momentum of the incident π^+. We would therefore expect to see a forward peak in the cross section for this reaction, and indeed there is a strong forward peak. For the reaction in Figure 2b, you simply change a π^+ to a π^- and simultaneously Σ^+ to Σ^-. However, the reaction $\pi^- p \to K^+ \Sigma^-$ requires the exchange of a doubly charged meson if the K^+ is to carry off the momentum of the incident π^-. We don't know of any light doubly charged meson resonances, and would not expect to see a strong forward peak in this reaction. This is consistent with the data. So far, there is nothing in this argument to indicate that the exchanges in question have anything to do with Regge behavior. We are simply applying an idea which we have had for many, many years, that the forces which produce scattering are associated with the exchange of particles or resonances. This idea seems to be roughly correct.

The distinctions between Regge pole and single particle exchange models (with or without absorptive corrections), are connected with the energy dependence of cross section. It is a striking fact that one can correlate the energy dependence of cross sections in one channel with the existence of resonances in another channel.

Let me simply remind you of the familiar Chew-Frautschi plot, Figure 3, in which one plots Re $\alpha(t)$ versus t. The

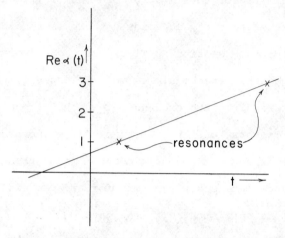

Figure 3. Typical Chew-Frautschi plot of Re $\alpha(t)$ versus t. The resonances in the t-channel occur for positive values of t such that Re $\alpha(t)=0, 2, 4, \ldots$ (even signature) or Re $\alpha(t) =1, 3, 5,\ldots$ (odd signature). The region $t \leq 0$ is the s-channel scattering region.

positive t region, of course, is the region of physical scattering in the t-channel, the resonance region. When a boson Regge trajectory passes through an integral value of the angular momentum, or when a fermion trajectory passes through a half-integral point, one expects to see resonances. Indeed, experimentally, one sees a considerable number of resonances.

The angular momentum of the resonances seems to be correlated with t, the square of the mass, in a particularly simple way. In fact, it is embarrassingly simple, namely a linear relation, Re $\alpha(t) \sim \alpha_0 + \alpha' t$. One should, of course, remember that only the positions of the resonances are observed, and it has not been possible to study the shape of the trajectory functions in between these points. One knows that there are threshold effects on the trajectory function, and there is undoubtedly structure $\alpha(t)$ between the resonances. However, in first approximation the known trajectories seem to be well described by straight lines. Incidentally, there is a remarkably astute speculation on the Regge trajectory functions in one of the first papers on the subject by Chew and Frautschi.[9] They knew, at that time, of only two resonances which should lie on the same Regge trajectory, the nucleon and the N*(1688). If you have two points, you can draw a straight line. Chew and Frautschi made the bold statement that the Regge trajectories might, in fact, be approximated by straight lines, that all of the trajectories could be more or less parallel. This seemed at the time to be the kind of crazy remark that was mentioned in the panel discussion yesterday. In fact, as far as we know now, the remark was basically correct. The trajectories, as observed, are more or less parallel and they are more or less straight.

The scattering region for the crossed (s-channel) reaction corresponds to negative t. If one extrapolates the Regge trajectory from the resonance region into the negative t region one can predict the energy dependence of the cross channel scattering amplitude, which is expected to vary as

$$A \underset{s \to \infty}{\sim} \beta(t) \, s^{\alpha(t)} (1 \pm e^{-i\pi\alpha(t)})/\sin \pi\alpha(t)$$

The energy dependence of a scattering amplitude in one channel is therefore strongly correlated with the existence of a set of resonances in the crossed channel. The observation that forward or backward peaks in cross sections are correlated with the existence of resonances, is quite old. If one resonance is observed, one can predict forward peaks or backward peaks, as the case may be. But when more than one resonance associated with a single Regge trajectory is observed, the form of the Regge trajectory can be guessed, and one can also make a guess about the energy dependence of the differential cross sections determined by these exchanges. This association of the energy dependence of the scattering amplitude with the cross channel resonances is characteristic of the Regge pole models. Note also that if the energy dependence of a scattering amplitude is given by $s^{\alpha(t)}$, and if the trajectory function α(t) continues to fall for large negative t, as in Figure 3, the cross section should fall more rapidly with increasing energy at larger values of -t, than it does near the forward direction. This is the famous "shrinkage" of the forward diffraction peaks.

In Figure 4 we show an example which is perhaps more atypical than typical. This is the famous πp charge exchange reaction, $\pi^- p \to \pi^0 n$, which from some points of view is the simplest reaction one can consider in a Regge pole model. Because of the quantum numbers involved, the only strong exchange that one expects is the exchange of the ρ meson, and one describes the reaction, typically, in terms of the ρ Regge trajectory. Figure 4 shows a recent fit to the data obtained by Barger and Phillips[18]. Their model actually includes a low-lying object, the ρ' in addition to the ρ. The ρ' may stimulate a cut rather than a pole. If the ρ trajectory determined in the fit is extrapolated back to positive values of t, it agrees quite well with the trajectory determined from the ρ and g resonances. The ρ' is not connected with any known resonances. Notice that Barger and Phillips get a very good fit to the cross section from 2.5 GeV/c to 18.2 GeV/c.

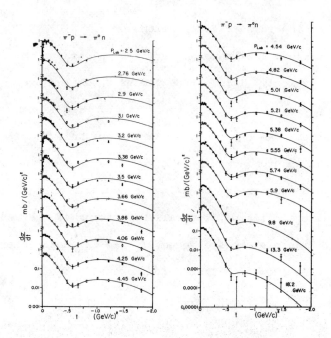

Figure 4. Regge pole fit to the data on the reaction $\pi^- p \to \pi^0 n$ from 2.5 to 18.2 GeV/c. The curves are theoretical fits by Barger and Phillips (ref. 18), which include ρ and ρ' Regge poles.

Note also that the cross sections show the typical Regge shrinkage, in this case by a factor of about twenty over the energy range considered. There is also a very considerable change in the value of the forward cross section. The model reproduces these energy dependent features of the data in a very natural way. These results certainly look very good. I should remark that one can obtain a fit which is almost this good using the ρ trajectory alone, provided only the cross section is considered. However, polarization is observed in the reaction, and one cannot explain this with a single Regge pole exchange, so something more is needed, in this instance, the ρ', which otherwise plays a minimal role in the fit.

A third characteristic of the Regge pole models is the correlation of the dips or structure in differential cross sections with those fine things, "wrong signature nonsense zeros" in the scattering amplitudes. If one looks at Regge pole models in the most simple minded way, it seems that if the scattering amplitude is not to have poles at unphysical integer (or half-integer) values of the angular momentum, the residue functions must vanish at those points (the so-called nonsense points). For example, the helicity-flip residue function for the charge exchange amplitude is dominated by ρ exchange and should vanish for $\alpha_\rho = 0$. That happens also to be a point at which the ρ signature factor $(1-e^{-i\pi\alpha})$ vanishes (a wrong signature point). Thus there is a zero in the residue function and a zero because of signature. The typical Regge denominator, however, contains the function $\sin \pi\alpha$ which also vanishes at $\alpha=0$. The net effect is a linear zero in the scattering amplitude at $\alpha=0$, and a corresponding dip in the charge exchange cross section.

Mandelstam and Wang[19] pointed out that these dips need not occur because the residue functions can have poles at wrong signature nonsense points, but the simplest interpretation of the data in Figure 4 is that the dip in the charge exchange cross section at $t \sim -0.5$ $(GeV/c)^2$ is due to a wrong signature nonsense zero. In particular, the energy dependence at the position of the dip is very accurately given by an amplitude which behaves as s^0, corresponding to angular momentum zero. Alternative interpretations of this dip phenomenon, which involve the introduction of cuts, have difficulty with this point since they tend to predict an incorrect energy dependence in the dip region. It seems to be difficult to get the dip in the right place and to get the correct energy dependence at the same time. In Figure 5, we see another example of the dip phenomenon, this time in backward πp scattering. The left hand curve shows the π^+p scattering cross section, and the theoretical fit obtained by Barger et al.[20] This cross section has an extremely sharp dip very close to the backward direction. The main contribution to backward π^+p scattering is probably given by

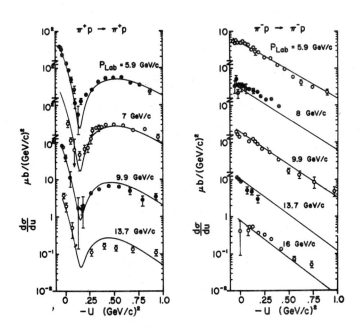

Figure 5. Fits to backward $\pi^{\pm}p$ scattering. The theoretical curves are those obtained by Barger et al., (ref. 20), in a model which includes only N (nucleon) and Δ_δ (N*(1238)) exchange.

exchange of the nucleon trajectory. If one extrapolates the nucleon trajectory from the resonance region to the position of this dip, one finds a value of α_N very close to 1/2. This value of α_N corresponds theoretically to a wrong signature nonsense point in the nucleon-exchange amplitude. It is correlated empirically with a very sharp dip in the $\pi^+ p$ scattering cross section. Again, the people who use cut models would contend that this dip can be explained in other ways and, in fact, it can be. Nevertheless, if one wants to take the simplest point of view, the dip represents a striking correlation of structure in a differential cross section in one channel with angular momentum structure (wrong signature nonsense zero) in another channel.

Figure 6 shows fits to the forward elastic $\pi^+ p$ and $\pi^- p$ scattering cross sections obtained by Barger and Phillips.[18] In this case, there are several poles involved in the Regge pole model. The energy dependence and the structure in the cross sections are both described very well. Figure 7 shows the structure at much larger momentum transfers. The smooth curves are theoretical fits to the cross section data. The fits look quite impressive. However, one should always recall when discussing Regge models that the theorists have not been

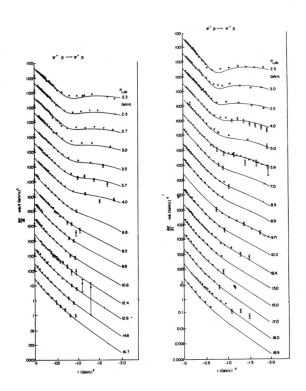

Figure 6. Fits to forward $\pi^{\pm}p$ elastic scattering cross sections obtained by Barger and Phillips (ref. 18).

able to produce any real information so far as to the way in which residue functions β(t) should behave. One can make educated guesses of one kind or another based on potential scattering, but that may be a very bad guide. The phenomenologists simply proceed empirically, parametrize the residue functions in some convenient way, and adjust the parameters to obtain a fit to the data. As a result, the energy dependence of the cross sections is much more significant as a

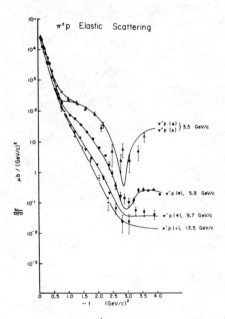

Figure 7. Fits to forward $\pi^{\pm}p$ elastic scattering cross sections obtained by Barger and Phillips, ref. 18.

test of the Regge-type models than the particular shapes of the curves which they produce. (There is one exception. The positions of the dips which have to do with wrong signature nonsense zeros are predicted by the models.) The energy dependence of the πp cross sections is in fact described quite well by the Barger-Phillips model.

Figure 8 shows the polarization observed in the $\pi^+ p$ and $\pi^- p$ scattering, and again Barger and Phillips[18] have provided us with careful fits. One can understand the approximate mirror symmetry in the $\pi^+ p$ and $\pi^- p$ polarization very easily. Most of the polarization has to do with the interference of the ρ exchange amplitude with the other amplitudes which are present. The ρ term changes sign between the two reactions (ρ exchange is analogous to heavy photon exchange) and the polarization consequently changes sign, a very simple interpretation of a feature of the data which might otherwise seem quite mysterious.

Problems with Pole Models

Up to now, I have told you mainly about the apparent successes of the Regge pole model. There are also lots of problems. These have become increasingly apparent in the past few years. One problem is associated with the so-called cross-over effect. In the forward direction, the π^-p cross section is larger than the π^+p cross section. However, the π^-p cross section decreases more rapidly than the π^+p cross section with increasing momentum transfer. The two cross sections become equal for $t \sim -0.2$ $(GeV/c)^2$, and switch roles as far as relative magnitudes are concerned for larger values of $-t$. One concludes from this that the amplitudes which give the difference between the cross sections, the odd charge conjugation amplitudes, must vanish at the point at which the two cross sections are equal. The same phenomenon occurs in the case of $\bar{p}p$ and pp scattering, K^-p and K^+p scattering, etc. The anti-particle cross sections always start out larger in the forward direction and fall more steeply than the particle cross sections. In the simple Regge pole models, the cross-over is described in terms of zeros in the odd-C Regge residues at the cross-over point. However, arguments based on the factorizability of the Regge residues lead at once to the prediction of sharp dips at the same point in the differential cross sections for various inelastic reactions, for example, $\gamma p \to \pi^0 p$. These reactions have now been studied and the predicted dips are not found. Factorization is one of the basic results that one should have in a pure pole model, so there is clearly something wrong.

There are other reactions which also cause problems in simple Regge pole models. A whole class of reactions, those which can be described dynamically by single pion exchange, turn out to be described very poorly by the simple pole model. Photoproduction reactions are also difficult to describe. On the other hand, the basic features of these reactions are easily accounted for by the old absorption model. In that model, one began with single particle exchange amplitudes in Born approximation and made rescattering corrections to take account of the absorptive effects in the initial and final states. There is lots of good physics here. The lower partial waves are strongly absorbed into the inealstic channels which are open, and one ought to correct for these absorptive effects in any model which begins with a basic single particle exchange. Indeed, for a very large number of reactions, the absorptive corrections to the single pion exchange amplitude allow one to reproduce the observed cross sections and spin dependence with essentially no free parameters.

This idea has been taken over in the last two or three

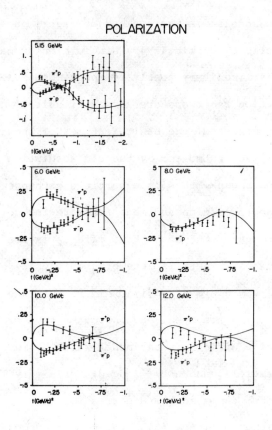

Figure 8. Fits to the polarization observed in forward $\pi^{\pm}p$ elastic scattering obtained by Barger and Phillips, ref. 18.

years by Regge phenomenologists who make the same kind of corrections to Regge pole models. When I was first considering making absorptive corrections to Regge exchange amplitudes some 5 years ago, the idea was very controversial. It was said that this procedure involves double counting of various Feynman graphs, which is undoubtedly true. My view was that the model was good physics, though not very good mathematics. Nevertheless, the results were highly model-dependent so they were not pushed very far. (The first calculations of this type that I know of were in a thesis of a student of mine--Dr. Y. T. Chiu--and were never published.[21]) The reason for the recent popularity of the Regge pole models with absorptive corrections is connected with the need to incorporate cuts in Regge type models, and the fact that one does not know how to calculate cut contributions from first principles.[6] It is possible to proceed naively and simply put in a free parametrization of the cut and fit the data, but one would like to constrain the parametrization as much as possible. It was observed several years ago that the absorptive model corrections generate cuts with the right general properties, and provide a simple way of estimating cut contributions. Unfortunately, the absorptive corrections generate the (presumably incorrect) Amati-Fubini-Stanghellini cuts, rather than Mandelstam cuts, but both have the same branch points, and their strengths may very well be similar.[6] This is a controversial subject, but the absorptive model at least provides a simple way of introducing some information on cuts without introducing too many free parameters.

There are basically two different ways of constructing a Reggeized absorption model. One can start with a Regge pole model which nearly fits the data and then make minor absorptive corrections. In this weak cut model, the cuts don't do very much, and change the initial amplitude only a little. Alternatively, one can start with an initial Regge pole amplitude which looks nothing like the final amplitude and rely on cancellations between the poles and rather strong cuts to give the dips that we have seen in the differential cross sextions. In these strong cut models, the wrong signature nonsense zeros are omitted in the initial pole terms and the absorptive corrections are quite large and tend to produce dips in about the right places.[22] In either the weak-cut or strong-cut models, the difficulties with factorization noted above are removed, and it is relatively easy to fit the data on pion exchange and photo-production reactions.

The following question remains: are the dips the result of interference between pole terms and cut terms or do they have to do with wrong signature nonsense zeros? This question is simply not answered at this point. Undoubtedly nature is trying to tell us something, since the two different ways of looking at the problem lead to similar results.

To summarize, the models which people use at the present time to fit data at energies from 2 to 30 GeV generally contain several poles or, as in the fits by the Michigan or Argonne groups, poles and absorptive or other types of cuts. As demonstrated by the foregoing figures, one can fit many reactions very well. There is an enormous literature on the Regge phenomenology, and practically all reactions on which data exist have been analyzed in this way, generally with fairly decent results.

Very High Energies

Figure 9 illustrates another problem which has been very embarrassing. This figure shows the Serpukhov data on total cross sections at very high energies,[15] and the theoretical predictions which were made using extrapolations of Regge pole models fitted to the lower energy data. You saw in Figures 6

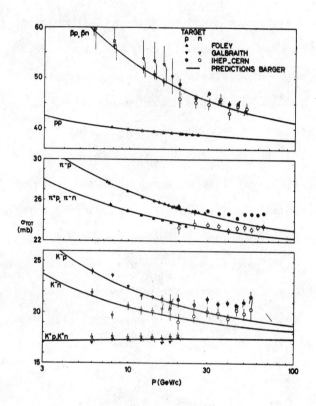

Figure 9. Comparison of the predictions of Barger et al., for high energy πN, KN, K̄N, NN and N̄N total cross sections with the new data from Serpukhov (J. V. Allaby et al., IHEP-CERN collaboration, in, ref. 15).

and 7 that the Barger-Phillips parametrization of πN scattering provides an extremely good fit to the πN data over a large range of energies. Moreover, the Barger-Phillips model fits not only the energy dependence of the differential and total cross sections, but also the momentum transfer dependence, and reproduces a great deal of detailed structure. As a first guess, one would attempt to fit the Serpukhov data by simply projecting this fit and similar fits to the other cross sections to higher energy. The solid lines in Figure 9 represent cross sections obtained in this extrapolation. The predicted cross sections fall very smoothly with increasing energy and do not agree with the Serpukhov points. We have been warned by experimentalists that this may be partly an experimental problem. If, however, we take the data seriously, there is clearly something wrong with the fits. Even if one shifts the normalization of the data, the pole model still does not provide a satisfactory fit.

Now, there is a characteristic of cut models which was observed several years ago by Gribov and various other people. In models with cuts, it is not quite a theorem, but it is generally the case, that the asymptotic limit of the cross section is approached from below. Barger and Phillips[23] have made a fit to the Serpukhov data using a model which takes this into account. They parametrized the forward scattering amplitude in the following way:

$$A(\nu) \sim i\nu \{\gamma_p - \gamma_c [\ln \nu + b - i\frac{\pi}{2}]^{-\beta} + \ldots\}, \text{ where } \beta \gtrsim 1$$

The constant γ_p represents the ordinary Pomeranchuk contribution, and would lead to a flat asymptotic cross section. The variable ν is defined by the relation $\nu = (s-u)/4M$, where M is the nucleon mass and s, u are the usual Mandelstam variables. The term proportional to γ_c is the cut contribution, and enters the foregoing expression with a negative sign. The parameter β is taken slightly larger than one, although that is not really terribly crucial. This parametrization represents the contribution of the kind of cut that one would generate in an absorptive calculation ($\beta=1$). Gribov has obtained similar results from his Reggeon calculus. It is not at all an unreasonable model. The constants γ_p and γ_c are positive, and the cut subtracts from the main term, so the asymptotic cross section is approached from below, but it is approached very slowly, namely, logarithmically in the energy. Figure 10 is the Barger-Phillips fit obtained using this model. The model fits the data quite satisfactorily, and leads to some interesting predictions. Note in particular,

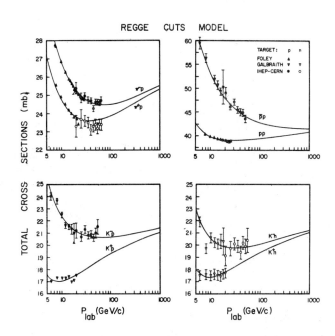

Figure 10. Fits to high energy total cross section data obtained by Barger and Phillips, ref. 23, using a model which contains a leading Regge cut as well as the usual pole terms. Note in particular the predictions for the K^+p and K^+n cross sections.

the K^+p cross section in Figure 10 c. There are no K^+p data from Serpukhov yet, but there is a very definite prediction in the cut model that the K^+p cross section will rise by an observable amount over the Serpukhov energy range. This model does not give exactly flat cross sections, by any means, but the levelling out of the cross sections occurs in the transition from the region in which the lower-lying poles determine the energy dependence of σ_T, to the region in which the cut is more important. Note also that this figure illustrates a phenomenon which is no doubt not peculiar to phenomenologists: if you plot data on a logarithmic scale, everything looks better. But the fits are indeed quite respectable.

In Figure 11, we see the predictions of this model for very high eenrgies. This is simply the same cut model pushed out to ridiculously high energies; the scale goes out to 10^{20} GeV/c. The highest energy cosmic rays observed are 10^{21} or 10^{22} eV, so one is certainly not going to get out into that region experimentally. This figure does, however, illustrate how very slowly a logarithmic term goes away. Thus, if cuts are important it will not be possible to test the Pomeranchuk theorem by seeing if cross sections become equal asymptotically. It will be necessary to use dispersion re-

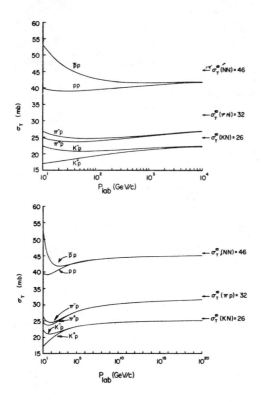

Figure 11. Asymptotic predictions for the Barger-Phillips cut model.

lations or some other means to test the theorem. However, the cross section differences are really very small at energies of 1000 GeV.

Another class of models used to fit the Serpukhov data is based on assumptions of unconventional asymptotics, that is, violations of the Pomeranchuk theorem. This is certainly a theoretical possibility, and one can clearly get fits in which the asymptotic π^+p and π^-p cross sections and the asymtotic K^+p and K^-p cross sections differ by constant amounts. Although I do not have a figure to illustrate the point, one can fit the data that way. This is one possible way in which the Pomeranchuk theorem may fail. I want to discuss, however, another model which is slightly more interesting. This model is, in fact, generated or at least suggested by some of the work on the summation of Feynman graphs, and involves a form with an extra logarithmic term but without the negative exponent. The expression for the forward amplitude is

$$A(\nu) \sim i\nu \{i\gamma_S + \gamma_D [\ln \nu + b - i\frac{\pi}{2}] + \ldots\}$$

(The positive exponent of the logarithmic term could exceed one, but it cannot exceed two without violating the Froissart bound.) The subscript D stands for dipole, for this is the form produced by a Regge dipole. If you would like to believe that some of the resonances are dipole resonances, this may be a natural form to consider. This is the simplest model which violates the Pomeranchuk theorem but which possesses correct analytic properties, crossing, and is consistent with the Froissart bound and the results of Roy and Singh for πN scattering. The high energy limit of the amplitude is just

$\nu \log \nu$. The factor ν is divided out when one calculates the total cross section, so the total cross section will increase logarithmically in this model. Figure 12 shows a fit to the data obtained by Barger and Phillips[24] using this model.

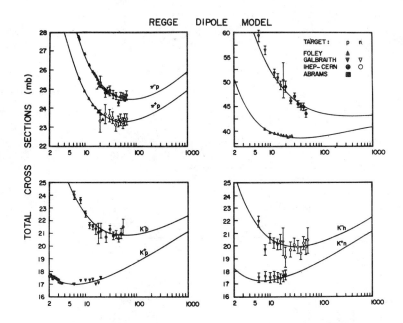

Figure 12. Fits to high energy total cross section data obtained by Barger and Phillips, ref. 24, using the Regge dipole model.

Clearly one cannot distinguish between the preceeding models using only the present total cross section data. At NAL energies, however, it will be possible to draw some conclusions, for there are some predictions of these models which are significantly different. For both models, however, the K^+N cross sections should increase very markedly at NAL energies, a very striking result, if true, considering the remarkable constancy of K^+N cross sections below 20 GeV.

Figure 13 shows the difference between the π^+p and π^-p cross sections for the dipole model. In both parts of the

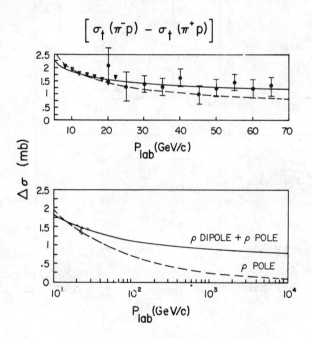

Figure 13. Fit of Barger and Phillips ref. 24, to the total cross section difference $\Delta\sigma(\pi N) = \sigma(\pi^-p) - \sigma(\pi^+p)$

figure the dashed curve is the prediction from pure ρ exchange and the solid curve is the prediction of the model including the term which produces the logarithmically increasing cross section. There is clearly not much difference at the Serpukhov energies, but at NAL energies there may be a measurable difference.

This figure illustrates something else which I think is important, namely, that Regge models work better for inelastic than elastic reactions. This cross section difference is related to the charge exchange amplitude, and the charge exchange amplitude was very well fitted by the Regge model. In general for the inelastic reactions the simple Regge pole models with a few exchanges fit the data successfully, and these models can make fairly successful predictions for cross section differences up to quite high energies. On the other hand, total cross sections, from the diffraction point of view, average over all of the inelastic intermediate states. It clearly is a very delicate question how the sum over many intermediate states adds up, whether it yields one over a logarithm, a logarithm in the numerator, or an exact constant. Accordingly, I think the models are much more reliable for the inelastic than for the elastic reactions, at least insofar as the overall behavior of the scattering amplitude is concerned.

Figures 14 a and b show some data which Professor Linden-

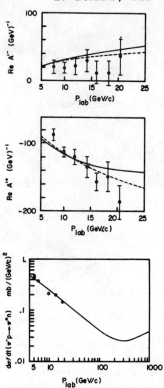

Figure 14. Fit to the real parts of the forward πN scattering amplitudes $A'^{\pm}$ and the forward charge exchange cross section using the Regge dipole model, ref. 24.

baum talked about, the data on the real parts of the forward scattering amplitudes. The dipole model gives quite respectable fits to the real parts, although there is some deviation from the measured values at the higher energies. Let me remind you that this model violates the Pomeranchuk theorem. It nevertheless gives acceptable results for the real part of the scattering amplitude, hence, acceptable results as far as tests of disperison relations are concerned. Again, it will clearly require measurements of the real part of the forward scattering amplitude at much higher energies than we have now, to distinguish among the various models. Figure 14c shows another interesting prediction of the dipole model. The forward $\bar{\pi}p$ charge exchange cross section decreases in this model to a minumum at about 300 GeV, and then begins to increase. This is a rather striking prediction, but one which will be harder to check than the general increases predicted in other cross sections.

I should like to discuss one more class of models in connection with the Serpukhov data, the models with complex Regge poles. It is well known in potential scattering that if two poles moving along the real axis collide, they will bounce off into the complex J-plane at right angles and form initially a complex conjugate pair of poles. They may eventually wander away from their conjugate positions.) Such complex conjugate Regge poles may lead to oscillations in total cross sections.[14] It is a delicate question how these poles are connected with Regge cuts. If cuts are present, the poles may be on the second sheet of the J-plane and one will not see the oscillations.[25] In any case, Chew and Snider[14] pointed out that multiperipheral calculations generate complex conjugate Regge poles which lead to oscillating total cross sections at high energies. Thus there is some theoretical indication that complex conjugate Regge poles exist in high energy scattering.

Barger and Phillips[26] have naturally considered this possibility also. They considered a model in which there is a normal Pomeranchuk term which produces constant asymtotic total cross sections, but in which the next lower lying trajectory (the P' trajectory associated with the f-meson) is replaced by a pair of complex conjugate poles. Figure 15 shows the results of their fit to the Serpukhov data on πN scattering using this model. The leveling off of the cross sections observed at about 30 GeV/c is actually associated with an upturn in the cross sections, which in this case is followed by strongly damped oscillations. Again, the result is very interesting because the first maximum should occur at about NAL energies, and at these energies one would begin to see interesting effects if the cross section is going to level off and decrease again.

There is clearly a great deal which we can learn in the

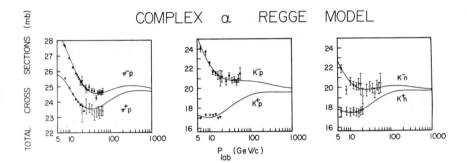

Figure 15. Fit to the πN and KN cross sections obtained using a model with complex conjugate P' poles (V. Barger and R. J. N. Phillips, ref. 26.)

future, but there is just not enough experimental information at this point to decide among the various types of models, nor enough theoretical information to say what is reasonable and what isn't. To summarize, the basic strength as well as the basic weakness of the Regge models is in their flexibility. One can clearly fit large quantities of data for a wide variety of reactions using a Regge model with cuts, and get a fairly consistent picture. The Regge models are quite successful in describing the energy dependence of cross sections, and in correlating resonance structure with the observed presence or absence of peaks in cross sections. However we do not have enough theoretical information with regard to the momentum transfer dependence of the Regge residues. We also do not have enough theoretical information on the properties of Regge cuts. Until this information is forthcoming, it will be very difficult to decide among the various possibilities.

Diffraction Models

I will finish with a few remarks on an entirely different kind of model. This is the diffraction model for elastic scattering cross sections. The version of the diffraction model which I will talk about is the one which Chou and Yang proposed a couple of years ago.[27] The history of this model actually goes back considerably farther to the work of Wu and Yang and Byers and Yang.[28] Other people have also been involved at one time or another. In this model, one regards the hadrons as extended objects with some kind of matter distribution. We can think of the hadrons as being made up of small objects, partons, quarks, or whatever, and consider a collision between hadrons as being something like a collision between two nuclei. Thus, we will consider a very simple, classical, almost optical picture in which two rapidly moving particles, Lorentz contracted (though that is not too essential) collide and pass through each other at some impact parameter b. One then makes the assumption that the absorption parameter, or in the optical sense, the imaginary part of the index of refraction, is proportional to the amount of overlap of the two objects. This model can be justified fairly directly for nucleus-nucleus scattering or in a parton description of the hadrons.

Although one knows nothing about the matter distribution of hadrons in general, one at least knows a great deal about the electromagnetic form factors of the proton, so one has some idea of what the charge distribution of a proton looks like. It is a plausible assumption, but not necessary, that the matter is distributed in a way similar to the charge. This goes back to the original Wu-Yang conjecture about the behavior of scattering amplitudes at very large momentum transfers, namely, that the pp cross section at large momentum transfers should behave as the fourth power of the charge form factor. Richard Lipes and I[29] therefore considered a model of this kind for pp scattering with the matter distribution taken as being approximately the same as the charge distribution. We used the "asymptotic" total pp cross-section to determine the strength of the absorption. (Of course, we used Barger's asymptotic cross sections, obtained before the advent of the Serpukhov data.) We also fitted the real part of the forward scattering amplitude, which you would expect in an eikonal model of this type to fall off roughly as p^{-1}, although other types of behavior are possible. Also, there was a little polarization data at very small momentum transfers available at the time, so we put in a simple spin-orbit interaction and tried to fit that, too. Now, as you should have gathered from this talk, you can fit almost anything, if you try hard enough. The amazing thing is that we hardly had to try at all. The whole model is tied down rather tightly. Once we fitted the asymptotic total cross section, and the real part of the forward scattering amplitude, we could predict the differential cross section and the polarization at large momentum transfers.

The solid curves a and b in Figure 16 are the predictions which we made before any of the data shown existed. Curve a is the fit with no real part, curve b includes a real

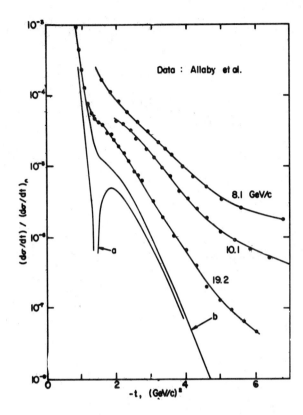

Figure 16. Durand-Lipes prediction (ref. 29) for the high energy pp scattering cross section compared to the CERN data (ref. 30) in the vicinity of the first dip. The curves labelled a and b correspond to models with a purely imaginary scattering amplitude, (a), and with the real part of the scattering amplitude included, 26 GeV/c, (b).

part of the forward scattering amplitude fitted to the observed real part at 25 GeV. We predicted that there should be a diffraction dip in the pp cross section at a momentum transfer of about 1.7 GeV. Exactly where the dip appears depends somewhat on the details of the model, but it should be in that region. We obviously did not spend any time fitting the data, because we did not have any very precise data to fit. But notice that the CERN data accumulated over the last two years apparently shows some structure in the dip region.[30] Figure 17 shows our prediction for the polari-

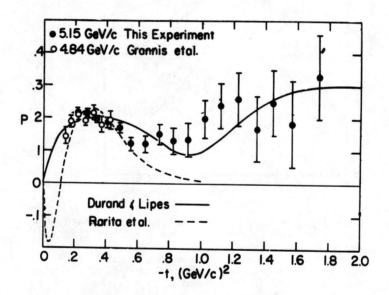

Figure 17. Comparison of the Durand-Lipes prediction for the pp polarization with the data of Booth et al., (ref. 31). The curve labelled Rarita et al., is the prediction of an old Regge pole model.

ization. We fitted the white points, which were all that existed at the time. The model predicts a nice double bumped structure similar to that observed in πN scattering. Subsequent measurements at Argonne seem to show structure of that type.[31] Undoubtedly, the fits could be improved using all this new data.

Diffraction structure in cross sections and polarizations is, of course, very familiar to nuclear physicists. The reason this is basically interesting for high energy phenomena, I think, is that the simple composite model of the hadrons is based on geometrical information and is related to the momentum transfer dependence of the cross sections. We are therefore able to predict the dip structure, that is, momentum transfer structure, even though we can say very little about the energy dependence of the scattering amplitude. In contrast, the Regge models predict the energy dependence of the scattering amplitude, but generally tell one rather little about the momentum transfer dependence. It would clearly be very nice to combine these ideas. Some people have proposed hybrid models in which the index of refraction or eikonal function is taken from Regge pole exchanges, and indeed one can build some energy dependence into the diffraction cross section in this way.[2] So far, however, no one has given a really convincing explanation of how or why one should do this. Chiu and Findelstein[2] took the work that Lipes and I had done, and put in energy dependent terms, and were able to fit all the pp data very nicely. Their hybrid model contains energy dependent terms which can be associated with Regge poles, and in the multiple scattering expansion, with Regge cuts. It is all very interesting and provocative, but at the moment it is all, also, very hazy. To summarize: there are two quite different, complementary ways of looking at the data, one in terms of momentum transfer dependence and diffraction type ideas, the other in terms of the Regge models and the energy dependence of cross sections. I think we are going to have to learn to combine these ideas in a more precise, less _ad hoc_ fashion, if we are going to get models in which we can make predictions of both.

The Regge models have been very successful over the last few years in some ways, but in some other ways they have been unsuccessful. When we get better theoretical input, and more high energy data to help tie down some of the possibilities, we can perhaps draw some fairly definite conclusions as to which of the possible models are reasonable and which are not.

The final remark I would make is that the Regge models, as I have already mentioned, seem to be more reliable in their main features for inelastic than for elastic processes. Experimentalists who wish to predict cross sections so that they can design experiments at 500 GeV should take this into account. If you want to know how large a charge exchange cross section is likely to be, go ahead and use the Regge model. It will give you a pretty good idea of what cross section you are going to have to measure. If you are considering total cross sections, the interesting features have to do with the small variations in the cross sections as functions of energy, not with the gross energy dependence. The Regge models are not much help in predicting these small variations. There are too many different possibilities which cannot now be distinguished. In fact, one hopes that the higher energy data will test these models and allow one to decide among the various possibilities.

Acknowledgment

I would like to thank Professor V. Barger for supplying most of the figures for this paper, and for many useful conversations on Regge phenomenology.

REFERENCES

1. G. Hite, Rev. Mod. Phys. $\underline{41}$, 669 (1969).
2. C. B. Chiu, Rev. Mod. Phys. $\underline{41}$, 640 (1969).
3. J. D. Jackson, Rev. Mod. Phys. $\underline{42}$, 12 (1970).
4. L. Durand, III, in Proceedings of the Boulder Conference on High Energy Physics, Colorado Associated University Press, Boulder, 1970.
5. G. Fox, in High Energy Collisions (Proceedings of the 3rd International Conference at Stony Brook), Gordon and Breach Science Publishers, New York, 1969.
6. Proceedings of the 1969 Regge Cut Conference, P. M. Fishbane and L. M. Simmons, Jr., editors. University of Wisconsin report, 1969.
7. Proceedings of the Regge Pole Conference, University of California, Irvine, California, December, 1969.
8. T. Regge, Nuovo Cimento $\underline{14}$, 951 (1959); $\underline{18}$, 947 (1960).
9. G. F. Chew and S. C. Frautschi, Phys. Rev. Letters $\underline{5}$, 580 (1960); $\underline{7}$, 394 (1961); $\underline{8}$, 41 (1962).
10. V. Gribov, JETP $\underline{41}$, 667 (1961) [translation: Soviet Physics - JETP $\underline{14}$, 478 (1962)].
11. H. Cheng and T. T. Wu, Phys. Rev. Letters $\underline{22}$, 666 (1969); Phys. Rev. $\underline{182}$, 1852, 1868, 1873, 1899 (1969); $\underline{183}$, 1324

(1969); <u>184</u>, 1868 (1969); <u>186</u>, 1611 (1969); Phys. Rev. Letters <u>23</u>, 670 (1969); Phys. Rev. D; <u>1</u>, 456, 459, 467 (1970).

12. H. D. I. Abarbanel and C. Itzykson, Phys. Rev. Letters <u>23</u>, 53 (1969); F. Englert et al., Nuovo Cimento <u>64A</u>, 561 (1969), M. Levy and J. Sucher, Phys. Rev. <u>186</u>, 1656 (1969).

13. S. J. Chang and S. K. Ma, Phys. Rev. <u>188</u>, 2385 (1969); S. J. Chang and P. M. Fishbane, University of Illinois preprints, (1970).

14. G. F. Chew and D. R. Snider, Phys. Letters <u>31B</u>, 75 (1970).

15. J. V. Allaby et al., Phys. Letters <u>30B</u>, 498 (1969).

16. G. F. Chew and A. Pignotti, Phys. Rev. <u>176</u>, 2112 (1968), G. F. Chew, M. L. Goldberger, and F. Low, Phys. Rev. Letters <u>22</u>, 208 (1969).

17. J. S. Ball and G. Marchesini, Phys. Rev. <u>188</u>, 2209 (1969), and paper to be published.

18. V. Barger and R. J. N. Phillips, Phys. Rev. <u>187</u>, 2210 (1969), Phys. Rev. Letters <u>21</u>, 865 (1968); <u>22</u>, 116 (1969). Phys. Letters <u>26B</u>, 730 (1968); <u>29B</u>, 503 (1969).

19. S. Mandelstam and L. L. Wang, Phys. Rev. <u>160</u>, 1490 (1967). C. E. Jones and V. L. Teplitz, Phys. Rev. <u>159</u>, 1271 (1967).

20. V. Barger, C. Michael and R. J. N. Phillips, Phys. Rev.

<u>185</u>, 1852 (1969). V. Barger and D. Cline, Phys. Rev. Letters <u>21</u>, 392 (1968).

21. Chiu and I made a rather thorough analysis of the charge exchange reaction $\pi^- p \to \pi^0 n$ using a Reggeized absorption model, but found the results to be highly model-dependent and not terribly satisfactory, mainly, it seems, because we did not take linearly falling Regge trajectories and exponential residues seriously. The results included a prediction of polarization in this reaction, then not measured, but were never published. Such is the effect of paying too much attention to one's theoretical biases about what is reasonable in a largely unexplored theory.

22. F. Henyey et al., Phys. Rev. <u>182</u>, 1579 (1969). M. Ross et al., and R. L. Kelly et al., University of Michigan preprints, and private communication from G. Kane.

23. V. Barger and R. J. N. Phillips, Phys. Rev. Letters <u>24</u>, 291 (1970).

24. V. Barger and R. J. N. Phillips, University of Wisconsin report COOO-262.

25. J. Ball and F. Zachariasen, Phys. Rev. Letters <u>23</u>, 346 (1969). J. Ball, G. Marchesini, and F. Zachariasen, Univ-

ersity of Utah preprint, 1970. R. Oehme, University of Chicago preprint, EFI 70-16.

26. V. Barger, private communication. V. Barger and R. J. N. Phillips, University of Wisconsin report COOO-281 (1970).

27. T. T. Chou and C. N. Yang, Phys. Rev. $\underline{170}$, 1591 (1968); $\underline{175}$, 1832 (1968); $\underline{188}$, 2469 (1969); Phys. Rev. Letters $\underline{20}$, 1213 (1968).

28. T. T. Wu and C. N. Yang, Phys. Rev. $\underline{137}$, B708 (1965). N. Byers and C. N. Yang, Phys. Rev. $\underline{142}$, 976 (1966).

29. L. Durand, III and R. Lipes, Phys. Rev. Letters $\underline{20}$, 637 (1968).

30. J. V. Allaby, et al., Phys. Letters $\underline{28B}$, 67 (1968).

31. N. Booth et al., Phys. Rev. Letters $\underline{21}$, 65 (1968).

DISCUSSION OF PROF. DURAND'S TALK

R. P. Feynman, California Institute of Technology: You gave the impression that the energy dependence of the photoproduction amplitudes was in accordance with Regge theory. The last time I looked, there was some difficulty.

Durand: The energy dependence in the models with cuts is okay. In the models with poles you are in trouble.

J. Reichert, University of Southern California: Several of these models predict a striking rise of the total cross-sections, but it seems to occur safely past the end of the present Serpukhov data. Are we being fooled by the logarithmic scale, or is the effect already present in the existing data?

Durand: The current interpretation is that the data come down and level off. The error bars, however, are wide and the normalization somewhat uncertain, so that one cannot be certain of this. Horn has made a model based on this idea that the data level off abruptly.

The other models, however, are based on very smooth functions of energy which produce curves that pass rather accurately through the data points. The effect arises because you have to have upward curvature in the region where there is data.

The really interesting question, of course, is what happens beyond this region where the models differ, and where NAL will be operating.

Reichert: If you put in a cut, it seems likely that you lose factorization. Do these models demand factorization asymptotically, or something?

Durand: Factorization doesn't enter into these models in any very important way. As soon as cuts are included, factorization is gone.

P. G. O. Freund, University of Chicago: You showed us that since the original Barger-Phillips fit disagreed with the Serpukhov data they have now improved it by including cuts and they chose the sign in accordance with a model by Gribov and also by Frautschi and Margolis. However, the Gribov-Migdal and Frautschi-Margolis models predict not only the sign of the cut contribution, but also the size. The Barger-Phillips fit uses a size about ten times the one predicted; moreover, it violates, for instance, exchange degeneracy and other properties of the usual Regge poles by violent factors. So I think that this fit is not to be viewed as of any theoretical significance.

Durand: No. I think it simply illustrates that once one

takes cuts seriously, the situation changes rather drastically. They can indeed fit the data in a number of ways as you saw, and there is no way, at the present time, of distinguishing between these fits. The dispersion relation type appraoch which Dr. Lindenbaum talked about yesterday simply doesn't give you a distinction. I agree that the Barger-Phillips cut is much stronger than you would expect in a Frautschi-Margolis type model, but there are so many papers which have calculated wrong Regge properties that the phenomenologists tend to be very suspicious of any calculations of this sort.

R. J. Yaes, University of Texas: Would you say that the Batavia data will provide a decisive test between the Horn conjecture that the usual Regge prescription can't be applied, say, over 30 GeV, and the models which Barger and Phillips have proposed. That is, will it be decisive to see whether the Weston total cross section data actually do turn up or whether they remain flat? And another question, in order to see how seriously to take these fits, how many free parameters are there?

Durand: In the fits to the total cross section data they have included sort of a minimum number of parameters. For example in fitting the πp, I don't think they adjusted the ρ and P'

contributions much. They simply took the usual intercepts, that is, the energy dependence, that had been given in the past, and adjusted the magnitudes. I think their fits are not to be taken as best fits, by any means. To produce a best fit, one has to include all the differential cross section data as well as the total cross sections and that reanalysis has not been done. They have only done the simplest part of it, including only the total cross sections and fit essentially just the parameters which appeared on the slides, three of four parameters. Unfortunately, the π^+p, K^+p, and pp cross sections have not yet been measured at the Serpukhov machine because of technical difficulties. When they are measured, then already at Serpukhov energies one should begin to see a rise in the K^+p cross sections, for example. At NAL energies one should certainly be able to distinguish between some of these possibilities. I think it is very premature to make any conjectures as to how well the models can be tested until full fledged fits have been done, putting whole bodies of data together. This has been done with the pole models, but not, so far, with models which explicitly include cuts.

<u>Yaes</u>: But if the Weston data is still flat, would you say

that Regge theory is dead?

Durand: Well, Regge theory will be very hard to kill.

BEHAVIOR OF INELASTIC NUCLEON-NUCLEUS CROSS SECTIONS FROM 10^2 to 10^5 GeV[†]

G. B. Yodh[‡], J. R. Wayland and Yash Pal[*]

Department of Physics and Astronomy

University of Maryland

College Park, Maryland

ABSTRACT

Recent cosmic ray measurements of inelastic cross sections of protons on carbon, primary proton spectra and unaccompanied hadron spectra at different atmospheric depths have been analyzed to derive the energy dependence of σ_{in} (p-air) from 100 to 10^5 GeV. We show that σ_{in}(p-air) would increase with energy above 30 GeV from 235 mb and reach a saturation value of $\sim$ 350 mb, somewhere between 3000 and 10^4 GeV, if the conventional primary proton spectrum ($AE^{-1.67}$) is used. If, on the other hand, the primary proton spectrum

---

[†]Supported in part by the Air Forces Office of Scientific Research number F-44620-69-C-0019.

[‡]Expanded version of an invited talk presented at the symposium "The Past Decade in Particle Theory" at the University of Texas at Austin on April 14-17, 1970.

[*]On leave from the Tata Insitute of Fundamental Research, Bombay, India.

steepens above 2000 GeV in accord with Grigorov's[10] recent measurements, σ_{in}(p-air) would rise to a broad maximum around 2000 GeV and then decrease to a plateau value of 280 mb around 3×10^4 GeV.

I. Introduction

Recent experiments at Serpukhov[1] show that meson baryon total cross sections have stopped decreasing between 30 and 60 GeV. These results have been accommodated in the present theoretical framework by either assuming violation of the Pomeranchuk Theorem[2], by using Regge cuts[3] or by invoking asymptotically rising cross sections[4]. The only information about hadron cross sections that we have above 60 GeV is from cosmic ray experiments. This is from two kinds of experiments: (i) direct measurements[5,6] and (ii) indirect measurements[7,8]. Grigorov, et al.,[5] have presented some new data on total inelastic cross sections of protons on carbon and hydrogen from 30 to 600 GeV based on direct measurements. The experiment measured transmission of cosmic ray protons through carbon and polyethylene targets. The energy was measured by a 3 interaction mean free path calorimeter and inelastic events were those which resulted in either no outgoing charged particles or 2 or more charged particles. The experiment was done in Proton Satellites. The basic result was that σ_{inel}(C) increased by about 20% from 20 to 200 GeV and σ_{inel}(H)

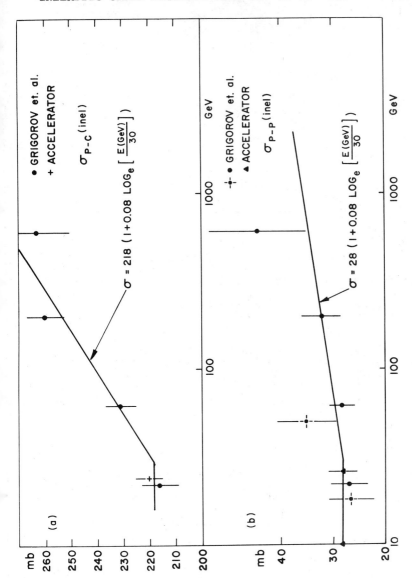

Figure 1 Experimental cross sections using direct measurements from Grigorov et al. (Ref. 5)'s direct measurements. The lines are empirical fits of the form $\sigma = \sigma_0 (1 + a \ln E/E_0)$.

also appeared to increase but less slowly. (See Figures
1(a) and 1(b)). The measured rise of $\sigma_{inel}(C)$ can be fitted by a form

$$\sigma = \sigma_o \left[1 + a \ln(E/E_o) \right] \quad (1)$$

where E is in GeV, $a \approx 0.1$, $E_o \approx 30$ GeV and $\sigma_o = 210$ mb. Preliminary results of the Echo Lake experiment[6] for hadron cross section in Fe are consistent with a 5% rise in a similar energy range although the results do not require it. Indirect measurements attempt to observe the spectra of "surviving protons" or unaccompanied hadrons at different depths. These measurements give information up to 10^5 GeV, about 1000 times higher energy than direct experiments.

In this paper we compare unaccompanied hadron spectra at different depths in the atmosphere to primary proton spectra and derive a lower bound to the proton-air inelastic cross section above 500 GeV. We show that σ_{in}(p-air) would increase with energy above 30 GeV from 235 mb and reach a saturation value of ~ 350 mb, somewhere between 3000 and 10^4 GeV, if the conventional primary protron spectrum ($AE^{-1.67}$) is used[9]. If, on the other hand, the primary proton spectrum steepens above 2000 GeV in accord with Grigorov's[10] recent measurements, σ_{in}(p-air) would rise to a broad maximum around 2000 GeV and then decrease to a plateau value of 280 mb around 3×10^4 GeV.

II. Analysis of Data

Experiments which measure unaccompanied hadron spectra usually consist of a total ionization calorimeter for the hadrons and an extensive air shower array to establish accompaniment of the hadrons. The calorimeter (usually ~ 3 to 7 interaction mean free paths deep) determines the energy of the hadron to an accuracy which can vary from ±100% to ±30%. The shower anticoincidence array, which is located close to the calorimeter, will not detect showers accompanying the hadron if they have a size less than some N_{min} (typically $N_{min} \sim 100$ to 500 shower particles). Therefore, the meaning of adjective "unaccompanied" varies from experiment to experiment. However, the measured "unaccompanied" hadron flux does provide an <u>upper limit</u> to the surviving proton flux. Furthermore, the efficiency of the shower array to detect accompanying showers increases as the energy of the hadron increases and, consequently, at sufficiently high energies this upper limit approaches the true value of the surviving proton flux.

Let us make a crude estimate of the number of showers that may be missed at Chacaltaya ($x = 530$ gr/cm^2) for hadrons of $E_o \sim 5 \times 10^4$ GeV. Conservatively, the shower array will detect 90% of all events where the shower size was ~ 500 particles[11]. If $N_e \sim 500$ then the energy in the shower is $E_e \sim 10^3$ GeV, which is about 2% of E_o. The probability of having a single collision in depth x (with interaction mean free path L) is $e^{-x/L}x/L$, the probability of having no collions is $e^{-x/L}$. As $x/L \sim 5$ at Chacaltaya, the number of protons which have interacted once is five times the number surviving. The question is: in what fraction of the interactions is energy transferred to secondaries less than 2%? A very rough approximation would be to say ~ 0.02. In that case percentage contamination due to once interacted protons will be $\lesssim 10\%$. In the same spirit the contamination at 10^4 GeV could be as high as 50%! We assert, therefore, that measurements at Chacaltaya above $\sim 20,000$ GeV are a close approximation to the true uninteracted or surviving proton flux and those at much lower energies are in the nature of upper limits to this flux.

Integral spectra for primary protons $f(>E; x=0)$ are shown in Figure 2. The units are m^{-2}, st^{-1}, s^{-1}, E is in GeV and x in g cm^{-2}. In Figure 3 we show experimental spectra of unaccompanied hadrons at depths of 698[7] g cm^{-2} and 530[8] g cm^{-2}. The latter correspond to unaccompanied omnidirectional bursts. We have not used the measurement at E = 3000 GeV of Grigorov et al.[7] and at 6×10^4 GeV from Chacaltaya since these are based on only 2 events and as the energy fluctuations in individual events can be very large.

Now we analyze these data to obtain lower bounds to σ_{in} (p-air). Using the spectra in Figures 2 and 3 we calculate an upper limit to the effective interaction length $L_{eff}(>E)$

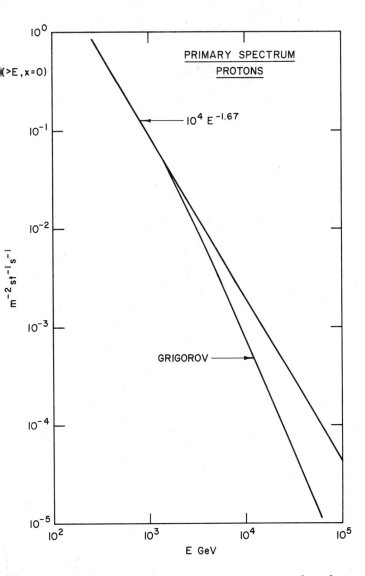

Figure 2 Primary proton spectra: Conventional spectrum (Ref. 9) and Grigorov et al.'s steeper spectrum (Ref. 10).

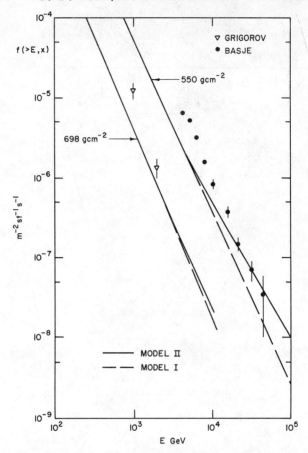

Figure 3 Unaccompanied "hadron" spectra at $x = 500$ g cm^{-2} (Ref. 7) and $x = 698$ g mc^{-2} (Ref. 8). The curves represent model fits using conventional primary spectra. Model I uses asymptotically increasing cross section and corresponds to formula (7) of text. Model II uses a rising cross section up to 10^4 GeV and then a constant cross section and corresponds to formula 8(a) and 8(b) of text.

for hadrons of energy greater than E (and the corresponding lower bound to the effective cross section $\sigma_{eff}(>E)$) by using the equation

$$f(>E,x) = e^{-x/L_{eff}} f(>E,0) \qquad (2)$$

The results for the conventional primary proton spectrum are shown in Figure 4. Also plotted in the same figure for comparison are the direct differential measurements $\sigma_{in}(E)$ of Grigorov et al. below 600 GeV (carbon cross sections were scaled to N_2). It is to be noted that $\sigma_{eff}(>E)$ is in general different from $\sigma_{in}(E)$. If $\sigma_{in}(E)$ increases with E then $\sigma_{eff}(>E)$ will be greater than $\sigma_{in}(E)$ though it will increase with energy in a manner similar to that for $\sigma_{in}(E)$; this displacement is due to the effect of the energy spectrum. This figure clearly shows that $\sigma_{in}(E)$ must increase with E between 30 and about 3000 GeV and then can either keep on increasing or saturate above 10^4 GeV. One particular energy dependence which also fits the direct measurements of Grigorov[10] below 600 GeV is given by

$$\sigma = \sigma_o(1 + a \ln(E/E_o)) \qquad 30 < E < 10^4 \text{ GeV}$$

$$\sigma = \sigma_o(1 + a \ln(10^4/E_o)) \qquad E > 10^4 \text{ GeV} \qquad (3)$$

where σ_o = 232 mb, a = 0.09 and E_o = 30 GeV. The upper curve represents $\sigma_{eff}(>E)$ calculated for this model of $\sigma(E)$ using

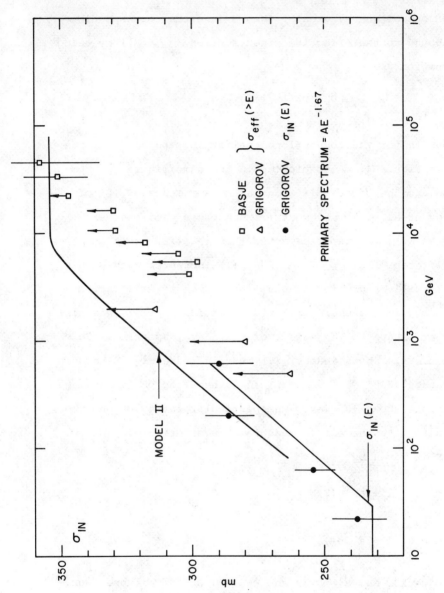

Figure 4 Lower bounds to $\sigma_{eff}(>E)$ derived from measurements at $x = 550$ and 698 g cm^{-2} (according to equation (9) of text) using conventional proton spectrum (Ref. 9). The solid dots are Grigorov's direct measurements of $\sigma_{in}(E)$.

the formulae given later on. The lower curve gives the $\sigma(E)$ in lower energy region.

For the newly reported spectrum of Grigorov et al.[10], similar analysis gives the results shown in Figure 5. The main difference from results in Figure 4 is that above 6000 GeV, $\sigma_{eff}(>E)$ has decreased and reached a plateau above 10^4 GeV of ~ 280 mb. Thus, there seems to be a maximum in the cross section around 3000 GeV.

Next, we calculate $F(>E,x)$ in terms of $f(>E,x=0)$ for this model. For the conventional primary proton spectrum the differential spectrum is given by

$$g(E,x=0) = BE^{-\gamma-1} \text{ particles } m^{-2}sec^{-1}st^{-1}GeV^{-1} \quad (4)$$

where $B = 1.67 \times 10^4$ and $\gamma = 1.67$. The surviving spectrum at depth x will be given by

$$g(E,x) = Be^{-x/L(E)}E^{-\gamma-1} \quad (5)$$

where $L(E)$ is the interaction mean free path of protons in air. Using Equation (1) to represent the increase of the cross section in air, one gets

$$\frac{1}{L(E)} = \frac{1}{L_o}\left[1 + a\ \ln(E/E_o)\right] \quad (6)$$

where $L_o = 100$ gr/cm^2. To compare with the experimentally measured integral spectra at depth x one integrates from E

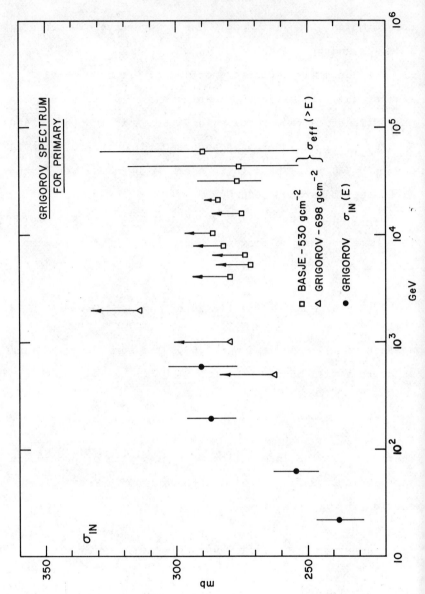

Figure 5 Lower bounds to $\sigma_{eff}(>E)$ derived from the measurements at x = 550 and 698 g cm^{-2} using Grigorov et al. spectrum (Ref. 10).

to ∞ to obtain (model I)

$$F_I(>E,x) = \frac{B \exp\left\{-\frac{x}{L_0}\left\{1 + a\ln(E/E_0)\right\} - \gamma \ln E\right\}}{\gamma + \frac{xa}{L_0}} \quad (7)$$

The new slope is $\gamma' = \gamma + ax/L_0$. For the case when the cross section saturates at $E = E_1$ (model II) the spectrum becomes

$$F_{II}(>E,x) = \frac{Be^{-x/L_2}}{\gamma'}\left\{E^{-\gamma'} - E_1^{-\gamma'}\right\} + \frac{B}{\gamma}e^{-x/L_1} \times E_1^{-\gamma} \quad (8a)$$

for $E < E_1$ and

$$f_{II}(>E,x) = \frac{Be^{-x/L}}{\gamma} \times E^{-\gamma} \quad (8b)$$

for $E < E_1$; where $L_2 = L_0/(1 - a \ln E_0)$ and $L_1 = L_0/(1 + a \ln(E_1 E_0))$. The effective cross sections $\sigma_{eff}(>E)$ (or L_{eff}) is obtained by

$$L_{eff}(>E) = x/\ln(f_I \text{ or } f_{II}/f(>E,x=0)) \quad (9)$$

The predicted spectra f_I and f_{II} are compared with experimental data at $x = 530$ and $x = 698$ g cm^{-2} in Figure 3. (In the calculations the values of x were increased to 550 g cm^{-2} and 740 g cm^{-2} respectively to take into account zenith angle distributions). As is apparent, we have relied on the highest energy points at Chacaltaya for absolute numbers and therefore the curves fall below lower energy points as expected. The last Grigorov point and the last Chacaltaya points are omitted because both of them are based only on a couple of events.

III. Discussion
Conventional Primary Proton Spectrum

Both Models I and II, combined with the conventional primary spectrum, are consistent with the data on unaccompanied hadron spectra at $x = 698$ and 530 g cm^{-2} (See Figure 3). As pointed out earlier, it is unlikely on experimental grounds that shower detection efficiency is independent of hadron energy. This conclusion is supported by independent estimates of MacKeown[11] based on Monte Carlo calculations for the BASJE set up. However, the highest energy (at 33,000 and 45,000 GeV) points should be a good approximation to the true surviving proton flux (better than 10%). The curve for model I (indefinately rising cross sections) at Chacaltaya, therefore, should be raised to go thru the two highest energy points without changing its slope. If this is done, it will go thru all the measurements down to 4000 GeV. The corresponding predicted spectrum at $x = 698$ g cm^{-2} will be raised also making it pass thru the measurements at 500 GeV and 1000 GeV, at the same time violating the upper limit to the surviving proton flux at E = 2000 GeV. The anticoincidence arrangement of Grigorov is definitely inadequate to reject showers accompanying 1000 GeV particles, hence the curve should stay below this point. Furthermore, such a fit implies that the cross-section σ_{in}(p-air) at about 200 GeV is no higher than $\sim$ 240 mb. This would contradict the fact that steady state fluxes of secondary cosmic ray hadrons (i.e. all hadrons) at a few hundred GeV require an interaction length 75-80 g cm^{-2} or equivalently σ_{in}(p-air) $\sim$ 300 mb[12] and also Grigorov's[5] direct measurements of σ_{in}(p-c). Thus, we conclude that model I is ruled out and that model II is favored by experimental data.

It is appropriate to point out the sensitivity of model II to the assumed form of σ_{in}(p-air) below 200 GeV. Although the saturation value of σ_{eff}(E=3 x 10^4 GeV) of about 350 mb is rather well determined, the manner in which the σ_{eff}(>E) approaches this value is not. For instance, we cannot rule out the possibility that this saturation value is reached already at approximately 3 x 10^3 GeV.

Grigorov et al.'s Primary Proton Spectrum

If the true primary proton spectrum above 2000 GeV steepens according to Grigorov et al.[10] and if the unaccompanied hadron fluxes as measured by Grigorov et al.[5] provide true lower bounds to σ_{in}(p-air), then the data require a broad maximum in the cross section at ~ 2000 GeV. New and independent measurements of the primary proton spectrum are necessary before considering this possibility.

Could the Rise in σ_{in}(p-air) be an Effect of the Nucleus?

Although there are a number of theoretical approaches which can accommodate asymptotically rising p-p cross sections[3,4], the direct measurements[5] of σ_{in}(p-p) do not necessarily require a rising cross section. It is interesting to speculate whether the observed rise in σ_{in}(p-A) could be a nuclear effect[13], in particular, could there be an energy variation in the A dependence? The observed rise σ_{in}(p-air) = $\sigma_o(1 + a \ln(E/E_o))$ would imply an energy dependence.

$$\alpha(E) = \ln\left[1 + .09 \ln(E/30)\right]/\ln(14)$$

for the exponent α in the relation

$$\sigma_{in}(p\text{-air}) = \sigma_{in}(p\text{-p}) A^{2/3+}$$

At $E = 10^4$ GeV α is equal to 0.17.

If $\alpha(E)$ is a universal function independent of A then the percentage increase in proton nucleus cross section σ_{in}(p-A) at a given energy will increase with A. For instance, at E=

600 GeV the percentage increase for $p-N_2$, p-Fe and p-Pb will be 27%, 43% and 62%. Such an A dependence is unlikely in view of measurements reported by several experimenters.[14] Thus it is probable that this rise is due to an increasing $\sigma_{in}(p,p)$ as pointed out by recent analysis of Grigorov's data by Balashov and Korenman.[15]

A final remark: if the ratio $\sigma_{in}(pp)/\sigma_{tot}(pp)$ does not change with energy, then such an energy dependence would imply an asymptotic $\sigma_{tot}(pp)$ of about 50 to 60 mb. This could rule out some of the models with Regge cuts.[3,4]

The analysis presented here clearly defines the need for further measurements of both primary proton spectrum and unaccompanied hadron fluxes at different depths above 1000 GeV. Such measurements, combined with the study of A dependence of σ_{in}(p-nucleus)[15] can give significant information as to the asymptotic cross sections.

We wish to thank Drs. R. W. Ellsworth, K. Kamata, V. K. Balahsubramanian, L. Jones, M. La Pointe and P. K. MacKeown for interesting discussions.

REFERENCES

1. J. A. Allaby et al., Phys. Letters 30B, 500 (1969).
2. I. Pomeranchuk, Zh. Exsperim. i. Theor. Fiz. 34, 725 (1958); Soviet Phys. J.E.T.P. 34, 499 (1958).
3. V. Barger, Review talk at the Regge Pole Conference, University of California at Irvine, Dec. 1969 (to be published); D. Horn, Phys. Letters 31B, 30 (1970).
4. V. Barger and R. J. N. Phillips, Phys. Rev. Letters 24, 291 (1970); H. Cheng and T. T. Wu, Phys. Rev. Letters 23, 670 (1969).
5. N. Girgorov et al., 11th Int. Conf. on Cosmic Ray Physics, Budapest, August 1969, Paper HE40 (to be published 1970).
6. L. W. Jones et al., Universtiy of Michigan preprint 03028-2-T (1969); 11th Int. Conf. on Cosmic Ray Physics, Budapest, August 1969, Paper HE33 (to be published); K. N. Erickson, University of Michigan preprint 03028-4-T, April 1970 (unpublished)(thesis); and L. W. Jones et al., Renewal Proposal for Echo Lake Experiment, submitted to N.S.F., Jan. 1970. Preliminary results on σ_{pp}(inel) show no rise up to about 800 GeV but the results are again consistent with the curve in Figure 1(b).
7. K. Kamata et al., 11th Int. Conf. on Cosmic Ray Physics, Budapest, August 1969, paper HE9 (to be published). A summary of hadron flux measurements in the atmosphere is

given in paper HE53 by R. W. Ellsworth and G. B. Yodh, 11th Int. Conf. on Cosmic Ray Physics, Budapest, August 1969 (to be published).

8. Kh. P. Babayan et al., Izv. Akad, Nauk. SSSR **31**, 1425 (1967); N. Grigorov et al., Proc. of Int. Conf. on Cosmic Rays, London, Vol. 2, 860 (1965).

9. P. K. Malhotra et al., Proc. of Int. Conf. on Cosmic Rays, London (1965) Vol. 2, p. 875 and Y. Pal and S. N. Tandon, Phys. Rev. 151, 1071 (1966).

10. N. Grigorov et al., 11th Int. Conf. on Cosmic Ray Physics, Budapest, August 1969, paper 0G99 (to be published).

11. P. K. MacKeown, University of Maryland Technical Report No. 7-123, July 1970 (unpublished).

12. Y. Pal and B. Peters, Mat. Fys. Medd. Dansk Vid. Selsk **33**, No. 15 (1966).

13. M. Gell-Mann and B. M. Udgaonkar, Phys. Rev. Letters **8**, 346 (1962).

14. Japanese and Brazilian Emulsion Groups, Akashi et al., Proc. of Int. Conf. on Cosmic Rays, London (1965) page 878; E. L. Andronikashvili, et al., Int. Conf. on Cosmic Rays, Calgary, Candian J. of Phys. **46**, S689 (1968), measure inelastic hadron cross sections in Fe to be 800 ± 100 mb at 800 GeV which would rule out a 43% rise and which is constant with $A^{2/3}$.

15. V. V. Balashov and G. Ya. Korenman, Phys. Letters **31B**, 310 (1970).

DISCUSSION OF PROF. YODH TALK

<u>J. A. Campbell, University of Texas at Austin</u>: There was a report about 2 years ago by the McCusker group that in some of the work they had done on measurement of extensive air showers with a large array of detectors in the Pilliga State Forest they had found that the apparent average transverse momentum of secondaries produced in showers started to go up again when the primary energy was above 10^{14} eV. There were about 18 such showers in the first published record. Firstly, is this a generally accepted effect? Secondly, has anybody else got any measurements in the energy range 10^{14} eV and up?

<u>Yodh</u>: I can answer the first question. That is certainly believed by McCusker himself. There are points in the interpretation of his widely separated cores which are hard to disentangle because you have to do a Monte Carlo cascade calculation through the atmosphere which involves many parameters. The fluctuation problem has to be carefully handled and I personally would feel that it would take some time and many more events before I will believe that large transverse momenta are indeed present. At energies above 10^{15} eV, he claims that you have an average of 10 GeV/c in transverse momentum rather than 1 GeV/c for these cores. Maybe this is so, but certainly, independent verification would be useful. The data which seem to corroborate higher transverse momenta

have to do with lateral spread of multiple muon events at very high energies. This occurs in the work of Keuffel and others. There is again a lot of interpretive theory going into the final answer. However it seems that again their distributions are wider than one would expect from the low transverse momenta we are used to.

<u>Campbell</u>: When his work was done or when the plots were made, as far as I know there was no Monte Carlo simulation of the electromagnetic cascades inside the showers. They counted the number of electrons and positrons that reached their ground detectors and they stated a rule, which is that you take that number, multiply it by 2×10^9 and the result is the primary proton energy in eV. The justification for that can be traced back to a technical report by two researchers at M.I.T. which was published in some obscure journal that I cannot find. Is there any wider circulation of folklore like this, and is that rule of thumb used in general to analyze these experiments and to get primary energies, or do you in fact use Monte Carlo calculations?

<u>Yodh</u>: Nowadays with large computers we in fact do Monte Carlo calculations. In view of the number of assumptions that you put into the calculations, you end up with an answer which I would believe to within a factor of 2 for many of these devices. In other words, 2×10^9 may be a nice conversation factor but 4×10^9 may not be a bad one either.

The errors are fairly large, and unless you see the whole shower and do the track-length integral you should worry about large fluctuations.

EXPERIMENTAL VERIFICATION OF DISPERSION THEORY AND HIGH ENERGY SCATTERING[*†]

S. J. Lindenbaum

Brookhaven National Laboratory, Upton, New York

Abstract: Dr. Lindenbaum presents a review of the status of experiments to test forward dispersion relations. After discussing the relevance of these measurements, he briefly describes the apparatus. This is followed by a description of exactly how the theory is tested. The second portion of Dr. Lindenbaum's talk is devoted to the subject of the Pomeranchuk Theorem and the recent total cross section data from Serpukhov. These results are compared with the extrapolations of data available from Dr. Lindenbaum's experiment.

I first want to point out that my talk will be divided into three parts: first, an introduction, second, where we were on this subject on testing dispersion relations and asymtotic behavior, general theorems, axioms, etc., prior to

---

[*]This work was performed under the auspices of the U. S. Atomic Energy Commission.
[†]Presented at the Symposium "The Past Decade in Particle Theory" at University of Texas in Austin, April 14-17, 1970.

Serpukhov, and finally what about the Serpukhov data. What do they do to our prior conclusions?

The subject that I am speaking about really started, in some form or other, five "past decades ago". I think that in the spirit of the Conference, this is a convenient unit to use in reviewing the situation. At that time the forward dispersion relations were first applied to a classical electromagnetic phenomenon. It actually was only one and a half "past decades ago" that these were formally derived, even for electromagnetic phenomena, from the point of view of quantum field theory. The charged $\pi^{\pm}$-p forward dispersion relations, which are the first forward dispersion relations for strong interactions that can be tested rigorously, were developed about that time by Goldberger, Oehme, and various others subsequently worked on it.[1]

And now the first point I would like to make is that the forward dispersion relations for ρp are in a unique position. They are the only common bridge between all of the basic structures that we have, and in fact, the only ones that give precise testable numbers. For a long time the proper experimental approach for a critical test was not developed, or even how to establish the forward dispersion relations theoretically, turned out to be quite a sophisticated problem. The original proofs got the right answers, but had some heuristic elements in them, and it was only subsequently that Bogolubov, and Symanzik derived them, from essentially the LSZ type of formalism, and recently there's a paper in this decade by Hepp, 1964, if I remember rightly, that establishes the analyticity properties required to deduce the forward dispersion relations from the Wightman axioms.

One can start from either the S-matrix, or you can start with various axiomatic field theories based on the LSZ formalism or on the Wightman axioms. In either case the goal is to conclude that the forward scattering amplitude is analytic (except for certain singularities) in the upper complex plane. These analyticity properties are assumed as "the gospel" in the S-matrix approach. In all the field theoretical approaches based on axioms, one derives the analyticity properties from the axioms. In fact, following the points of view in the newer material presented at this Conference, in particular Salam's and Sudarshan's work, one would also obtain forward dispersion relations.*

Now although I am speaking about forward dispersion relations, all of these approaches give a domain of analyticity that goes off the forward direction, and yields the so-called fixed-t dispersion relations. The minimum domain is the well-known Lehmann ellipse. This has been extended by Martin recently

---

*However, these may be somewhat modified in form since somewhat different singularities may be present. The reader is referred to the papers by Salam and Sudarshan, this Conference.

using unitarity to go somewhat further, and then, of course, there is the exotic domain of Mandelstam, which has never been demonstrated, but you can go as far as you like. But certainly, even the most rigorous purists in this business, and almost anybody, eventually agree on dispersion relations in the forward and near forward direction in the π-p case. I will endeavor to review briefly what has happened in the past that has been published, and then go on to the unpublished part of the talk, in the latter part.

In the π-p case, our attempt was to do something like what Salam mentioned at the roundtable, which is to put these things in a very tight box, and then say, yes they're right, or they're wrong, but not maybe. In other words, conclusively and reasonably prove* that they are correct or they are incorrect. Now, at the very least, if you wish to do that, you need to use them in a practical form. In order to derive forward dispersion relations, you need analyticity in the properly cut complex energy plane, unitarity, and crossing symmetry. So if they do work, you have at least demonstrated that in this special case of $\pi^{\pm}$-p, these requirements are met by the forward scattering amplitude. That does not mean, for example, that they are correct

---

*As the author has pointed out in the past, there are no absolute experimental proofs possible. The best one can do is to present a reasonable proof.

everywhere for all possible scattering amplitudes, but at least in the one case that can be checked critically, the predictions would be shown to be correct. One can always imagine a high order of accident, but it would have to be an unusually high order of accident as you will see shortly.

Now the other point is that the forward dispersion relations give you a "crystal ball" to study asymptotic behavior of forward scattering amplitudes simply because they contain integrals over total cross sections to infinity. So in checking them for validity, you want to desensitize them to these integrals to infinite energy (via subtractions), but once you have checked that the structure is correct, you want to sensitize* them as much as possible to these infinite energy integrals to allow one to draw conclusions about asymptotic behavior of total cross sections.

Figure 1 shows the conventional (π-p) forward dispersion relations in the least subtracted form. One obviously, for convenience, forms the symmetric and antisymmetric combinations of the real part of the amplitude, writes a once subtracted dispersion relation for the symmetric and an unsubtracted dispersion relation for the antisymmetric combination of the real part of the amplitude. The latter would only converge if the

---

*Minimize the number of subtractions to the level required just to ensure convergence.

It is convenient to employ the symmetric and antisymmetric forms:

$$D^+(\omega) = \frac{1}{2}\{D_-(\omega) + D_+(\omega)\}$$

$$D^-(\omega) = \frac{1}{2}\{D_-(\omega) - D_+(\omega)\}$$

where $D(\omega)$ is the real part of the forward scattering amplitude.

$$D^+(\omega) = D^+(\mu) + \frac{f^2 k^2}{M\{1 - (\mu/2M)^2\}\{\omega^2 - (\mu^2/2M)^2\}}$$

$$+ \frac{k^2}{4\pi^2} P \int_\mu^\infty \frac{d\omega'\,\omega'}{k'} \frac{\sigma_-(\omega') + \sigma_+(\omega')}{\omega'^2 - \omega^2}$$

(Once subtracted forward dispersion relation)

$$D^-(\omega) = \frac{2f^2\omega}{\{\omega^2 - (\mu^2/2M)^2\}} + \frac{\omega}{4\pi^2} P \int_{+\mu}^\infty \frac{d\omega'\,k'(\sigma_-(\omega') + \sigma_+(\omega'))}{(\omega'^2 - \omega^2)}$$

(Unsubtracted forward dispersion relation)

Figure 1: π-p forward dispersion relations in the least subtracted form.

Pomeranchuk theorem works, essentially if the difference $\sigma(\pi^-+p)$ $-\sigma(\pi^++p)$ converges faster than the log of the energy. The total cross sections in the symmetric case could go up somewhat more slowly than linearly with the energy (i.e., as $\omega^{1-\epsilon}$), and D^+ would still converge, and obviously the Froissart bound makes it converge quite well. In addition, when we checked these relations, we also took another subtraction at the highest energy point (as you will see later) to ensure that there was no sensitivity to any reasonably conceivable asymptotic behavior of the total cross sections.

Figure 2 shows that the π^-p scattering amplitude can be reduced to one complex function and, that if you allow for Coulomb scattering, the resultant Coulomb interference, with the real part of the nuclear amplitude, will allow you to measure the real part. The imaginary part can be obtained by measuring the total cross section and then using the optical theorem. There is the well known phase angle (β) and other effects which modify the simple addition of nuclear and Coulomb amplitudes. This is the subject of a paper published by West and Yennie, and in the interest of speed, I will refer you to it.[2]

The differential scattering cross section at small $|t|$ is essentially the sum of a Coulomb part and an interference part. Therefore, one can measure the ratio (α) of the real to the imaginary part of the forward scattering amplitude by observing these interference effects. One can then measure the imaginary part by total cross section measurements utilizing the

$$f(x,t) = f_{ord}(s,t) + \underbrace{g\,\vec{\sigma}\cdot\vec{n}}_{\text{spin flip}} \to 0 \text{ as } t \to 0$$

where $\vec{n} = \vec{k}_{inc} \times \vec{k}_{fin}$

Thus
$$\frac{d\sigma}{dt}\left\{\begin{array}{c}\pi^{\pm}\\ \pi\end{array}+p\right\} = |F_{total}|^2 = \left|\mp\frac{F_{coul}\,e^{2i\delta}}{|t|} + F_{N_{re}} + iF_{N_{im}}\right|^2$$

where
$$\delta = \frac{e^2}{\hbar c}[2\ln(kb'\theta) + \lambda]$$

(See West and Yennie reference.)

F is the point charge constant times the equivalent form factor.

Defining $\alpha = \text{Re}\,F_N/\text{Im}\,F_N$

$$\frac{d\sigma}{dt}(\pi^{\pm}+p) = \frac{F_{coul}}{|t|^2} \mp \frac{2F_{coul}}{|t|}\,\text{Im}\,F_{N\pm}\{\alpha_{\pm}\cos 2\delta \pm \sin 2\delta\}$$

$$+ \{1+\alpha_{\pm}^2\}\{\text{Im}\,F_{N\pm}\}^2 + \text{multiple scattering correction}$$

Figure 2: $\pi^{\pm}+p$ (near) forward scattering amplitudes.

optical theorem, and thus reconstruct the entire forward scattering amplitude. Our group did this experiment[3] between 8-29 GeV/c, and we got complete data on both $\pi^{\pm}$-p to 20 GeV/c. The idea was to check these relations critically in that domain.

I will say very little about the apparatus (Figure 3a & b). These are two 40' x 10' trailers (in the background) which contain a computer. This trailer (in the foreground) contains electronics. The system of a counter hodoscope, which was used to run the experiment, extended for about 350 feet (the total beam extended for about 500 feet), and it had the equivalent of 10^{12} counter combinations, and gathered several billion events with resolutions in angle and momentum which were higher than ever attained before in this type of experiment. In fact, it is this kind of new technique -- Telegdi referred to the advantages of techniques -- which really makes this kind of physics doable.

The on-line computer analyzes the data and prints out relevant information like, for example, the following. For 17.96 GeV/c π^{+}+p, it gives you the elastic differential cross section (Fig. 4), and it gives you what would happen if we subtracted out the Coulomb amplitude squared term -- we referring to my co-workers.[3] If you had no real amplitude, you would have this dotted line; that's the best fit for any reasonable parameterization, such as $e^{a/2+bt/2+ct^2/2}$ for the nuclear amplitude. The interference effect is obviously a very sizeable construc**tive inter**ference effect. This gives you an idea of the sensi-

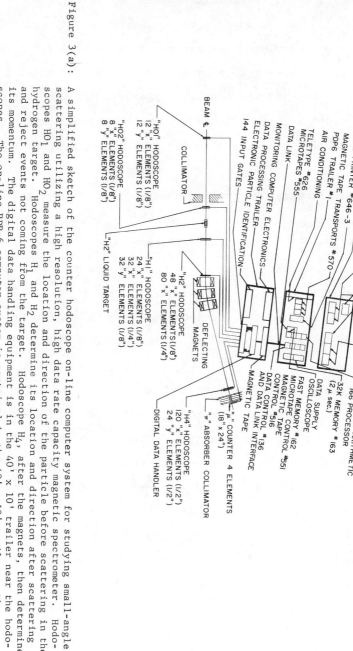

Figure 3(a): A simplified sketch of the counter hodoscope on-line computer system for studying small-angle scattering utilizing a high resolution, high data rate capacity magnetic spectrometer. Hodoscopes HO_1 and HO_2 measure the location and direction of the particle before scattering in the hydrogen target. Hodoscopes H_1 and H_2 determine its location and direction after scattering and reject events not coming from the target. Hodoscope H_4, after the magnets, then determines its momentum. The digital data handling equipment is in the 40' x 10' trailer near the hodoscopes. The on-line PDP-6 computer system is contained in the 40' x 10' trailers. The apparatus is about 400' long.

EXPERIMENTAL VERIFICATION OF DISPERSION THEORY

Figure 3(b): A photograph of part of the apparatus.

tivity at one of the higher energy π^+p points, and demonstrates that the real amplitude is negative (i.e., repulsive).

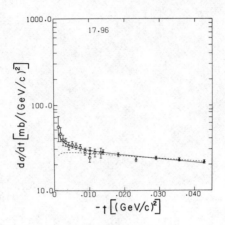

Figure 4: The elastic differential π^+-p scattering cross sections after subtraction of single Coulomb scattering. The dashed line represents the best fit with $\alpha = \text{Ratio} \dfrac{(F_N)_{\text{Re}}}{(F_N)_{\text{Im}}} = 0$. There is obviously considerable constructive nuclear-Coulomb interference in the small $|t|$ region, thus indicating that α is a negative quantity.

Figure 5 shows the highest π^-p point, and you notice the change in sign. The interference now is destructive, thus the real amplitude is also shown to be negative (i.e., repulsive).

Now with essentially the same apparatus, not using the magnet, we measured the total cross sections, which are shown in Fig. 6. The differential cross sections were measured to an absolute precision of 1%. These total cross sections were

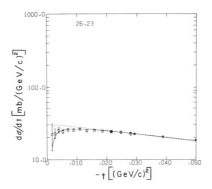

Figure 5: The elastic differential π^--p scattering cross sections after subtraction of single Coulomb scattering. The dashed line represents the best fit with $\alpha = \text{Ratio} \dfrac{(F_N)_{\text{Re}}}{(F_N)_{\text{Im}}} = 0$.

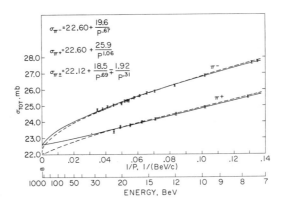

Figure 6: The predicted values of $\sigma_{\text{total}}(\pi^{\pm}\text{-p})$ above 8 BeV/c plotted versus a linear plot of 1/P in the lab system, with a nonlinear energy scale (BeV) also shown. The solid lines represent a fit of Type I. The dashed lines represent a fit of Type II.

measured to an absolute precision of 0.3%. That's about an order of magnitude better, roughly speaking, than before. We used an interesting plot, which is very bad, from the point of view of impressing funding agencies, but is very nice from the point of view of theorists. It's a 1/p plot, and zero, the origin, then becomes infinite energy. Essentially, instead of always pointing to something miles off the graphs, we are always able to reach infinity (which is at the origin), but this graph leaves very little spacing for new accelerators. However, I think it really shows you what we are facing -- it's an honest appraisal of the situation. In spite of the prevalence of the Regge theory, we fit this total cross section data first, with the minimum number of parameters, just by putting two simple power laws, a constant plus another constant, over a power of incident momentum p, and we let these parameters vary, and they came out the same, within errors, so we made them the same, in agreement with the Pomeranchuk theorem. Then we also fit the sum and difference separately, again using five parameters. This would be equivalent to a three pole Regge fit. Now you see, both fits to the data fit extremely well in the momentum region we worked in. They depart somewhat from each other at higher energies. So with five parameters, or you might say with the equivalent of three Regge poles, we were able to fit these $\pi^{\pm}$-p total cross sections.

Well, now having fitted the total cross sections, we

simply plugged these fits and the known lower energy data into the dispersion relations and got the predicted D^+ (Figure 7). It is most appropriate to look at D^+ and D^- separately, and therefore, we formed the appropriate combinations from the data. This black line is the predicted D^+. These are the experimental points. Now that impressed us as a remarkable degree of agreement. You can notice that the D^+ is a number like 0.27 in this region. It is not a special number like zero, and there is no reason these points could not have fallen, in principle, at zero, which many theorists thought would happen before the experiments. Also they could, in principle, fall any other place on the graph, so this is quite a striking confirmation. But, of course, one could ask, what about the unknown integral to infinity? Well, in order to handle that question, we put in another subtraction at the high energy end point, thus even drastic changes in high energy behavior of the total cross sections go up somewhat faster than linearly, or even drop to zero and stay there above 35 GeV/c -- these are obviously unphysical behaviors that we now experimentally know (after Serpukhov) do not happen. Yet we could barely make a dent in the predictions. So we really have locked the D^+ doubly subtracted dispersion relation predictions in, independent of asymptotic behavior, and all the low energy cross sections and parameters are known, and variations of them, by many times the errors, do not appreciably affect the predictions. So there is no question that these results reasonably demonstrate that using the doubly subtracted D^+, up to 20 GeV, you have the appropriate analyticity, you have the appropriate unitarity, and you have crossing symmetry of the forward scattering amplitude. These are the only precise number checks that I know of, of any of those three quantities.

It seems also that our fits to infinity were not completely wrong either. Otherwise, the D^+ with one subtraction would not have fit so well. In fact, changing the constant by a few mb's leads to trouble. We just put in the second subtraction to make sure that no matter what happened (within reason) at higher energies, our conclusions would still be valid.

Now let us look at D^- (Figure 8). This lowest solid line is the calculated results for a fit of the Type I, where we used individual power laws. This highest solid line is the calculated result for a three pole Regge fit (Type II), and this is a fit where we used individual power laws in the $T_{3/2}$ and $T_{1/2}$ cross sections separately (Type III). You will notice that the displacement of the dashed line from Fit I is the systematic error. In D^+, the individual systematic errors in $\pi^+ + p$ and $\pi^- + p$ cancel out almost entirely because of the difference in sign of the Coulomb interference effect. In D^-, on the other hand, we do get a sizeable systematic error, because for the same reason, they add. But when we shifted the data by

EXPERIMENTAL VERIFICATION OF DISPERSION THEORY 271

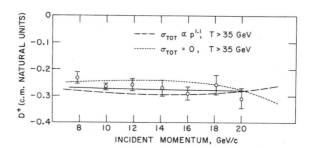

Figure 7: Values of D^+ in c.m. natural units. The solid line is the prediction of the singly subtracted D^+ using Fit I for high energy cross sections. The broken curves are the results obtained from the doubly subtracted dispersion relation integral using the assumed total cross section behavior shown on the figure.

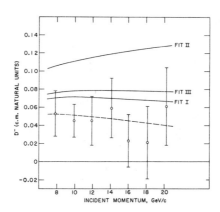

Figure 8: Values of D^- in c.m. natural units. The curves are the results of the dispersion relation integrals discussed in the text. The dashed line is Fit I displaced downward by the systematic error.

this systematic error uncertainty, we get an excellent fit for Types I and III. Now I say that this is certainly not a proof, since as I previously mentioned, there are never ever experimental proofs. However, I say that this is about as reasonable a demonstration as you could expect, of the validity of the Pomeranchuk theorem, because if it were not valid, then this $\bar{D}$ calculation should have diverged, and it would be a high order of accident that we got just these numbers, which fit the predictions. In fact, you see for a slightly different fit with a slightly different convergence; this Fit II has a convergence of the difference of cross sections by $1/p^n$, where power $n \simeq 0.3$, the predictions move out of agreement with the experiment. So if we wanted to use the Regge mechanism, three poles were probably not enough. We would need a ρ', or if we were so bold at that time, perhaps a cut. We never took that seriously enough to bother with introducing more parameters, since we got very satisfactory fits with just five parameters.

No one has ever been able to obtain the analyticity properties of the scattering amplitudes from basic axioms without including local commutativity, and furthermore, they rapidly convince themselves it has to be global. If it fails anywhere, it fails everywhere. So we really were fishing for big game when we went after this one. Now Oehme, a long time ago, and

more recently in a paper he's written, for a Festchrift for
Gregor Wentzel,[4] analyzes this problem of what would happen if local commutativity failed at small distances. It is well known that one cannot ever do this rigorously in a consistent manner. I refer you to his paper, which discusses various physical models for estimating what such effects might do to the forward dispersion relations. If local commutativity failed at $\sim 10^{-15}$ cm, you would mix up the D^+ and A^+, according to Oehme's model. Essentially, the analytic function gets multiplied by $e^{i|\ell|\omega}$, and you would have gotten horribly violent oscillations so that if it failed at distances $\sim 10^{-15}$ cm, the effects would be expected to be clearly detectable. However, three dimensional distance concepts are obviously not rigorously tenable in relativistic concepts. Furthermore, attempts to violate local commutativity anywhere generally require that it be violated everywhere or unacceptable consequences occur. Therefore, estimates along the lines I just mentioned are at best justifiable on "dimensional grounds" and are obviously model dependent.

So it looks like analyticity is good as far as we can see. It's hard to see how it couldn't be good theoretically or in a unique way anyway, but it's nice to know that this wasn't the case. If these numbers didn't fit, I think that modern theory would have been confronted with its greatest catastrophe, but I am equally happy that the data fit. The thing that would have made me unhappy is if it didn't decide the question one way or another.

Where Would the Pomeranchuk Theorem Come True?

Fits I and III, which give good fits to the $\sigma(\pi^{\pm}+p)$ total cross sections and both the once-subtracted D^+ and the unsubtracted D^- would predict a cross section difference within 0.1% (the best experimental precision one could foresee) to occur at about 25,000 GeV. One reasonable operational definition of where "Asymptotia" lies is at those energies beyond which the Pomeranchuk theorem comes true. In that sense, "Asymptotia" lies beyond ~ 25,000 Gev.

Fit II (which does not fit the D^- data) would predict "Asymptotia" as lying beyond 1,000,000 GeV.

Now as a by-product, we also were able to calculate what the charge exchange cross section at t=0 should be, from having the complete amplitudes. In comparing with the data, we found that charge independence is correct, in the sense that if there are charge dependent terms in the scattering amplitudes these are $\lesssim$ 1%.

The next figure (Figure 9) is the total cross section for p+p and $\bar{p}$-p. The analysis of p+p and $\bar{p}$+p is not as precise as $\pi^{\pm}$+p. We fit the world integral of total cross section data for p+p and $\bar{p}$+p using a five parameter fit of the Type I previously discussed for π-p. For the p+p, $\bar{p}$+p case the fit has the Pomeranchuk theorem coming true at ≳ 35,000 which in these uncertain extrapolations is the same number as in the $\pi^{\pm}$-p case.

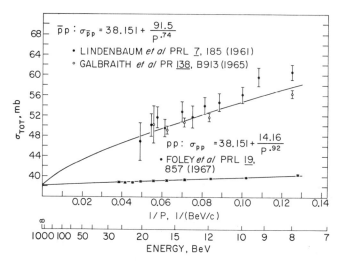

Figure 9: The predicted values of σ_{total} (p-p) and $\sigma_{total}(\bar{p}$-p) by a fit of Type 1 to the experimental data, versus a linear plot of 1/P in the lab system with a non-linear energy scale (BeV) also shown.

There were several small-angle scattering experiments for p+p. The first investigation was done by a group of USSR scientists.[5] There were also the Nimrod[6] experiment, and the CERN experiment[7], etc.[8] and our most recent investigations,[9] which are the most accurate. In the p-p, $\bar{p}$-p case, one cannot make precise predictions such as in the π-p scattering case, since there are three non-vanishing amplitudes in the forward direction. In order to make a sensible analysis, one must assume spin independence of the interaction -- a questionable assumption, in view of the existence of sizeable polarizations. There is a large nonphysical region which also introduces considerable uncertainty. These (Figure 10) are the

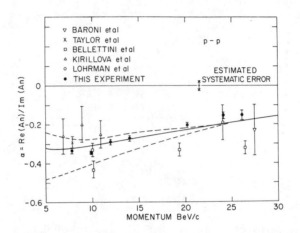

Figure 10: α, the ratio of the real to the imaginary part of the forward p-p scattering amplitude (assuming spin independence) compared to other data, and forward dispersion relation calculations.

forward dispersion relation predictions assuming spin independence and making some reasonable assumptions about what the nonphysical region might contribute. These outer boundary lines are estimates of how wrong these nonphysical region estimates could be, and with spin independence assumed, you see, one does get a good agreement even in the p+p, p̄+p case. I consider this merely of comfort and not of any great intrinsic value because it is not really a critical test. You can poke holes in this analysis almost anywhere you like.

Well, we were in a happy state, and we thought we knew what would happen at Serpukhov at this point, and now we go on to perhaps the most interesting part--what did happen at Serpukhov? This is from the paper of the Serpukhov-CERN team, Allaby et al,[10] which has been published. As you can see, (Figure 11) they extended the cross sections essentially up to

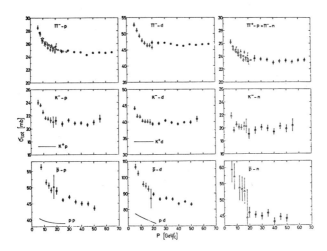

Figure 11: The Serpukhov total cross section measurements compared to the lower energy data.

65 GeV for π^--p, k^--p, $\bar{p}$+p, π^--d, k^--d, and $\bar{p}$-d. There were various momentum cutoffs, depending on the particle. Now we will see this in more detail shortly, but one notices immediately that the new data seem to fit in the overlap region with the older data (~ 20 GeV), but the monotomic tendency to decrease seems to disappear, and in fact, there is an apparent

flatness, but notice that there is quite a bounce in the points. The errors on these points are not statistical errors; they are systematic errors, and this, to me, implies that the experiment has possibilities of systematic uncertainties. I'm not surprised by this, because the experiment employed conventional counter telescope transmission techniques, with another added cross to bear, which was not a scientific choice but an operational one, in that they could not use a liquid hydrogen target due to the early stage safety limitations. This in itself led to an uncertainty of about a percent. In my opinion, telescope techniques could be uncertain by 1-2% in this energy range. I do not want to imply any disagreement. I must say, that the authors in writing their paper have been very suitably cautious and have pointed out the limitations in their conclusions. Personally, I think the odds are in favor of this kind of effect happening, although I do not think it has been demonstrated yet, in a conclusive fashion.

Well the first thing we did was to try and take our old data and the Serpukhov data and fit them together in the old way (using our five parameter Fit I), and you see, we could not do that successfully (Figure 12). However, so far we took no account of the fact that these authors said there was a 1% scale uncertainty.

We took advantage of that in the next slide* (Figure 13).

---

*Unpublished calculations of our group. The parametric fits were performed primarily by W. A. Love.

EXPERIMENTAL VERIFICATION OF DISPERSION THEORY

We just took this 1% uncertainty and said, all right, they say they're uncertain by 1%, let's move their cross sections down by 1% (in the direction to accommodate a fit) and see if we can still use the same kind of parametric fit. I should mention that another experimental difficulty which they had no control over is that there was no positive beam, so the only way they can get π^+-p data is by also measuring π^--d and then using Glauber-Wilkins' corrections to deduce π^--n, and then, assuming charge symmetry, assume π^+-p = π^--n. Now this fit looks pretty good except for the few points (π^--p) at the highest momentum. You notice, that after the Serpukhov data is lowered by 1%, you have one, two, three points below, three points on the Fit I line, and four points above. Now if you do not attach any significance to the sequence of these points, you'll come back saying it's not improbable that this is the right line. But obviously, these are four points at high p are consistent with the line, in that (at high momentum) there is one point that fluctuates down from the line by about as much as these last (highest momentum) four points fluctuate up. This, to me, indicates that the systematic errors are capable of producing this type of fluctuation, and whether it is a real effect or such a fluctuation, I don't know, but I'm not convinced at this point that it is necessarily a real effect. If you look at the χ^2, we get 37 with 44 points, but that's fooling you unfairly because our older data

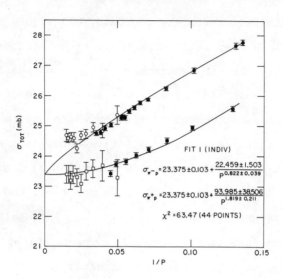

Figure 12: A fit of Type I to the Serpukhov and lower energy data combined.

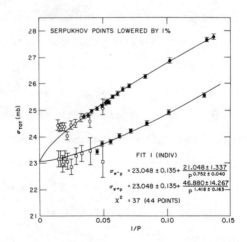

Figure 13: A fit of Type I for the Serpukhov data lowered by 1% and the lower energy data combined.

points give practically no χ^2 by themselves due to the point to point smoothness of the systematic errors in our data. So this is definitely not a good fit to the Serpukhov data, but factoring in everything I know about it or I can think of, if I had to put betting odds, I would give it less than a three standard deviation rating. It's very hard for me (or anyone else) to do this (considering the uncertainty in systematic error estimates) in a logically unique manner. Therefore, this is merely my own personal estimate, and I don't put any more weight on it than that.

Just to show that we are not prejudiced against European and Soviet installations, I have shown what happened when we did our last high precision experiment in our own laboratories. This slide (Fig. 14) compares our data with the previous inves-

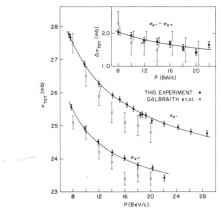

Figure 14: The measured total cross sections versus laboratory momentum for the lower energy data. Also shown are the data. The lines are Fit I. The inset shows the difference $\sigma_{\pi^-} - \sigma_{\pi^+}$ with Fit II. The fits are made to the data of this experiment only.

tigation done at BNL by another BNL group.[11] They had done about the best that could be done with the counter telescope techniques in our energy range, and you see that our points measured subsequently differed from theirs by quite a bit,* and in some cases, the differences being enough to bring the Serpukhov points down to our old fit, almost. And now, you say, maybe there is an absolute normalization error. This, in fact, appears to explain most of the difference, but in the $\sigma_{\pi+}$ data the point at 16 BeV/c dips down, and the data subsequently dip back up again, and almost overlap, except for the last point, which in our case also dipped. No one is immune to such a thing. There seems to not be any reason to believe that this is a real structure. These are not statistical errors, but definitely systematic. On the difference of total cross sections, of course, we agree very well. And that, by the way, is true with the Serpukhov data -- generally, for the differences, there is no problem. The errors in the lower precision experiments are so large that you have no problem accommodating D^-. Therefore, we have no problem whatsoever with satisfying the Pomeranchuk theorem. The only problem we have is with what happens to the sum of the cross sections.

---

*Reference 11 was more accurate than an older 1961 measurement of our group and also correctly differed from some fluctuations in our earlier measurements.

What happens to the D^+, perhaps, or what happens to the asymptotic limits of the sum.

Well, we also tried the other fit -- the three pole Regge fit. Without even moving the Serpukhov points, it accommodated them quite well (Figure 15), and if we move them one percent,

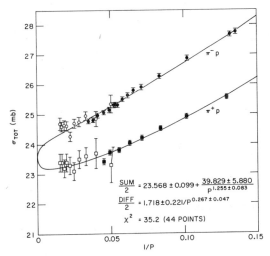

Figure 15: A fit of Type II (three pole Regge fit) to the combined Serpukhov and lower energy data.

it would be an excellent fit, but it would disagree with our D^- prediction, just as it did with our points alone. This fact hasn't changed.

Now this shows a recalculation of the D^+ and the D^- with these fits of Type I of our data and the Serpukhov data that I just described (Figure 16). Now you see, in the case of D^+, the Fit I, including the 1% lower Serpukhov data, does not make any difference to speak of. In fact, any reasonable fit

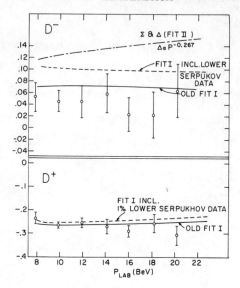

Figure 16: Recalculation of D^+ and D^- using Serpukhov data and lower energy data.

to the data, for example, Barger and Phillips fit,[12] would give a predicted D^+ which agrees reasonably well with the data. The small cut term that these authors put in has very little effect. In the D^- case, the Fit I of our type, including the lowered Serpukhov data, and taking account of the systematic error, which is of this order, could bring the prediction for the data down close to the old Fit I and thus in reasonable agreement with the data. For the three pole Regge fit type (Fit II), the systematic error uncertainty is not enough to give agreement, and you need one more parameter to fit D^- reasonably. Now whether you call that a cut or another pole or another parameter, I don't care. In any event, the old script repeated itself; you do a new measurement, and you find you need another parameter. Whether this is of great intrinsic significance or not, it's certainly very interesting, but I don't get very excited about that because we found, even in our own work in our own energy range, many examples of where another measurement introduced another parameter. When one measured the polarization of the charge exchange, one immediately needed another parameter. So the fact that you measure total cross sections at a higher energy, and it introduces one more parameter does not personally make me think that this is necessarily the key to all our problems.

Now this slide (Figure 17) shows that we have no

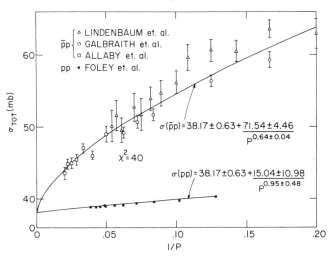

Figure 17: A fit of Type I to the p-p and p̄-p Serpukhov and lower energy data.

trouble accommodating the p̄+p and the p+p data from Serpukhov with our own, and we get an excellent χ^2, and in fact, the p-p data showed none of this levelling off and was the one case which did not indicate any of this effect at all.

The next slide (Figure 18) shows the famous K case. As I said,

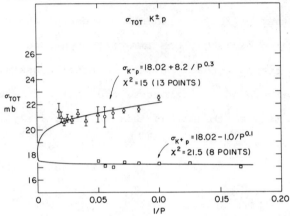

Figure 18: A fit of Type I to the $K^{\pm}$-p Serpukhov and lower energy data.

we would not show a violation of the Pomeranchuk theorem because without even adding any other parameters, we get a fit (which is, of course, bad) that accommodated the Pomeranchuk theorem. Considering the wobble of the points, the poorness of the fit can be partially understood. If the points were smooth, it would not be so bad. Obviously, if we put in another parameter, whether you call it a cut, or another pole, or simply another parameter, we could accommodate the Pomeranchuk theorem beautifully. The real question is, where is the Pomeranchuk theorem satisfied? I do not think you can estimate that without real amplitude measurements. There aren't any in the K case. There is absolutely no reason from the data to conclude, in my opinion, that there is a violation of the Pomeranchuk theorem. Now it is true that these cross sections seem to have flattened off with a constant difference, but if you look at the scatter of the points, you could almost convince yourself that there is a linear rise of the K^+-p cross section from below. I think that one clearly needs more accurate experiments to answer questions as to whether the Pomeranchuk theorem is violated. Furthermore, you must at least have the real amplitude measurements, and in the K case, to make the analysis conclusive is much more difficult -- it's not as bad and indeterminate as the p+p and $\bar{p}$-p case, since the nonphysical region is smaller. By the way, the $\pi^\pm$-p case would satisfy the Pomeranchuk theorem from weaker arguments like the Pomeranchuk-Okun rule.

Here are the Barger and Phillips fits (Figure 19) using a Regge cut term which rises from below as the energy increases toward infinity. They had no problem whatever fitting all of these with the usual poles. I refer you to their paper[12] but this just shows that by adding essentially one more parameter, there is no problem.

And of course, they had no difficulty fitting the $\pi^\pm$-p real part. They just considered the sum, but I am sure looking at their fits, that the difference would have fitted.

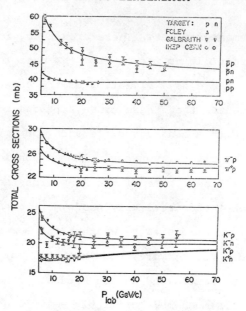

Figure 19: Barger and Phillips fits to the Serpukhov data. They also have no difficulty doing this in the p+p, and the $\bar{p}$+p cases.

The last point worth mentioning is that these authors then, using this Regge cut, get sizeable increases towards asymptotic constants. Now my own feeling is, that whether you take a cut from below which approaches an asymptotic constant, or you think of including a positive cut term, which continually rises slower than the Froissart bound, is almost a matter of taste. Whether the cross section at infinity is a constant or it's logarithmically rising with s by a power of $2-\varepsilon$, just so that we do not violate the Froissart bound, which I think probably is correct, does not particularly excite me, and I would not use that as a definition of "Asymptopia". I would still use the definition of where the Pomeranchuk theorem comes true, and that has not changed whatsoever, because the convergence of differences of incident particle and antiparticle cross sections both, in the Serpukhov data and our own, are in agreement, and if that convergence had changed, then we would have expected to have seen it in our $\bar{D}$ measurements at lower energy. So what has happened is, that even if you take the Serpukhov data at face value, the major change is that the sum of the cross sections either goes up to a constant logarithmically, or flattens out and stays constant, or goes up logarithmically indefinitely. These details have very little effect on the basic checks of the forward dispersion relations and Pomeranchuk theorem, etc., which we were talking about, and the accuracy is so weak that they do not even seriously reflect on it.

So, in summary, I would say that basically the situation is the same as it was when we published our paper, with the reservations I just mentioned, and I am looking forward to seeing more accurate results with time. Of course, the authors (of the Serpukhov work) themselves are trying very hard to utilize a hydrogen target. I understand that they have now gotten approval. The next set of data is bound to be better from that point of view. I want to summarize by saying that I think this Serpukhov investigation was a heroic undertaking. I think that the physicists involved did pretty well in these preliminary investigations considering the conditions. They had to work at a new accelerator with more primitive techniques and various limitations connected with the early stages of operations.

I would, in retrospect, want to make one more comment. One other thing, besides the general analytic structure of scattering amplitudes, which impresses me, is the success of unitary symmetry.[13] So let us ask how does one estimate asymptotic energies from that point of view? As I have previously mentioned, if you take a perhaps fictional quark triplet as representing the observed symmetry for a 10 GeV quark mass, the threshold production energy for a quark pair is 250 GeV. If you want to make sure that we do not have a repeat of the π case, where investigations of the then postulated fundamental new particle led to a whole new particle spectroscopy eventually then you have got to go another factor of 10 in the center of mass energy to make sure that the quark is unique. You then again get to numbers like 25,000 GeV minimum. So I think numbers like 25,000 GeV seem to be the minimum for "Asymptopia" and if anything, the new evidence points suspiciously toward even much higher energies. I therefore think that one should push hard on all work at present at the available accelerators and not be surprised if more accurate techniques in subsequent experiments help solve the basic problems, but also vigorously push for much higher energies as soon as practically attainable.

NOTE: BeV and GeV have been used interchangeably to avoid changes in various existing slides or figures which used one or another of these equivalent units.

REFERENCES

1. For a general review of the early history and a complete set of references, see S. J. Lindenbaum, "Forward Dispersion Relations -- Their Validity and Predictions", Pion-Nucleon Scattering, pp 81-113, Edited by G. L. Shaw and D. Y. Wong, John Wiley & Sons, New York, 1969.

2. G. B. West and D. R. Yennie, Phys. Rev. 172, 1413 (1968).

3. K. J. Foley, R. S. Jones, S. J. Lindenbau, W. A. Love, S. Ozaki, E. D. Platner, C. A. Quarles, and E. H. Willen Phys. Rev. Letters 19, 193 (1967); Phys. Rev. 181, 1775 (1969).

4. R. Oehme, "Forward Dispersion Relations and Microcausality". in "Quanta", Essays in Theoretical Physics Dedicated to Gregor Hentzel, Ed. by P. G. O. Freund, C. J. Goebel, and Y. Nambu, University of Chicago Press 1970, p. 309.

5. L. Kirillova, P. Markov, Kh. Tchernev, T. Todorov, and A. Zlateva, Phys. Letters 13, 93 (1964).

6. A. E. Taylor, A. Ashmore, W. S. Chapman, D. F. Falla, W. H W. H. Range, D. B. Scott, A. Astbury, F. Capicci, and T. G. Walker, Phys. Letters 14, 54 (1965).

7. G. Belletini, G. Cocconi, A. N. Diddens, E. Lillenthun, J. Pahl, J. P. Scanlon, J. Walters, A. M. Wetherell, and P. Zanella, Phys. Letters 19, 705 (1966).

8. E. Lohrman, H. Meyer, and H. Winzeler, Phys. Letters 13, 78 (1964); G. Baroni, A. Manfredini, and V. Rossi, Nuovo Cimento 38, 95 (1965); K. J. Foley, R. S. Gilmore, R. S. Jones, S. J. Lindenbaum, W. A. Love, S. Ozaki, E. H. Willen, R. Yamada, and L. C. L. Yuan, Phys. Rev. Letters 14, 74 (1965).

9. K. J. Foley, R. S. Jones, S. J. Lindenbaum, W. A. Love, S. Ozaki, E. D. Platner, C. A. Quarles, and E. H. Willen, Phys. Rev. Letters 19, 857 (1967).

10. J. V. Allaby, Yu. B. Bushnin, S. P. Denisov, A. N. Diddens, R. W. Robinson, S. V. Donskov, G. Giacomelli, Yu. P. Gorin, A. Klovning, A. I. Petrukhin, Yu. D. Prokoshkin, R. S. Shuvalov, C. A. Stahlbrandt, and D. A. Stoyanova, Phys. Letters 30B, 500 (1969).

11. W. Galbraith, E. W. Jenkins, T. F. Kycia, B. A. Leontic, R. H. Phillips, and A. L. Read, Phys. Rev. 138B (1965).

12. V. Barger and R. J. N. Phillips, Phys. Rev. Letters 24, 291 (1970).

13. Y. Ne'eman, Fields and Quanta, 1, 55 (1970)

DISPERSION THEORY AND THE COMPLEX J-PLANE

Stanley Mandelstam

University of California, Berkeley

Abstract: In a paper based on a talk at the symposium "The Past Decade in Particle Theory," Prof. S. Mandelstam reviews some of the basic ideas of dispersion theory and presents examples of the successful applications of the theory in understanding elementary particle reactions. He then discusses the principles of Regge theory including phenomenological applications and the composite nature of Regge recurrences. Finally, he discusses the prospects for dynamical calculations based on dispersion relations and Regge theory.

In this talk I would like to give a summary of the part which dispersion relations have played in the physics of the past decade. The approach to physics by dispersion relations essentially studies the analytic properties of the scattering matrix. One might think, off hand, that such an approach could only give very general results, and that one would have to put in more specific assumptions in order to get a complete theory. This is by no means obviously the case. It is, in fact, plausible that dispersion relations provide us with a complete theory. There is, however, no general agreement among physicists in that respect, and the successful applications of dispersion relations have mostly been semi-phenomenological in nature and do not depend on whether dispersion relations provide us with a complete theory.

After outlining the principles of dispersion relations I shall give a summary of the types of semi-phenomenological applications which have been fairly successful. I shall then go on to discuss the complex J-plane and its relation to the asymptotic behavior of scattering amplitudes. Finally, at the end of the talk, I shall say something about the more ambitious program which attempts to use dispersion relations to provide us with a complete dynamical theory. In particular I shall discuss the so-called dual resonance models which, to my mind, seem to provide us with the best hope at the moment of constructing a theory from dispersion relations.

To begin with, let us discuss kinematics. Suppose we have an elastic reaction in which the particles are labeled as shown is Figure 1.

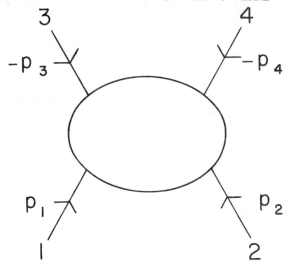

Figure 1: Kinematics of Two Body Scattering

I will define ingoing momenta as p_1 and p_2 and outgoing momenta as $-p_3$ and $-p_4$. I take the s- reaction to be $1+2 \rightarrow 3+4$, and define $s=(p_1+p_2)^2$, $t=(p_1+p_3)^2$. Then s is the square of the center of mass energy in this reaction, and t is the square of the momentum transfer. The variable t is linearly related to $\cos \theta$, where θ is the s-channel center-of-mass scattering angle. In the physical region for this reaction, $-1 \leq \cos \theta \leq 1$, s is always time-like, and t is always space-like.

The scattering amplitude A(s,t) is given by a dispersion relation of the following form.

$$A(s,t) = \frac{1}{\pi} \int_{-\infty}^{\infty} \frac{\text{Im } A(s',t)}{s'-s}$$

As you see, we always keep one variable constant, and integrate over the other variable. There are various types of dispersion relations depending on which variable or which combination of variables one keeps constant, but this is the simplest type.

Such dispersion relations are by no means confined to elementary particle physics. As a matter fact they were first proposed as early as 1926 by Kramers and Kronig in the study of the scattering of light by a dispersive medium; hence the name dispersion relations. At the moment this name is really of historical rather than descriptive significance. Dispersion relations were also applied to other branches of physics, for example potential scattering, before they were applied to elementary particle physics. They are always the result of some type of causality postulate which one applies to the system.

Now you may question whether one gets a complete theory from casuality plus other rather general requirements. The reason why this is at least plausible is that in elementary particle physics one applies causality in the form of the locality of fields. One assumes that there exist local fields and, from that assumption, one obtains the dispersion relations. Furthermore, one knows that the assumption of local fields is an extremely strong one; so strong that so far no-one has constructed any non-trivial example of such a theory. (By a non-trivial example, I mean an example other than that of free fields.) So, we see that the causality postulate required for the proof of dispersion relations is actually a fairly strong assumption.

In fact, models based upon perturbation theory seem to indicate that dispersion relations do provide us with a complete theory to within a finite number of coupling constants, provided we say which particles in the theory are elementary. In strong interaction physics it may even be that this freedom of stating which particles are elementary is absent, and that the general principles do provide us with a complete theory. I will say more about that later, but as I mentioned the practical applications of dispersion relations do not really depend on such speculative considerations.

It follows from our dispersion relation, that we have one relation between A and ImA, and we need a second equation between these two quantities. This equation is provided by the unitarity relation, but when s is above its lowest possible physical value, say, $s > s_0$. However, you notice from the dispersion relation that the integral is over all values of s. The unitarity equation is thus insufficient; we still have to specify how we can find the imaginary part of A for $s < s_0$. This further information is provided by the so-called crossing relation, which is again a consequence of the general principles of quantum field theory or of S-matrix theory.

The crossing relation says that, in addition to the reaction $1+2 \to 3+4$ which I have just written above, the same amplitude describes the reaction $1+\bar{3} \to \bar{2}+4$. For instance the same set of Feynman diagrams describes both reactions, or, if you wish to state the result without reference to Feynman diagrams, the same analytic function describes the two reactions. As a matter of fact, it also describes a third reaction, $1+\bar{4} \to 3+\bar{2}$, but we shall not consider this reaction.

For the original reaction $1+2 \to 3+4$, s is the square of the energy and t the square of the momentum transfer. For the crossed reaction, $1+\bar{3} \to 2+\bar{4}$, the square of the energy is t and s is the square of the momentum transfer. Since I have already stated that, in a physical region for a reaction, the square of the energy has to be positive or time-like and the square of the momentum transfer has to be negative or space-like, you observe that the combination of values of s and t for which the first reaction is physical, and that for which

DISPERSION THEORY AND THE COMPLEX J-PLANE

the second reaction is physical are disjoint from one another. Therefore, we cannot really say that the two amplitudes are represented by the same function; we should really say that the amplitude for one function is an analytic continuation of the amplitude for the other.

Now, if we are given the scattering amplitude for the first reaction with experimental errors, one cannot use it to obtain the amplitude for the second reaction, because analytical continuation of a function which is subject to experimental errors is in practice impossible. However, the crossing relation is just what we need for the dispersion relation, since it enables us to calculate the imaginary part of A for values of s below threshold in terms of the imaginary part of A in the physical region for the crossed reactions. This means, that in trying to construct a theory, one has to consider the equations for the direct reaction and for the crossed reactions simultaneously. Crossing, analyticity, and unitarity are the three inputs that are needed.

We now turn to the discussion of semi-phenomenological applications of dispersion relations. More such applications depend on the fact that the amplitude is dominated by single poles. Suppose that the first reaction can go through a single-particle intermediate state of mass m. Then the imaginary part of A contains a term $\text{Im}A = \pi g^2 \delta(s-m^2)$. When this term is inserted in the dispersion integral, we obtain the following contribution to A:

$$A = g^2 \frac{1}{s-m^2}$$

so that A has a pole at $s=m^2$. Thus, scattering amplitudes have poles associated with stable particles. Of course, there are not many stable particles in a strongly interacting system. The pion and the nucleons are the only ones. On the other hand, there are a large number of approximately stable particles, for when one plots the imaginary part of A against the energy, one often obtains curves of the form shown in Figure 2.

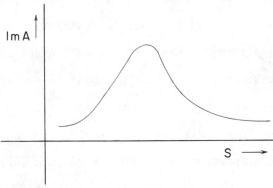

Figure 2. Imaginary part of the amplitude in a resonance region.

That is, there are narrow maxima. Now, as I said a while ago, dispersion relations essentially tell us that the scattering amplitude is analytic in the s-plane with a cut along the real axis. The imaginary part of A is related to the discontinuity across the cut, and if the imaginary part contains such a narrow maximum, one may analytically continue through the cut to obtain a pole, not on the real axis, but just a little below the real axis. That is, A contains a term of the form

$$A = g^2/[s-M_R^2-iM_I^2]$$

which is familiar as the Breit-Wigner formula for a scattering amplitude near a resonance.

We thus observe that the situation with a stable particle and a pole on the real axis is similar to that with an unstable particle, when the pole is slightly off the real axis. The modern tendency is not to distinguish, in principle, between stable and unstable particles. It is merely a matter of detailed dynamics whether a certain particle has high enough energy to decay, in which case it would be an unstable particle, or whether it does not, in which case it would be a stable particle. Thus, the Breit-Wigner formula itself is the simplest application of dispersion theory, though it was originally obtained long before the use of dispersion relations.

When one uses the fact that both the direct and the crossed reactions are dominated by single poles, one does obtain non-trivial examples of dispersion theory. Let me stress that it is by no means obvious (as a matter of fact it is surprising) that the imaginary part of A has this narrow resonance shape. Nevertheless, it does seem to be an experimental fact that, in many reactions at not too high energy, the scattering amplitude is dominated by narrow resonances. If one assumes that the direct and the crossed reactions are dominated by narrow resonances, one can calculate the imaginary part of A and can do the dispersion integral. Thus, one can use dispersion theory to obtain a semi-phenomenological description of the scattering amplitude.

As the first example, let me discuss the electromagnetic structure of nucleons, which was one of the early successful applications of dispersion theory. The actual experimental reaction studied is shown in Figure 3. Of course, we know the electron-photon vertex so the only unknown is the nucleon-photon vertex, $\gamma+N \rightarrow N$. By crossing, that is equivalent to the reaction $\gamma \rightarrow N+\bar{N}$. Let us assume that this reaction is dominated by narrow resonances. (See

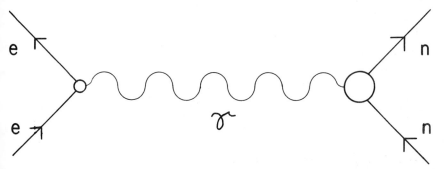

Figure 3. Electron nucleon scattering by virtual photon exchange.

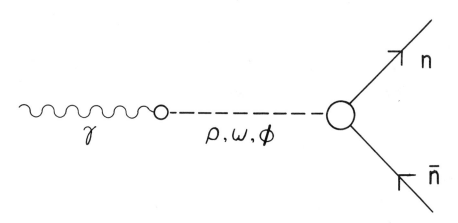

Figure 4. Resonance dominance of nucleon form factors.

Figure 4.) The quantum numbers have to be the same as the gamma, so the resonances have to be vector particles. The obvious such resonances are the ρ, ω, ϕ. By the pole-dominance formula quoted above, we find that the form factor has the form:

$$f(t) = \frac{c}{t - m^2}$$

where c is the coupling constant and m is equal to the mass of the ρ, the ω, or the ϕ. If we include all of them, there will be three such terms. Of course, we would not expect such a crude approximation to be quantatively accurate, but it does give a rough qualitative description of the electromagnetic structure of nucleons.

It is worth remarking that if the ρ, ω, and ϕ did not exist, one would get a totally wrong form factor from this sort of consideration. It would fall off much too slowly as a function of momentum transfer. In fact, it was from considerations such as these that the ρ and ω were first predicted, before they were discovered experimentally.
This then was one of the first successes of dispersion theory. The dispersion theoretic approach to quantum field theory is not the only possible one, and these particles were also pre-

dicted by non-dispersion-theoretic methods.

As a second application, let us discuss nucleon-nucleon scattering. The process is shown in Figure 5.

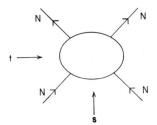

Figure 5. Channel designations for nucleon-nucleon scattering.

The direct reaction is N+N→N+N and the crossed reaction is N+N̄→N+N̄. Suppose that the crossed reaction is dominated by single particle intermediate states, or in other words, that the scattering amplitude is dominated by poles in the crossed channel. This is shown in Figure 6.

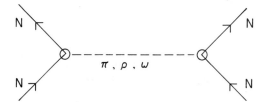

Figure 6. Meson resonance dominance of nucleon-nucleon scattering.

The nucleon-antinucleon system has zero baryon number, so these poles will be meson poles, such as ρ, π, ω, and several others.

Again, this type of diagram is present in approaches other than dispersion theory. For example, it is often said that the nuclear force is caused by exchanges of ρ, π, and ω. In dispersion theory, however, one does not generally think of one type of reaction as being caused by another. Instead one has a set of self consistency equations that one has to satisfy, and no reaction is, a priori, more fundamental than any other. The difference is really a matter of language; to say that the crossed reaction is dominated by single particle intermediate states is the same as saying that the direct reaction is caused by the exchange of single-particle intermediate states.

We know that the approximation of single-particle dominance is not correct and that there are also other diagrams. How would we minimize the effect of these diagrams? One way is to look at the states of higher angular momentum in nucleon-nucleon scattering. The reason is that the higher the mass of the exchanged object, the shorter the range of the associated force. The multi-particle intermediate states will generally have a higher mass than a single-particle state, so the single particle diagram will produce forces of longer range than the other diagrams. Short range forces are unimportant for high angular momentum because they are masked by the centrifugal barrier. Therefore, for high angular momentum, one would expect the pole diagram to dominate. One can test this hypothesis by making a phase shift analysis in which only the lowest few phase shifts are experimental parameters, and all of the phase shifts associated with the high angular momentum-states are calculated from pole diagrams. In spite of the fact that one has far fewer parameters, one can get a better fit to the data by means of such a procedure than by taking all the phase shifts as phenomenological parameters. This is another success of the theory.

At one time, it was said that meson theory gave completely wrong descriptions of nuclear forces. There are two reasons why we can now reverse that statement. First, we are being slightly less ambitious since we are not trying to calculate all partial waves. We leave out the S and P waves, say, which are dominated by short range forces which we cannot calculate. Second, we also include in our calculation the ρ, the ω, and other particles whose existence was not known when one originally tried to calculate nuclear forces from the exchange of mesons. With this modified approach,

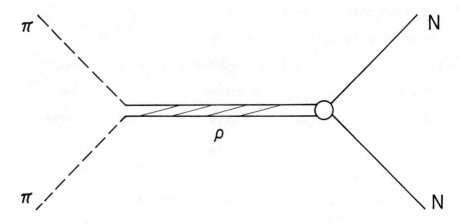

Figure 7. Rho dominance of πN scattering.

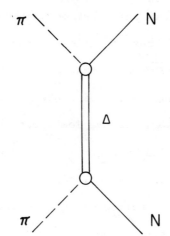

Figure 8. Direct channel resonance in πN scattering

meson theory does give an adequate model for nuclear forces. In fact, if one only calculates the very high partial waves, even the forces due to the ρ and ω can be considered short range, because the ρ and ω are much heavier than the pion. For the very high partial waves one gets good results by considering the exchange of the pion alone.

Although I have taken nucleon-nucleon scattering as an example of a case where one can use disperison theory to get reasonably good results, other elastic scattering amplitudes have also been considered; pion-nucleon scattering for instance. In the t- channel one would again have ρ exchange, whereas in the direct channel one would have the Δ, plus other nucleon resonances. (See Figures 7 and 8.) For various reasons, the calculations of pion-nucleon scattering have not been as good as those of nucleon-nucleon scattering, but some successful results have nevertheless been obtained.

In addition to elastic reactions, inelastic reactions have also been treated. One may consider a high energy process in which two particles go to many, as shown in Figure 9, and one may make the approximation that the cross channels are dominated by a single particle intermediate state, and particularly by the pion. This is the so-called multiperi-

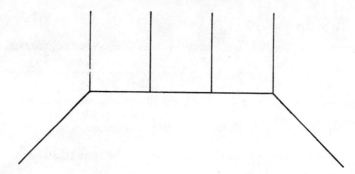

Figure 9. Many particle final state with single particle exchanges.

pheral approximation. One can get better results by taking competing, non-multi-peripheral processes into account phenomenologically. One considers the diagrams shown in Figure 10. This is the so-called absorption model, which has been very successfully applied by Gottfried and Jackson, by Dar, and others. In many cases, but not in all cases, such models do provide us with a reasonably successful semi-phenomenological

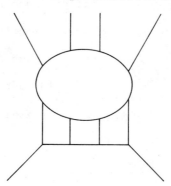

Figure 10. Diagramatic representation of the absorptive multi-peripheral model.

approach to high energy reactions.

These are just examples of methods by which one can attack various elementary particle reactions phenomenologically by means of dispersion relations. There is another completely different phenomenological approach to dispersion relations, namely, the testing of the dispersion relations themselves, we saw that the scattering amplitude is of the form

$$A(s,t) = \frac{1}{\pi} \int_{-\infty}^{\infty} ds' \, \frac{\text{Im}A(s',t)}{s'-s}$$

For forward scattering $t=0$, and one can show from the unitarity equation that ImA is related to the total cross-section, which is something one can measure experimentally. Therefore, one can insert the experimental values of $\text{Im}A(s,o)$ in this equation, and one can test whether it gives the real part of A correctly. In most cases there is, unfortunately, a complication, due to the fact that the range of integration usually includes a region below threshold where ImA is not given by any experimentally measurable quantity. Therefore, part of the integrand on the right-hand side is usually unknown. In two cases, pion-pion scattering and pion-nucleon scattering, there is no unphysical region and, in principle, $\text{Im}A(s,o)$ is susceptible to experimental measurement. This is not very helpful for pion-pion scattering, but for pion-nucleon scattering it is, and the equation has been tested, in particular by Lindenbaum and his co-workers, and has been found to be correct to within the experimental errors.

I mentioned that dispersion relations are consequences of locality, and Lindenbaum stated his result in terms of the amount of non-locality which is allowed by his experimental errors. Of course, with one equation one cannot test all possible types of non-locality, but as far as the types of non-locality which would effect this equation, the theory is local to within 10^{-14} cm.

Leaving general dispersion theory, we turn to the complex j-plane. The only physical points at which the partial wave amplitude is known are the integers. When I say the integers I shall always mean the integers for a boson system and the half-integers for a fermion system. Most of you are familiar with the mathematical probelm of analytically continuing a function which is defined for a continuous range of values of the variable However, there is also another problem of analytically continuing a function when it is defined at an infinite series of integers. That problem is not always soluble, but when it is, we know that the solution is unique, provided one makes certain neccessary assumptions about asymptotic behavior. This is a mathematical problem which had been considered long before it was applied to this kind of physics. From a study of either potential theory models or other simple models of elementary-particle physics one assumes that the analytic continuation is possible, and once one assumes that, the analytic continuation is unique.

Therefore, the amplitude is defined for general complex values of j. (Actually, in the simplest applications one only makes the analytic continuation into the right half j-plane.) In particular, one can ask questions about bound states for complex values of j. For instance, in a potential theory model, one can simply solve the Schrödinger equation, putting in a non-real value for j. The procedure is quite straightforward and one can examine the results by plotting j against $E^2 = s$.

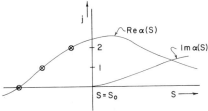

Figure 11. Regge trajectory with bound state poles.

Let the threshold be $s = s_0$. Considering first integral values of j, one may get a bound S-state; at some higher energy one might get a bound P-state, and at a still higher energy one might get a bound D-state. One then has the three points as shown in Figure 11, representing the values of j at which a bound state occurs corresponding to particular values of s. Since we are now considering non-integral values of j, we can join these points to obtain a continuous curve which we label $\alpha(s)$. At any combination of values of j and $s(< s_0)$ on this curve we have a bound state. Above threshold a bound state never occurs at a real value of the angular momentum, but we can get a bound state at a complex value of the angular momentum, as represented by the continuation of the curve to $s < s_0$. Above the threshold, α is complex, and we plot two separate curves, Re $\alpha(s)$ and Im $\alpha(s)$. The curve $\alpha(s)$ is known as a Regge trajectory.

I have already mentioned the fact that bound states correspond to poles of the scattering amplitude. If we consider the scattering amplitude, at fixed s as a function of complex j, we shall get poles at values of j corresponding to this curve $j=\alpha(s)$. Conversely, one can consider the scattering amplitude at some fixed complex value of j as a function of s, and again one gets poles corresponding to the points of this curve.

This is all very interesting mathematically, but you may ask why one worries about complex angular momentum. The answer is that it is related to the asymptotic behavior of scattering amplitudes. As I said, we have already made the assumption, which has not been proved, but which is probably true, that one can analytically continue the scattering amplitude into the complex j-plane. Let us strengthen this somewhat and make the assumption that when one does so the only singularities which one finds in the j-plane are these complex poles. That is, we assume that there are no branch cuts or other singularities in the j-plane. (We will see that we shall have to modify this assumption later.) Under this assumption, one gets results about the asymptotic behavior of the scattering amplitude from purely kinematic considerations. The relation between the analytic properties in the j-plane and the asymptotic behavior of scattering amplitudes is not conjectural. It does not depend on any assumptions, but is based on quite firm reasoning, and as a matter of fact it is true in potential theory as well.

Consider a reaction where t is the energy and s is the momentum transfer. In general, as s goes to infinity:

$$A(s,t) \to \frac{\beta(t) \, s^{\alpha(t)}}{\sin \pi \alpha(t)}$$

where $\beta(t)$ is related to the residue of the pole in the j-plane, and $\alpha(t)$ is just the Regge trajectory that I sketched. (I sketched it as a function of s, but of course if one considers the t-channel partial-wave amplitudes the function is $\alpha(t)$.) It is straightforward kinematics that, given the analytic properties in the j-plane, one can get the asymptotic behavior of the amplitude as the momentum transfer tends to infinity. In potential theory this is completely uninteresting, because the momentum transfer is related to $\cos \theta$ and, in the physical region, $|\cos \theta| < 1$, so we are not interested in what happens when $\cos \theta \to \infty$.

In the elementary-particle problem for the reaction where t is the energy, we are again uninterested in what happens when s, the momentum transfer, goes to infinity. But there now exists the crossed reaction in which s is the energy, and we are interested in what happens when the energy goes to infinity.

Let us again consider the diagram shown in Figure 1. By the s reaction I will always mean the reaction where s is the energy, by the t reaction I will mean the reaction where t is the energy. We recall that $s = (p_1 + p_2)^2$. We have just seen that the asymptotic behavior of the scattering amplitude in the s-channel as s goes to infinity is related to the Regge poles in the crossed channel, the t-channel. One can determine points on the Regge trajectory at positive values of t, corresponding to the bound states or the resonant states in the t-channel. One can also study the asymptotic behavior of the s reaction, in which t, the momentum transfer, is negative, and determine $\alpha(t)$ in the $t<0$ region. Generally, one finds that when one plots the Regge trajectory, $\alpha(t)$, for $t<0$ and extrapolates to the $t<0$ region, the bound-state poles fall on the trajectory $\alpha(t)$. This is the impressive success of Regge theory. It is this success which resurrected it from a premature death which it was supposed to have met a few years after it was proposed.

The most spectacular example is pion-nucleon charge exchange scattering. The s-channel reaction is $\pi^+ + n \to \pi^0 + p$, and we consider $s \to \infty$.

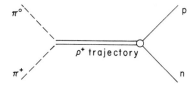

Figure 12. π-N charge exchange scattering dominated by the ρ^+ trajectory.

The asymptotic behavior is related to the Regge trajectories in the t-channel. In this channel the Regge trajectory has to be a meson-like trajectory of positive change, and the obvious trajectory is that which contains the ρ^+ (see Figure 12). From the asymptotic behavior of this reaction, one can measure the ρ Regge trajectory in the $t<0$ region, and it seems to be almost a straight line. If one extrapolates it into the $t>0$ region it does go clean through the mass of the ρ at $\alpha=1$.

As another example, consider backward pion-nucleon scattering, $\pi N \to N\pi$. The particles exchanged will now be particles with baryon number 1 instead of baryon number 0. Again, when we plot the Regge trajectory for negative t and extrapolate it into the positive t region, it does pass through the masses of the baryons at the corresponding angular momenta.

Thus, the most striking successes of the Regge theory have been the quantitative correspondence between the positions of the Regge trajectories as measured in the asymptotic behavior of the scattering amplitude and the positions of the resonances in the crossed channel. Two completely different kinds of experiments produce results which are correlated with one another. There is also the question of parametrizing the scattering amplitude by a term with the Regge asymptotic form, or more generally by the sum of a number of such terms. With regard to this, there is less universal agreement among physicists about how successful the parametrization has been. One usually can fit the experiments, but the question usually asked is, "Is it significant, in view of the number of parameters one has to take?" Probably in some reactions it is, and in others it is not.

Perhaps the reason why the success in quantitative fitting of asymptotic behavior by means of Regge trajectories has been less than spectacular is that our fundamental assumption, that the only singularities in the j-plane are poles, is wrong. It leads to inconsistencies and, in the elementary particle problem, there must be cuts as well as poles in the j-plane. Furthermore we know that, for very high values of the energies, the terms produced by the cuts must dominate over the terms produced by the poles. Therefore, Regge-pole terms alone cannot give the correct asymptotic description of the scattering amplitude at very high energies. However, they may well give a qualitatively reasonable description of the scattering amplitude at moderately high energies if the cut terms are not too big. For many reactions they seem to do so.

The reactions where Regge fitting has been least successful are the elastic reactions. Consider a reaction A+B→A+B, shown in Figure 13. In this reaction the exchanged object has quantum number zero, because the crossed reaction is A+$\bar{\text{A}}$→B+$\bar{\text{B}}$. So this reaction will be dominated by Regge poles, or cuts, corresponding to zero quantum number. We can determine the position of this j-plane singularity from the asymptotic behavior of A+B→A+B, and if we plot the position of the singularity, whatever it is, against t, we find that it is almost flat (see Figure 14).

By using the symbol α, I am not implying that the singularity is a pole; we do not really know what it is, but its trajectory is much flatter than the other trajectories. So far no one has seen any resonances at t>0 corresponding to points where the trajectory goes through integers. That may be because it is so flat that it never goes through an integer in the resonance region, or it may be because it is not a pole at all, but some other sort of singularity such as a cut. One

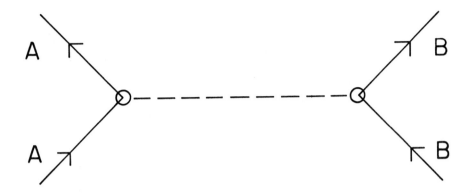

Figure 13. Regge exchange in elastic scattering.

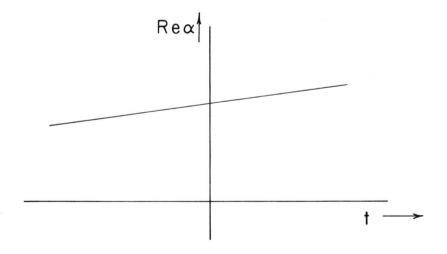

Figure 14. Chew-Frautschi plot of Pomeranchuk trajectory.

calls the singularity the Pomeranchuk singularity, because it is related to the asymptotic behavior of the scattering amplitude as given by the Pomeranchuk theorem and as discussed in Durand's lecture.* The nature of the Pomeranchuk singularity is one of the mysteries in Regge pole theory.

I would now like to leave the semi-phenomenological applications of Regge theory, and go on to discuss the relation between Regge trajectories and the elementary or composite nature of particles. Until we have a complete theory, we cannot say rigorously what we mean by an elementary particle, but we know roughly what we mean. In quantum electrodynamics the electron, the photon, and the proton are elementary particles. One criterion which we can give for the elementary nature of a particle is whether it forms part of a rotational sequence or not. We know that in the hydrogen atom, for given principal quantum number, there exists an S-state, a P-state, a D-state, and so on. These states are composite particles, and they do form part of a rotational sequence. On the other hand, there is no comparable rotational sequence associated with the electron or the photon. The electron has spin 1/2, and there are no corresponding particles with spin 3/2 or higher; the situation is similar for the photon. Therefore

---

*See Prof. L. Durand's talk, page-7, this issue.

we can tentatively say that a particle is composite or not depending on whether it forms part of a rotational sequence.

Furthermore, we have generalized the notion of rotational sequence to Regge trajectories by generalizing the angular momentum from integral to nonintegral values. Therefore we say that composite particles are those which lie on Regge trajectories and which exist when Regge trajectories pass through the integers. There may be other particles in the theory which have nothing to do with Regge trajectories, those would be the elementary particles. In terms of analytic properties in the j-plane, one says that there exists an analytic function in the j-plane which is equal to the scattering amplitude at most integers. However, if at one particular integer, say $j=1$, there is an elementary particle then analytic continuation of the amplitude is not equal to the true $j=1$ partial-wave amplitude at this particular integer. The integers at which elementary particles occur have to be treated separately.

In strong interactions a large number of particles have been seen experimentally, and so far we have no reason to separate them into two groups; elementary and non-elementary. Furthermore, we have seen how one can make a rough determination of Regge trajectories, and it is at least consistent with the available evidence that all particles lie on Regge trajectories. If we accept this fact, there are no elementary particles in strong interaction physics; all particles are composite. In a theory with elementary particles one first builds up some composite particles out of elementary particles, then one builds up other composite particles out of the first set of composite particles, and so on. However, without any elementary particles there is nowhere to start and one just has to say that any particle can be regarded as a bound state of other particles. This is sometimes called "bootstrap hypothesis;" it is also sometimes called "nuclear democracy" because it implies there is no preferred class of elementary particles.

Finally, I come to my last point of discussion, namely, the question of using dispersion relations to perform dynamical calculations. There is no absolute boundary between dynamical and phenomenological calculations; it is more or less a case of the philosophy of our approach. The question is, "Do we use dispersion relations to predict some experimental information from other experimental information, or do we actually use dispersion relations to construct a complete theory, like theories based on the Schrödinger equation in non-relativistic quantum mechanics?" At the moment it is an open question whether one can do so from dispersion theory. How much one tries to do depends on whether or not there exist elementary particles. For example, consider again quantum electrodynamics. The states of the hydrogen atom are composite particles and one is able to calculate their energies, or in other words, their masses. On the other hand the electron and proton are elementary particles and, within the framework of quantum electrodynamics, one does not try to calculate their masses. Thus, in a complete theory, one would expect to be able to calculate the masses and the coupling constants associated with composite particles, but not those associated with elementary particles, if any exist.

Very crude calculations done in this spirit, based on dispersion relations, unitarity, and crossing, have been reasonably successful. They are, however, very qualitative, and they have been successful in the sense that they imply that particles exist in those channels where attractive forces exist. Particular examples are the following: We know that there is an attractive force in the P-wave state of two pions and the ρ is a bound state of two pions. There is also an attractive force in the 33 state of the pion-nucleon system and experimentaly one sees the Δ in that state. Very crude calculations give the right order of magnitude (within a factor of two or three) of the masses and the coupling constants associated with such particles.

When one tries to improve the calculations one is faced with the difficulty that the integral equations are singular at high energies. The reason is almost certainly that the approximations give an incorrect representation of the asymtotic behavior of the scattering amplitude. The obvious thing to do is to try and combine such calculations with Regge theory, which we believe gives a reasonably good representation of the asymptotic behavior.

Incidentally, I should say that in the crude calculations I just mentioned, one gets over this difficulty by inserting a cutoff. One only has a rough qualitative idea of the position of the cutoff, and one cannot obtain results with any degree of accuracy.

One would like to perform calculations without such a cutoff, making use of Regge asymptotic behavior, but then they become fairly complicated. However, Collins and Johnson have recently done such a calculation for the two-pion system, by neglecting interactions between the two-pion system and any other system. They were successful in obtaining the ρ as a bound state of the two pion system. They could not calculate the mass of the ρ relative to that of the π, but they could calculate the width of the ρ with reasonable accuracy. Furthermore, and this is probably the most impressive feature of their calculation, they were also able to calculate the S-wave pion-pion scattering length and to get results which agree with experiments.

Now, in any bootstrap calculation, it is often difficult to distinguish between the imput and the output. For instance, Collins and Johnson certainly showed that they could get a consistent solution in which the ρ appeared as a bound state of two pions, but they did not rule out the existence of a consistent solution which does not have a ρ. They did not use any input connected with S-wave pions, however. After they had finished their calculations, and obtained the width of the ρ, they looked at the S-wave and found a result which agreed with experiments. This was definitely a prediction of their calculation which was independent of the input.

The calculation of Collins and Johnson is quite an impressive beginning, but it was pretty complicated, and at the moment it is just an open question whether it will be feasible to do such calculations where one includes interactions between the two-pion system and other systems. Another approach to such calculations has been suggested and developed by Chew, Goldberger, Low, and their co-workers. Their approach starts with the kind of multi-peripheral calculation. This approach has also been reasonably successful so far, but again no one knows whether it provides us with a complete theory.

To conclude I would like to outline still another approach. The Regge trajectory in my previous diagram rose to some finite value and then turned over. It was motivated by potential theory, and also by models in elementary particle physics where one neglects interactions with multi-particle states. However, the experimental Regge trajectories appear to be almost straight lines. As an example, let us take a fermion trajectory where we have most experimental information. For reasons which I have not gone into, resonances on a trajectory are spaced by two units of angular momentum. The nuclear trajectory will have resonances when $\alpha=1/2, 5/2, 9/2$ and so on. The first few have been observed, the energy and spin have been measured and have been found to correspond to a straight-line trajectory.

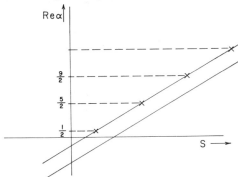

Figure 15. Chew Frautschi plot showing linearly rising trajectories and first daughter.

If one extrapolates this straight line to the point where it crosses the next half-integer, 13/2, a resonance does occur at almost exactly the predicted value of s; similar resonances are found at values of s corresponding to higher spin recurrences on the linear trajectory. The spins of these higher resonances have not been measured, but they are known to be high because these resonances are narrow and yet have a large Q-value for their decay. It is tempting to identify these higher resonances with the higher members of the Regge trajectory. Thus evidence is consistent with Regge trajectories which are linear to within the limits of experimental error, surprising actually if they were exactly linear.

Furthermore, the resonances are very narrow, a fact which

I mentioned at the beginning of my talk. These two features are related. If resonances are infinitely narrow, the Regge trajectory must be linear. One may enquire whether such a behavior is plausible? The simple models give Regge trajectories which look more like the potential-theory trajectories which I showed earlier. But the reason why the Regge trajectories tend to turn over is that, in all these models, there is an energy above which interactions tend to become weak. Once one exceeds this energy the Regge trajectories turn over and approach some finite value at infinite energy. In the physical elementary-particle problem, however, where there exist interactions with multi-particle systems of arbitrarily high complexity, there is probably no energy above which the interactions tend to become weak. It is thus not implausible that one would obtain infinitely rising linear Regge trajectories.

These facts suggest a new kind of weak-coupling approximation. Narrowness of resonances does mean that the interactions are weak; they take a long time to decay. This new type of weak coupling approximation assumes that all narrow resonances are on linearly rising Regge trajectories. It agrees with the conventional weak coupling approximation in that resonances are narrow, but it disagrees with the conventional weak-coupling approximation in that the resonances are composite and not elementary. That is, they do lie on Regge trajectories. This, if you like, is an extreme nuclear democracy where any angular momentum has resonances and no angular momentum is distinguished from any other. Whereas we know that the conventional weak coupling approximation is inadequate for strong coupling physics, this kind of weak coupling approximation may well be a reasonable approximation because resonances in nature are observed to be narrow. Effects due to finite width and due to non-linearity of the trajectories are regarded as corrections.

In this kind of weak coupling approximation, unlike the conventional weak coupling approximation, there are consistency requirements to satisfy even in the lowest approximation, simply because all particles are composite. The amplitude, viewed from the s-channel, is equal to the sum of terms representing scattering through a number of resonances. In the t-channel, the amplitude has a similar representation. This feature is shown graphically in Figure 16. We emphasize that we do not add s-channel diagrams and t-channel diagrams. The whole amplitude must be expressed as the sum of resonances in

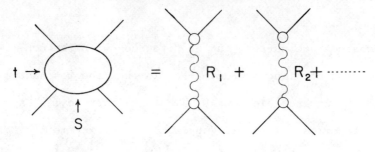

Figure 16. Resonance dominance of the scattering amplitude in both s and t channels.

the s-channel, or alternatively as the sum of resonances in the t-channel. This is due to the fact that these resonances are composite and not elementary, and this requirement is the so-called requirement of duality. Other self-consistency equations arise from the requirement that **any** resonance should also be able to take part as an external (scattered) particle.

There are thus a large number of self-consistency conditions to satisfy, and it surprised me to find out that these self-consistency problems could all be satisfied analytically. This has been done in the dual resonance models. The simplest model was first written down by Veneziano and it has been generalized by many people. The Veneziano model itself only solved the problem with scalar particles on the outside. It does not attack the further self-consistency problems where the higher-spin resonances play the role of external particles. (See Figure 17). However, the Veneziano model was indepen-

Figure 17. High-spin states as external lines.

dently extended from four particles to five by Bardacki and Ruegg and by Virasoro, and was then extended to six and higher numbers of particles by a number of people. Thus, one can also get amplitudes like that shown in Figure 18. Now consider an eight point function with scalars on the outside.

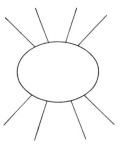

Figure 18. Eight-line process with scalar particles.

By looking at intermediate resonances one can get a four point function with higher-spin resonances on the outside and one can satisfy all the self-consistency problems. (See Figure 18

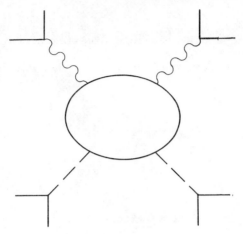

Figure 19. Four-line process with higher spin external states.

and 19.)

Perhaps I have been a bit too glib, because all the models which have been proposed so far are not free from defects. Either they have ghosts (particles with negative decay probability) or they have particles with imaginary mass. But I think they do offer a promising start.

In particular, having constructed a four-point amplitude from an eight-point amplitude, we no longer need the external scalar particles used in the construction. They were just a mathematical aid in getting the four-point amplitude for particles with spin. The reason this is important is that by a slight generalization from scalar external particles to particles of spin 1/2, we could regard these outside particles as quarks. Therefore the quarks might exist as mathematical entities but not as actual particle in the complete theory, because the real particles in the theory are those which appear on trajectories. By this method we can understand why the meson spectrum or the baryon spectrum that we observe in nature is the spectrum that one would get by combining two or three quarks, even if quarks as such do not exist as real particles. The trouble at the moment is that these quark-like models suffer particularly from ghost diseases which have not been cured as yet. We therefore do not yet have consistent theories satisfying all our requirements. Nevertheless, the Venziano type models, I believe, have so many attractive features that they provide the most promising starting point at the moment for constructing an elementary particle theory.

DISCUSSION OF PROF. MANDELSTAM'S TALK

A. O. Barut, University of Colorado: I have a question about one of your basic assumptions, namely unitarity. The successful applications of dispersion theory are based on the single-particle dominance of the amplitudes in question. The relation $\text{Im} A = g\delta(s-m^2)$ does not really follow from unitarity when s is below threshold. In a complete dynamical theory you would like to get these poles from unitarity in the physical region, and then reinterpret them in terms of single-particle Feynman diagrams, is that correct?

Mandelstam: In a conventional calculation of the sort performed before the Veneziano model, one essentially computed the amplitude from the discontinuity across the cut. Then one would analytically continue through the cut and find a pole (on the second sheet) which would correspond to the position of the unstable particle. The bound state poles also really follow from unitarity. Without going into details let me indicate the method in solving the dynamical equations one usually writes the scattering amplitude as the ratio of two functions, which are determined by integral equations which follow from the unitarity conditions. When one finds a zero in the denominator function, that zero corresponds to the position of a pole, which may occur below threshold.

S. Okubo, University of Rochester: As you have explained, Regge poles have a simple physical interpretation. Does the Regge cut have a similar simple physical meaning?

Mandelstam: No, the interpretation is not as simple as in the case of poles. Regge cuts arise from the unitarity condition, which involves integration over intermediate states. In particular, cuts follow from multiparticle intermediate states such as this:

Now as these intermediate momenta vary during the integration, the parameters associated with the two parts of the diagram vary and, in particular, the positions of the Regge poles vary. If you perform the integration, cuts appear in the final amplitude.

L. C. Biedenharn, Duke University: Doesn't the existence of a cut eliminate any predictive power? The pole is a very simple object, but the discontinuity of the cut is a function, and hence requires infinitely many parameters.

Mandelstam: If you want to make truly quantitative predictions, I agree. But if cuts account for only 25% of the asymptotic amplitude at moderate energies then at least we have some predictive power for 75% of the amplitude. This is not as much as one would like, but it is not zero.

M. Nauenberg, University of California at Santa Cruz: I would like to pursue this discussion of cuts a bit farther. If, as phenomenological analysis seems to indicate, cuts do play an important role, then they should be included in the duality program. You showed us how one can obtain a Veneziano model by assuming Regge pole dominance. But how do you implement the notion of duality if both poles and cuts are important?

Mandelstam: First, as I pointed out, the Veneziano model is just a weak coupling approximation, and in this approximation one does not have any cuts, either in the angular momentum or in the energy plane. We know that this is an approximation, and one has to take into account the finite width of resonances. This introduces cuts into the energy plane. Now at the moment the only way anyone knows how to use finite widths in a dualistic model is to regard the Veneziano amplitude itself as the Born term. Hopefully, when one understands more about this semi-perturbative method, higher order corrections will produce cuts, both in the energy and the angular momentum plane. Some work has been done on this question in simple models. That is the simplest way I know to calculate the cuts. I don't know if this program is possible, but it is at least less impossible than the program of calculating the cuts from the extremely complicated Feynman diagrams in which they occur.

J. W. Moffat, University of Toronto: One of the reasons for constructing the Veneziano amplitude was to get a crossing symmetric model. Yet, if one uses a K-matrix method or other methods to introduce absorptive effects, crossing symmetry will be lost. In fact, this is an embarassing feature of all phenomenological calculations which use cuts: the backward region does not like what you do in the forward region.

Mandelstam: I agree fully with this comment, and in fact, I was not thinking of K-matrix methods but rather of the semi-perturbative methods such as that originally proposed by Kikkawa, Sakita, and Virasoro. In semi-phenomenological applications, K-matrix methods are useful, but for a fundamental improvement in the theory one wants to retain crossing symmetry, and should therefore make unitarity corrections by these semi-perturbative techniques, rather than by the K-matrix approach.

R. Yaes, University of Texas: Does duality have an unambiguous meaning in situations other than these narrow resonance dual models? In other situations, it seems to me that everyone has his own duality.

Mandelstam: I don't think it has an unambiguous meaning.

R. L. Warnock, Illinois Institute of Technology: I noticed that in your talk you used only fixed-t dispersion relations. Could you comment on the existence of double dispersion relations with a finite number of subtractions?

Mandelstam: I don't believe that one has any reason to expect that the amplitude should not increase exponentially as both s and t go to infinity, because this is a completely unphysical region. In fact, if the Veneziano type amplitudes have any relation to the real world, then the amplitudes will increase exponentially as both s and t go to infinity.

Warnock: So far the most ambitious and realistic bootstrap calculation, namely the model of Collins and Johnson, seems to assume a Mandelstam representation with a finite number of subtractions. Is there some way in which this can be reinterpreted in terms of fixed-t dispersion relations?

Mandelstam: In some ways the Collins and Johnson calculation, of course, is completely alien to the Veneziano approach, in the sense that they only allow for interactions with two-pion systems, and therefore, one would not expect these calculations to produce infinitely rising trajectories. However, I don't think there is any reason why this model could not be a reasonable approximation to the real world at low energies. In particular, Collins and Johnson write down dispersion relations at fixed s and dispersion relations at fixed t, but I don't think they really ever have to assume that if they were to write down dispersion relations valid for all values of s and t the number of subtractions would be independent of s and t. Indeed, they might not be free from the necessity of using some kind of a cutoff, though their calculations are much less dependent on the cutoff than are those in the cruder models of which I spoke. I think they obviate the necessity of having to examine the question whether a finite or an infinite number of subtractions is necessary in the real world.

THE CONSTRUCTION OF CROSSING-SYMMETRIC UNITARY SCATTERING AMPLITUDES*

Robert L. Warnock

Illinois Institute of Technology, Chicago, Illinois 60616

and

Argonne National Laboratory, Argonne, Illinois 60439

Abstract: For the symposium "The Past Decade in Particle Theory", Prof. Warnock reviews the question of how to construct scattering amplitudes which satisfy the postulates of S matrix theory. The review traces the development of the theory from the early formulation of the Chew-Low equation to the recent proofs of existence of amplitudes with Mandelstam analyticity, crossing, and unitarity. The problem of constructing amplitudes is viewed as having two parts. The first part is to solve the partial-wave dispersion relation with given left-hand cut and given elasticity. The solution of this part by the N/D method, and the associated matter of CDD poles, is briefly summarized. The more difficult second part, which has to do with imposing crossing symmetry and Mandelstam analyticity, is considered as a problem in non-linear functional analysis. Elementary tools of the latter subject (fixed point theorems, implicit function theorems) are introduced and applied to the Low equation and to the Mandelstam crossing-unitarity equation. The article ends with some comments on the outlook for S matrix dynamics.

---

*Work partially supported by the National Science Foundation, and performed in part under the auspices of the U.S. Atomic Energy Commission.

I will discuss attempts to make S-matrix theory into a more or less complete dynamical scheme. This is the boldest part of S-matrix theory, and the part that some people find most inspiring. By combining Mandelstam analyticity, crossing symmetry, and unitarity with certain physical simplifications and assumptions of good asymptotic behavior, one derives singular, non-linear integral equations for the scattering amplitudes.[1,2,3] Up to now, these equations have been neither clearly successful nor clearly unsuccessful as a basis for dynamics, but they have at least been historically influential. The bootstrap philosophy, which is still actively cultivated in one guise or another, grew up in this context.[3,4] Furthermore, the introduction of Regge's complex angular momentum into relativistic theory was a direct result of work on the strip model equation.[5,6]

At the present time, S-matrix dynamics as originally conceived appears to be relatively quiescent, apart from local enthusiasms. A good example of the latter is the revival of the strip approximation by Collins and Johnson in Durham.[7] Part of the reason for decreased interest may be frustration about the problem of many-particle amplitudes. For a precise dynamical theory it would certainly be necessary to include all the many-particle amplitudes in the dispersion relations and in the unitarity condition. Since the analyticity properties of even the five-line amplitude are quite complicated, such a direct approach to a complete theory is not likely to succeed in the near future. For the time being, one relies on special models of inelasticity: the strip approximation, the multi-peripheral model, or a model in which all channels are approximated as two-body channels. Such models do not carry a great deal of conviction at present, but it should be realized that they are neither well developed nor fully evaluated. The strip approximation, for instance, has not been worked out with sufficient clarity so that one can either accept it or reject it. It is conceivable that with further improvements in many-particle models, but without a real solution of the n-body problem, S-matrix theory could lead to an understanding of particle dynamics comparable in depth to present day nuclear physics. In any case, S-matrix theory has produced more reasonable looking ideas on the n-body problem than any other approach. Also, it lends itself to restricted studies of a sort not usually possible in field theory. One can study in detail a part of the S-matrix in a restricted domain of energy and momentum transfer, while treating the rest in less detail through some phenomenological description.

Aside from the many-particle problem, there is another current doubt in S-matrix dynamics; namely, the question of existence of Mandelstam's double dispersion relations, with a finite number of subtractions. The analyticity conjectured by Mandelstam is usually accepted as a working hypothesis, but the matter of asymptotic behavior is in question. Doubt was expressed by Mandelstam himself in 1963.[8] His theory of Regge branch points seems to imply that if a fixed-t dispersion relation is written with N subtractions, the number N must grow indefinitely as t is increased in the positive direction. There are ways to escape from this argument, but no way of deciding whether it is wrong or right. This leaves us with three choices. First, we may suppose that the argument is wrong, and that a Mandelstam representation holds with some large number of subtractions. That many subtractions are involved should not bother us, since Froissart[9] and Martin[10] have shown that most of the subtraction constants and single-spectral functions are determined, in principle, by the double spectral functions. This is the view I will take for the purposes of this paper. As a second possibility, one can

try to get by with fixed-t dispersion relations. In such
an approach it is still possible to derive restricted S
matrix equations, and much of what I have to say will apply
to that scheme. Third, one can abandon ship entirely, and
try the generalized Veneziano model, which doesn't use dispersion relations at all. Although this approach aspires
to solve the many-particle problem, and boasts some elegant
formulas for many-particle amplitudes, it seems to be stuck
in some difficulties of principle at the moment.

A serious impediment to progress in S matrix dynamics
is the difficulty of the mathematical problems involved. The
integral equations are not well understood. We have not
given enough attention to the question: "Exactly what sort
of creature is a non-linear integral equation with Cauchy-type singularity?" How many solutions does it have, what
are their qualitative properties, how can they be computed
numerically? This is important, since physical intuition
often proves to be merely physical prejudice in problems of
this type.

During the past three years, there has been a modest
movement afoot in which a direct attack is made on these non-linear problems[11-15]. Instead of trying to circumvent the
equations, or to solve them numerically in a blind fashion,
one uses methods of non-linear functional analysis to prove
existence theorems, to study qualitative properties of solutions, or to provide convergent iterative methods and general

guidance for numerical solutions. A similar approach is used by applied mathematicians working in non-linear mechanics, where one finds an older tradition and more sophistication.

I will outline the progress which has been made in this line of work. The outline will serve as an indication of the present status of S matrix dynamcis as it would appear to a rather dour observer, one who would agree with Gertrude Stein that "There is no use in telling more than you know, no not even if you do not know it."

I apologize for saying nothing about the ingenious and persistent work of André Martin and others of the same school who have proved many valuable qualitative results about the S matrix on the basis of very economical assumptions.

The grandfather of S matrix equations is Low's equation for scattering of p-wave π mesons from a static nucleon source[16]. For complex meson energy z the equation reads as follows:

$$f_\alpha(z) = \frac{\lambda_\alpha}{z} + \frac{1}{\pi} \int_1^\infty \frac{d\omega' \, \rho(\omega') |f_\alpha(\omega'+i0)|^2}{\omega'-z} +$$

$$\sum_\beta c_{\alpha\beta} \frac{1}{\pi} \int_1^\infty \frac{d\omega' \, \rho(\omega') |f_\beta(\omega'+i0)|^2}{\omega'+z} , \qquad (1)$$

$$f_a(\omega+i0) = \frac{\sin \delta_a(\omega) \, e^{i\delta_a(\omega)}}{\rho(\omega)}, \quad \omega \geq 1;$$

$$\lambda_\alpha = -g_\alpha^2 + \sum_\beta c_{\alpha\beta} g_\beta^2. \qquad (2)$$

$$\rho(\omega) = \frac{k^3(\omega) \, v^2(k)}{12\pi}; \quad k = (\omega^2 - 1)^{\frac{1}{2}} \qquad (3)$$

The function $v(k)$ provides a high-energy cut-off, $-g_a^2$ is the residue of the nucleon pole in channel a, and $c = [c_{\alpha\beta}] = c^*$ is the crossing matrix, which satisfies $c^2=1$. Eq. (1) becomes a singular integral equation for $f_a(\omega+i0)$, upon taking the limit $z \to \omega+i0$, $\omega \geq 1$. This equation has some of the typical features of later, more complicated equations. It has the Cauchy singularity $1/(\omega'-\omega)$, and it is **strongly** non-linear at the physical value of the coupling constant. Although the equation is not likely to provide more than a vague account of low-energy pion-nucleon scattering, it is a good example on which to practice non-linear techniques. It is still not well understood in the physical domain of large coupling, and its properties for small coupling were only recently elucidated.[15]

Soon after Mandelstam's proposal[1] of simultaneous analyticity in s and t, Chew and Mandelstam[2] proposed a partial-wave equation which may be regarded as a relativistic generalization of the Low equation. If $A_\ell^I(s)$

is the partial-wave amplitude for π-π scattering with angular momentum ℓ and isospin I, the equation reads

$$A_\ell^I(s) = B_\ell^I(s) + \frac{s-s_o}{\pi} \int_4^\infty \frac{\rho(s')|A_\ell^I(s'+i0)|^2 ds'}{(s'-s_o)(s'-s)}, \quad (4)$$

where

$$A_\ell^I(s+i0) = \frac{\sin \delta_\ell^I(s)\, e^{i\delta_\ell^I(s)}}{\rho(s)}, \quad \rho(s) = \left\{\frac{s-4}{s}\right\}^{1/2}, \quad (5)$$

$$B_\ell^I(s) = R_\ell^I(s) + \frac{s-s_o}{\pi} \int_{-32}^0 \frac{\mathrm{Im}\, A_\ell(s'+i0)\, ds'}{(s'-s_o)(s'-s)}. \quad (6)$$

The discontinuity over the left cut in the interval $[-32,0]$ is determined by crossing as follows:

$$\mathrm{Im}\, A_\ell(s+i0) = \frac{1}{s-4} \sum_{I'} \beta_{II'} \int_4^{4-s} dt\, P_\ell\left(1+\frac{2t}{s-4}\right) \sum_{\ell'=0}^\infty (2\ell'+1) P_{\ell'}\left(1+\frac{2s}{t-4}\right) \mathrm{Im}\, A_{\ell'}^{I'}(t+i0). \quad (7)$$

The remainder of the left-cut discontinuity is not easily determined by crossing, since the partial wave series of (7) does not converge if $s < 32$. The contribution of this remainder is represented by $R_\ell^I(s)$ in Eq. (6). At least for the purposes of low-energy dynamics, it should be permissible to represent R_ℓ^I by a rational function or some other empirical formula with adjustable parameters. The parameters may be determined, in principle, by imposing constraints of crossing symmetry and correct threshold behavior, which are very powerful and might serve to determine a large number of parameters. The determination of the parameters is a non-linear problem, however, which has not yet been studied systematically. Recently developed techniques for imposing crossing on partial-waves may be helpful in approaching the question.[17] For a fixed and given R_ℓ^I, the Chew-Mandelstam equation may be analyzed by the same techniques that have been used on the Low equation.

A more sophisticated approach, in which crossing and threshold behavior are treated more clearly and completely, is based on Mandelstam's crossing-unitarity equation. For the case of π-π scattering with neglect of isospin, and with an unsubtracted Mandelstam representation for the amplitude $A(s,t)$, Mandelstam's relation is as follows:

$$\rho^{e\ell}(s,t) = \iint_{K^{-2}>0} K(s,t,t_1,t_2) d^*(s,t_1) d(s,t_2) dt_1 dt_2, \quad (8)$$

$$K = \left\{ \frac{4}{\pi s(s-4)} \right\}^{1/2} \left\{ t^2 + t_1^2 + t_2^2 - 2(tt_1 + tt_2 + t_1 t_2) - \frac{4tt_1 t_2}{s-4} \right\}^{1/2}, \quad (9)$$

$$d(s,t) = \frac{1}{\pi} \int_4^\infty ds' \left\{ \frac{1}{s'-s} + \frac{1}{s'-u} \right\} \rho(s',t). \quad (10)$$

Eq. (8) as it stands is merely a definition of $\rho^{el}(s,t)$, the part of the double-spectral function $\rho(s,t)$ which is "elastic" with respect to the s-channel. That is, $\rho^{el}(s,t)$ receives only contributions from two-particle intermediate states of the s-channel. It is equal to $\rho(s,t)$ in a restricted range of s:

$$\rho(s,t) = \rho^{el}(s,t), \quad \frac{4t}{t-16} \leq s \leq 16 \frac{t-4}{t-16} \quad (11)$$

Eq. (8) becomes an integral equation for ρ^{el} if we introduce the following approximation of Mandelstam,[1] which is the starting point of the strip model:

$$\rho(s,t) = \rho^{el}(s,t) + \rho^{el}(t,s). \quad (12)$$

This Ansatz is crossing symmetric, and it provides an interesting, specific model of inelastic processes as they affect the elastic scattering. The term $\rho^{el}(t,s)$ has only 2-particle states in the t-channel, but many-particle states in the s-channel.

For a realistic strip model, it is necessary to put subtractions in the Mandelstam representation, so that there may be Regge poles in the right half ℓ-plane. The corresponding subtracted version of the integral operator (8) then requires a "strip cut-off" function if it is to be well-defined as an operator in a reasonable space of functions.[18]

One may avoid approximations and still obtain an integral equation for $\rho^{e\ell}$ by regarding the difficult part of the double spectral function as given. This is the viewpoint of Atkinson,[13] who writes

$$\rho(s,t) = \rho^{e\ell}(s,t) + \rho^{e\ell}(t,s) + v(s,t), \quad (13)$$

with $v(s,t)$ regarded as a given "input force." Under appropriate conditions on $v(s,t)$, Atkinson shows that Eq. (8) has a solution, from which an unsubtracted Mandelstam representation for the amplitude may be constructed. This demonstrates, for the first time, the existence of functions which satisfy crossing symmetry, unitarity in the elastic region, and which have an unsubtracted Mandelstam representation. Under additional restrictions on $v(s,t)$, the unitarity inequality Im $A_\ell \geq \rho |A_\ell|^2$ may be met at all energies. The proof can be extended to include subtractions, but then the unitarity inequality cannot be enforced with present techniques.

One step in analyzing the equations just described is to solve a partial-wave dispersion relation with a given left-hand cut term. Of course, the left-hand cut term is not given, but it is helpful to pretend at first that it is. The partial-wave dispersion relation is then a singular, non-linear integral equation, but of such a special type that it can be solved by a clever procedure of Chew and Mandelstam, the N/D method. The N/D theory has been worked out rigorously to a fair degree of completeness. I will review this briefly, since the generality of the technique as it stands now may not be appreciated, even though the main idea is very well known. Suppressing reference to the angular momentum state and channel indices, we write the single channel partial-wave dispersion relation for π-π scattering as

$$A(s) = B(s) + \frac{s-s_o}{\pi} \int_4^\infty \frac{\text{Im } A(s'+i0) \, ds'}{(s'-s_o)(s'-s)}, \quad (14)$$

$$\text{Im } A(s+i0) = \rho(s)|A(s+i0)|^2 + \frac{1-\eta^2(s)}{4\rho(s)}. \quad (15)$$

Both the left-cut term $B(s)$ and the elasticity $\eta(s)$ will be regarded as given. These functions are determined, in principle, as functionals of all the partial-wave amplitudes. The elasticity is assumed to be Hölder-continuous in any finite interval:

$$|\eta(s) - \eta(s')| \leq K|s-s'|^\mu; \; K, \mu > 0. \quad (16)$$

The "force function" $B(s)$ is assumed to be analytic in at least a small neighborhood of the right-hand cut $[4, \infty)$. The width of this neighborhood may contract to zero at infinity. Thus, we do not require that (14) be a partial-wave dispersion relation in the strong sense, in which B is a convergent integral over a left-hand cut. The B function may have an essential singularity at infinity, for instance.

The amplitude is written as $A(s) = N(s)/D(s)$, where $D(s) = D(s^*)^*$ is meromorphic in the s plane with a cut $[4,\infty)$, is bounded by a power at infinity, and satisfies the Hilbert boundary condition

$$D(s+i0) = e^{-2i\delta(s)} D(s-i0). \qquad (17)$$

Here $\delta(s)$ is the real part of the phase shift. Provided that δ is bounded, and Hölder-continuous on any finite interval, the general D function with these properties has the representation

$$D(s) = R(s) \exp\left\{ -\frac{s}{\pi} \int_4^\infty \frac{\delta(s')ds'}{s'(s'-s)} \right\} \qquad (18)$$

$$= 1 + (s-s_0) \left\{ \sum_{i=1}^n \frac{c_i}{s-s_i} + \frac{1}{\pi} \int_4^\infty \frac{\text{Im } D(s'+i0)ds'}{(s'-s_0)(s'-s)} \right\}.$$

$$(19)$$

The rational function $R(s) = R(s^*)^*$ is arbitrary, so that the residues c_i and positions s_i of the Castillejo-Dalitz-Dyson (CDD) poles are free parameters at this point. Now if $A(s)$ is an amplitude with bounded and Hölder-continuous phase-shift, then a necessary consequence of Eq. (14) is that the imaginary part of $D(s+i0)$ satisfy the following linear integral equation:[19]

$$\eta(s)\phi(s) = \frac{C(s)}{s-s_0} + \sum_{i=1}^{n} c_i \frac{C(s)-C(s_i)}{s-s_i} +$$

$$\frac{1}{\pi} \int_{4}^{\infty} \frac{C(s)-C(s')}{s-s'} \rho(s')\phi(s')ds', \qquad (20)$$

$$\phi(s) = \frac{-\mathrm{Im}D(s+i0)}{\rho(s)(s-s_0)};$$

$$C(s) = B(s) + \frac{s-s_0}{2\pi} P \int_{4}^{\infty} \frac{1-\eta(s')}{(s'-s_0)(s'-s)} ds'. \qquad (21)$$

Conversely, if B and η are subject to reasonable conditions of asymptotic behavior, and $\eta(s)$ has no zero for $16 \leq s < \infty$, then Eq. (20) is a Fredholm equation in L^2, and a solution of (20) leads to a solution of (14), via (19) and (17), provided $D(s)$ has no spurious zero (no "ghost") in the complex plane. If the asymptotic behavior of partial waves is estimated from Regge theory with a moving Pomeranchuk pole (with or without associated branch points), it is found that δ is bounded, and B and η have the asymptotic behavior required for Fredholm theory to apply to Eq. (20).[19] It seems that the mathematics applies to the case of physical interest, and one need not truncate the equations at some finite energy, as is often done.

The general solution of (14) depends on the hidden CDD parameters, c_i and s_i. The existence of hidden parameters (i.e., the existence of infinitely many solutions) is a pervasive feature of equations with Cauchy singularities. We shall find that this feature persists in the full, relativistic equations with crossing symmetry. There is, therefore, a serious challenge to find the exact physical interpretation of the CDD poles. One way in which CDD poles could arise seems quite clear, if one is willing to imagine variations in strength of interactions or of interchannel couplings. Suppose that as interaction strengths are varied, a zero of the partial wave S matrix, $S=\eta e^{2i\delta}$, decides to pass through the physical cut above the inelastic threshold. If $\eta(\hat{s})=0$, say, the character of the N/D equation suddenly changes. It is no longer a Fredholm equation, because we cannot divide through by $\eta(s)$ to put it in the standard form. In general, it no longer has a square-integrable solution, but acquires instead a generalized function solution of the form

$$\phi(s) = \alpha\delta(s-\hat{s}) + \Psi(s), \qquad (22)$$

where $\Psi(s)$ satisfies a Fredholm equation, and the coefficient α of the δ function is determined uniquely by the solution of the latter equation. Inserting (22) in (21) and (19), we see that a CDD pole has appeared at the position $s=\hat{s}$, with a <u>determined</u> residue $c=\rho(\hat{s})a$. Such zeros of S can be associated with resonances in particular models, and they seem to occur very close to the real axis in certain partial waves of pion-nucleon scattering. Another origin for CDD poles was suggested by Dyson soon after they were discovered. They could be manifestations of discrete spectrum eigenstates of the unperturbed Hamiltonian; i.e., elementary particles. This interpretation is valid in soluble models of the Lee type, but whether it would hold up in a properly formulated relativistic field theory is a question for the future. At any rate, such "non-dynamical" CDD poles are regarded as poison by those committed to a nuclear democracy, in which no particle has the privileged status of elementarity. Chew and Frautschi proposed to rule them out by the principle of analyticity in angular momentum. Elementary particles were presumed to cause non-analytic terms in the scattering amplitude of the form $\delta_{\ell J}$, where J is the spin of the particle. It is difficult to find any strong experimental evidence for nuclear democracy, but it is still interesting to look for dynamical models with ℓ-plane analyticity. Besides being pretty, the latter principle is a very strong constraint which will certainly help to narrow down the class of possible theories.

Before leaving the N/D method, it should be noted that it generalizes nicely to the case of n coupled two-body channels, with a matrix elasticity function.[20] To establish this generalization rigorously, one has to prove that a matrix N/D representation exists for a sufficiently wide class of scattering matrices. This has been done under conditions weak enough to allow Regge asymptotic behavior (again with a moving Pomeranchuk pole).[21] Mandelstam has proposed an N/D method which will accomodate three-body channels, but it is rather complicated, and some points of rigor are not settled.[22]

The N/D method by itself will not produce any physics, but it is useful, as we shall see, in analyzing the full non-linear equations with crossing. We now turn to the full non-linear problem. The S matrix equation will be written as

$$y = A(y) , \qquad (23)$$

where y represents the quantity we wish to determine: a scattering amplitude, a double-spectral function, or whatever. The mapping A is a non-linear, singular integral operator on some function space, and a solution of (23) is called a fixed point of the mapping. The operator A depends on certain data: coupling constants, masses, CDD parameters, elasticity functions, etc. We denote these data collectively by x, and write (23) in the form

$$F(x,y(x)) = 0 \qquad (24)$$

where

$$F(x,y) = y - A(y;x). \qquad (25)$$

We wish to find the dependence of the solution y on the data x; i.e., to determine the "implicit function" y(x) in (24). There exist elementary answers to the fixed point problem (23) and the implicit function problem (24), which I will now review.

For the fixed point problem we may apply the <u>Contraction Mapping Theorem:</u>[23] Let K be a complete metric space, and let A be a mapping of K into itself; A : K → K. Suppose that the mapping is <u>contractive</u>; i.e.,

$$d(A(y), A(z)) \leq \beta \, d(y,z), \qquad (26)$$

$$\beta < 1, \text{ all } y, z \in K.$$

Here d(y,z) is the distance between y and z, and β is independent of y and z. In other words, A brings every pair of points closer together. Then the equation

$$y = A(y) \qquad (27)$$

has a unique solution in K, which is the limit of the convergent sequence $\{y_n\}$,

$$y_{n+1} = A(y_n), \qquad (28)$$

where y_0 is any element of K. At the n-th iteration, the error is bounded in terms of the error at the first iteration:

$$d(y,y_n) \leq \frac{\beta^n}{1-\beta} d(y_1,y_0) . \qquad (29)$$

Thus, we have existence, uniqueness, a method for calculating the solution, and an estimate for the error in the calculation. In applications, we normally take the metric space K to be a closed bounded subset of a Banach space, using the norm as the distance: $d(y,z) = ||y-z||$. A Banach space, you recall, is a linear space which is complete in the sense that every Cauchy sequence has a limit with respect to the distance provided by the norm: if a sequence $\{y_n\}$ has the Cauchy property,

$$||y_n - y_m|| \leq \varepsilon , \quad n,m > N(\varepsilon), \qquad (30)$$

then there is a y such that $||y_n - y|| \to 0$, $n \to \infty$.

Schauder's fixed point theorem is also available. This is a topological theorem, which is an infinite-dimensional generalization of Brouwer's theorem. The latter is as follows: A continuous mapping of a closed ball in n-dimensions into itself has at least one fixed point. By such a theorem one obtains mere existence of at least one solution, and no method of constructing the solution. Schauder's theorem has been applied to S-matrix equations, but the resulting existence theorems are not a great deal better than those from the contraction mapping principle.

For the implicit function problem we have the ordinary implicit function theorem of advanced calculus, generalized to Banach spaces. (See Dieudonné, Ref. 23.) Let $F(x,y)$ be a mapping from $B_1 \times B_2$ into B_3, where B_1, B_2 and B_3 are Banach spaces. Here $B_1 \times B_2$ denotes the Cartesian product; i.e., the set of all pairs (x,y), $x \epsilon B_1$, $y \epsilon B_2$. The <u>Implicit Function Theorem</u> is as follows. Suppose that

(i) $F(x,y)$ is continuous in a neighborhood ω of (x_o, y_o)

(ii) $F(x_o, y_o) = 0$

(iii) The partial Fréchet derivative $F_y(x,y)$ is continuous in ω, and $F_y(x_o, y_o)$ has a bounded inverse. (31)

Then there is a unique solution $y(x)$ of $F(x, y(x)) = 0$ in a neighborhood δ of x_o such that $y(x)$ is continuous in δ and $y(x_o) = y_o$.

The Fréchet derivative is familiar to physicists as the variational derivative in the case of integral operators. The Fréchet derivative $A'(x_o)$ of an operator $A : B_1 \to B_2$ is a <u>linear</u> operator such that

$$\lim_{||\delta x||\to 0} \frac{||A(x_o+\delta x) - A(x_o) - A'(x_o)\delta x||}{||\delta x||} = 0. \tag{32}$$

For instance, suppose that A is the integral operator

$$A(x) = \int_0^1 K(s,t,x(t))\,dt,$$

on the Banach space of real continuous functions $x(t)$ on $[0,1]$, $||x|| = \sup |x(t)|$. If the real function $K(s,t,x)$ is continuous in s and t, and continuously differentiable in x, then the Fréchet derivative $A'(x_o)$ is defined by

$$A'(x_o)\delta x = \int_0^1 \frac{\partial K}{\partial x}(s,t,x_o(t))\delta x(t)\,dt. \tag{33}$$

Under appropriate conditions the implicit function theorem gives us continuity of the scattering amplitude with respect to the data x, and local uniqueness. If more differentiability of $F(x,y)$ is assumed, then the solution $y(x)$ will possess derivatives (Dieudonné, Ref. 23). If $F(x,y)$ has a power series in both variables about the point (x_o,y_o), then $y(x)$ will have a power series about x_o. In some of the physical cases, F is highly differentiable or even analytic, so these are welcome facts.

It is important to know how far the curve y(x) extends; i.e., how big is the neighborhood δ for a given neighborhood ω? There are continuation theorems to the effect that the continuous, locally unique solution curve y(x) extends from (x_0,y_0) at least to the first point (x_1,y_1) where the derivative $F_y(x_1,y_1)$ fails to have a bounded inverse. See, for instance, Theorem (2.4) of Ref. 24. We shall be able to find solutions by the contraction mapping principle for small coupling constants. Whether these solutions can be continued into the large coupling region will depend on the location and nature of singularities of $F_y(x,y)$.

The study of phenomena in the neighborhood of the singular points goes by the name of <u>Bifurcation Theory</u>. At a singular point, the curve y(x) might split continuously into two or more branches, or it might turn back on itself, or simply come to an end (cf. Figure 1) Exactly what happens depends on the details of the operator.

A Schematic Representation of Possible Behaviors of y(x) at a Singular Point (x_0,y_0)

Figure 1

In favorable cases, all continuous curves passing through the singular point can be determined by solving a set of finite-dimensional, non-linear equations. For instance, a special case of a theorem of Vainberg and Trenogin[25] is as follows. Let $F(x,y)$ be as described in the implicit function theorem, except that $F_y(x_0,y_0)$ is singular. Suppose that $F_y(x_0,y_0)=I+C$, where C is completely continuous (i.e., C is a Fredholm operator). Then all continuous solutions $y(x)$ of $F(x,y) = 0$ for which $y(x_0) = y_0$ are determined in a neighborhood of (x_0,y_0) by r equations in r real variables

$$B_i(a_1,a_2,\ldots,a_r;x) = 0, \quad i=1,2,\ldots,r. \quad (34)$$

Here r is the dimension of the null-space of $F_y(x_0,y_0)$. The equations (34) are called "branching equations." Each solution of (34) gives a solution of the form

$$y(x) = y_0 + h(x) + \sum_{i=1}^{r} a_i(x)\phi_i, \quad (35)$$

where the ϕ_i form a basis for the null-space of $F_y(x_0,y_0)$, and $h(x)$ lies outside that null-space. Solving (34) for the a_i is a non-linear analog of determing the "right" linear combinations in degenerate perturbation theory. In the case of a one-dimensional null-space there are well-developed techniques for solving the branching equation.[24,25]

Bifurcation points are often physically interesting. In mechanics they might correspond to buckling of beams, to the onset of asymmetric flows, etc. In particle physics, the "spontaneous" symmetry breaking discussed by Baker and Glashow,[26] Cutkosky and Tarjanne,[27] and others, is associated with a bifurcation point. In such a scheme, one has non-linear equations for particle masses of the form

$$m_i = G_i(m_1,\ldots,m_n;\lambda), \quad i=1,2,\ldots,n; \qquad (36)$$

where λ represents an adjustable parameter or several such; for instance, coupling constants or cut-offs. It is assumed that there is a symmetric solution $m_i = m_o$, $i=1,\ldots,n$.

For most λ this symmetric solution $m_i = m_o(\lambda)$ will be locally unique, because the Fréchet derivative, which is the finite matrix

$$\left[\delta_{ij} - \frac{\partial G_i}{\partial m_j}(m_o,\ldots,m_o;\lambda)\right], \qquad (37)$$

is non-singular. At isolated values of λ this matrix may become singular, and a new, asymmetric solution $m_i = m_o + \delta m_i$ may branch off from the symmetric one. If

λ_o is the point of singularity, then we have $m_i(\lambda) = m_o(\lambda_o) + \delta m_i(\lambda)$, where $\delta m_i(\lambda)$ tends to zero as $\lambda \to \lambda_o$. Studies of such models could be systematized and clarified by applying bifurcation theory. One should do a careful study of the branching equations, using the well-developed methods which are available. At a more sophisticated level, it is possible to discuss spontaneous symmetry breaking in an infinite-dimensional context, still with the advantage of finite-dimensional branching equations. For instance, we could look for bifurcation points of the coupled Dyson equations for the vertex and propagator functions. This seems to be an interesting direction for future work.

Finally, in our summary of non-linear techniques, we should give some attention to methods of numerical computation, since we are not likely to find even approximate solutions by analytic means. Also, it is difficult to get even existence theorems in the realistic region where scattering amplitudes are large and the non-linearities very large. The contraction mapping theorem and Schauder's fixed point theorem are generally useful only when the problem is fairly close to linearity. At present, the only general systematic procedure I know of to reach the strongly non-linear regime is to begin in the near-linear regime, where solutions are obtained by iteration, and continue numerically, hopefully all the way into the regime of interest. According to the continuation theorem mentioned above, it is possible to increase the coupling strength (or some other parameter controlling the non-linearity), until a singularity of the Fréchet derivative is encountered. This suggests that the <u>minimum</u> tools required for a successful program of solution will be

(a) an efficient numerical scheme for solving the equations with successive increments of the parameter,

(b) a sufficiently complete analytic understanding of the linear Fréchet derivative operator.

Without (b), it will be difficult to know what is happening in a numerical caluclation, since it is almost certain that singularities of the derivative will be encountered. If the derivative is well understood, it may be possible to continue the solution curve right through its singularity, but not necessarily uniquely. For item (a) there may be many possibilities, but one standby which is often recommended in the mathematical literature is the Newton-Kantorovich iteration.[28] This is a straightforward generalization to Banach space of the familiar Newton method. The iteration, which amounts to successive linearizations of the operator equation $F(x) = 0$, is as follows:

$$x_{n+1} = x_n - F'(x_n)^{-1} F(x_n). \qquad (38)$$

Here F' denotes the Fréchet derivative. Kantorovich has given conditions which guarantee the success of this method. In the cases of interest to us, his conditions are met provided $F'(x_o)^{-1}$ exists, and $||x_1-x_o||$ is sufficiently small. If we have a solution (x_o,λ_o) of $F(x,\lambda)=0$, or a numerical approximation thereto, and $F'(x_o,\lambda_o)$ is non-singular, then we may expect that (x_o, λ_1) will be a suitable starting point for a successful Newton iteration provided λ_1 is sufficiently close to λ_o, because of the continuity in λ ensured by the implicit function theorem. Better yet, we can use a linear extrapolation in λ to get a starting point of the Newton method. An improved starting point, ξ_o, is

$$\xi_o = x_o + \frac{dx}{d\lambda}\bigg|_{\lambda_o} (\lambda_1-\lambda_o). \tag{39}$$

The derivative $dx/d\lambda$ may be computed using $F_x(x_o,\lambda_o)^{-1}$ which is necessary for Newton's method in any case. Since

$$F_x \frac{dx}{d\lambda} + F_\lambda = 0, \tag{40}$$

we have

$$\frac{dx}{d\lambda}\bigg|_{\lambda_o} = F_x^{-1}(x_o,\lambda_o) \cdot F_\lambda(x_o,\lambda_o). \tag{41}$$

Under Kantorovich's conditions, the convergence of the

iteration is very rapid. The error at the n-th iteration is bounded as follows:

$$||x_n - x|| = O(2^{-n} a^{2^n - 1}), \quad 0 < a < 1. \tag{42}$$

With less rapid convergence one can use the computationally easier iteration

$$x_{n+1} = x_n - F'(x_o)^{-1} F(x_n). \tag{43}$$

This concludes my primer of non-linear analysis. Let us now see what has actually been done with these methods.

The simplest case to analyze is the Low equation, which we study in an N/D formulation.[15] It is also possible to approach the Low equation directly, without N/D, or to work with equations for the inverse of the scattering amplitude.[15] The results are best in the N/D formulation. We regard the N function as the unknown, so the equation has the form

$$N = A(N), \tag{44}$$

where a channel subscript a is suppressed, and

$$A_a(N) = B_a(\omega) + \frac{1}{\pi} \int_1^\infty \frac{B_a(\omega) - B_a(\omega')}{\omega - \omega'} \rho(\omega') N_a(\omega') d\omega', \tag{45}$$

$$B_a(\omega) = \frac{\lambda_a}{\omega} + \sum_\beta c_{a\beta} \frac{1}{\pi} \int_1^\infty \frac{N_\beta^2(\omega')}{\left| 1 - \frac{1}{\pi} \int_1^\infty \frac{\rho(\omega'') N_\beta(\omega'') d\omega''}{\omega'' - (\omega'+i0)} \right|^2} \frac{\rho(\omega') d\omega'}{\omega'+\omega}.$$

(46)

We consider A as an operator on a Banach space B, which consists of all real, n-component, continuous vector functions $[N_1(\omega), N_2(\omega), \ldots, N_n(\omega)]$ defined for $1 \leq \omega < \infty$, such that the following quantity, identified as the norm in B, exists:

$$||N|| = \sup_{a,\omega} |N_a(\omega)| + \sup_{a,\omega,\omega'} |N_a(\omega) - N_a(\omega')| \cdot \left| \frac{\omega\omega'}{\omega-\omega'} \right|^\mu,$$

(47)

$$0 < \mu < 1.$$

The second term in (47) is introduced so that the principal value integral in the denominator of (46) may be given a bound proportional to $||N||$. For the complete metric space K of the contraction mapping principle, we choose a closed ball in B; i.e., all $N \epsilon B$ such that

$$||N|| \leq R.$$

(48)

The cut-off is restricted so that

$$|\rho(\omega) - \rho(\omega')| \leq k \left|\frac{\omega-\omega'}{\omega\omega'}\right|^\mu , \qquad (49)$$

$$\rho(1) = \rho(\infty) = 0. \qquad (50)$$

From (49) and (50) it follows that $\rho(\omega) = O(\omega^{-\mu})$, $\omega \to \infty$. A finite number of CDD poles may be introduced in the physical region in (45) and (46). The crossing matrix $c_{\alpha\beta}$ and the number n of channels are arbitrary.

The integrals in (45) and (46) may be bounded in terms of $||N||$ in such a way as to show that A is a contractive mapping of K into K, provided that $|\lambda_a|$, R, and CDD residues are sufficiently small.[15] It then follows from the contraction mapping theorem that for fixed CDD poles of small residue, and small coupling constant, there is a solution of N = A(N), unique in K. This solution gives a solution of the Low equation, provided $D(\omega)$ is free of zeros. It is possible to show that $D(\omega)$ indeed does not vanish, with appropriate additional restrictions on K and on the signs of the CDD residues. Since the CDD poles are in the physical region and have small residues, they produce narrow resonances. We conclude that there is a weak-coupling solution of the Low equation for a nearly arbitrary choice of narrow resonances.

The stable states in this solution are also associated with CDD poles, but poles of what I like to call the "canonical" D function. The canonical D function has a zero for each stable particle, no other zeros, and the number of poles required to make it 1 at infinity. Our solutions of Low's equation obey Levinson's relation in the form $\delta(\infty)/\pi = -n_b + n_c$, where n_b is the number of stable states in the channel in question, and n_c is the number of CDD poles of the canonical D function (which is not the D function used in (46)). For pion-nucleon scattering in the nucleon channel, our solutions have $n_b = 1$, $n_c = 1$. In a purely elastic theory, both our stable states and our resonances would be regarded as "elementary," rather than composite. If the Low equation is regarded as describing only one channel of a theory with inelastic effects, however, the CDD poles of both stable and unstable states could be manifestations of inelastic effects, and some or all of these states could, conceivably, be composites with a many-channel structure.

By attacking the Low equation directly without N/D, one gets only a single non-resonant solution from contraction mapping. This solution is analytic in the coupling constant g^2 in a circle $|g^2| \leq g_0^2$, as one proves by Vitali's theorem using the fact that every iterate (28) is a polynomial in g^2. The nearest singularity is probably quite a distance beyond the circle $|g^2| = g_0^2$, since g_0^2 is limited by technical considerations of the proof, not by actually finding a singularity.

The implicit function theorem may be applied to show that any solution of the N/D equation (not just the one we have found) is <u>normally</u> locally unique and continuous in the coupling constant and the CDD parameters. An abnormal situation can occur only at a singularity of the Fréchet derivative $1-A'(N_o)$. The operator $A'(N_o)$ turns out to be a well-behaved Fredholm operator under reasonable conditions on N_o, as long as N_o and D_o do not develop a simultaneous zero. If $A'(N_o)$ acquires a unit eigenvalue, or N_o and D_o acquire coinciding zeros, then $1-A'(N_o)$ becomes singular.

Since the solutions we have found are not of the bootstrap type, we must look further for bootstrap solutions. One definition of a bootstrap solution is that there should be no CDD poles in the canonical D function. For example, for π-N scattering one requires

$$\delta_{11}(\infty) = -\pi, \quad \delta_{13}(\infty) = \delta_{31}(\infty) = \delta_{33}(\infty) = 0 \qquad (51)$$

This definition may be unduly restrictive. It corresponds to the heuristic picture of "every particle a bound state" in a purely elastic theory, but in a theory with inelasticity other phase shift behaviors could be consistent with the same heuristic picture. Huang and Mueller[29] have proved that the (unsubtracted) Low equation has no solution obeying (51) provided the cut-off has the form

$$\rho(\omega) = k^3 \left\{ \frac{K^2}{k^2 + K^2} \right\}^n, \quad n = 0,1,2,\ldots \qquad (52)$$

The argument depends explicitly on the analyticity properties of the cut-off, so the question remains of whether there could be solutions with some other class of cut-offs. No existing numerical work throws any light on the question, as far as I know, and the approach via fixed point theorems is not yet up to the task. A numerical study of the Low equation, using Newton-Kantorovich iteration of the N/D equation, is now being undertaken. Presumably this will clarify the question of existence of bootstraps, and also determine whether there are ghost-free solutions of any kind at the physical value of the coupling constant. It will also be interesting to look for possible bifurcation points. It will not be straight-forward to generate a bootstrap starting from small coupling. One could try it by taking only the crossed channel nucleon pole as input, and turning up the coupling constant. This should produce a resonance in the (11) state, which moves down to form the nucleon as a bound state. As it crosses threshold, however, the operator $A'(N_o)$ becomes singular (the kernel acquires a pole, since N_o and D_o have a common zero) and it is not clear how to follow the solution N through this singularity.

Although the Low equation has not been put to rest, I will now turn to the much more complicated case of the Mandelstam crossing-unitarity equation as studied by

Atkinson. As was pointed out in Eq. (13), Atkinson writes the equation in the unsubtracted case as

$$\rho^{el} = A(\rho^{el}; v), \qquad (53)$$

where

$$A(\rho^{el}; v) = \int K(s,t,t_1,t_2) d^*(s,t_1) d(s,t_2) dt_1 dt_2, \qquad (54)$$

$$d(s,t) = \frac{1}{\pi} \int_4^\infty ds' \left[\frac{1}{s'-s} + \frac{1}{s'-u} \right] [\rho^{el}(s',t) + \rho^{el}(t,s') + v(s',t)]. \qquad (55)$$

We work in a Banach space C consisting of all real functions $\phi(s,t)$ with domain $4 \leq s,t < \infty$ such that

(i) $$\sup_{4 \leq s_1, t_1, s_2, t_2 < \infty} \frac{|\phi(s_1,t_1) - \phi(s_2,t_2)|(\log\Lambda\bar{s}\ \log\Lambda\bar{t})^2}{\left|\frac{s_1-s_2}{s_1 s_2 \bar{t}}\right|^\mu + \left|\frac{t_1-t_2}{t_1 t_2 \bar{t}}\right|^\mu}$$

$$= ||\phi|| < \infty, \qquad (56)$$

$\Lambda > 1$, $0 < \mu < \frac{1}{2}$, $\bar{s} = \min(s_1, s_2)$, $\bar{t} = \min(t_1, t_2)$

and

(ii) $\quad \phi(s,\infty) = \phi(\infty,t) = 0$, all s,t. $\qquad (57)$

Conditions (i) and (ii) together imply that $\phi(s,t)$ has the bound

$$|\phi(s,t)| \leq ||\phi|| \, (st)^{-\mu} \{\log \Lambda s \, \log \Lambda t\}^{-2}. \tag{58}$$

The input function $v(s,t)$ is restricted as follows:

(i) $v \in C$
(ii) $||v|| \leq V$
(iii) $v(s,t) = 0$, $s \leq \sigma_1(t)$, $4 \leq t < \infty$. (59)

The boundary curve $\sigma_1(t)$ of v should satisfy

$$16 \leq \sigma_1(t) \leq 16 \, \frac{t-4}{t-16}, \quad 16 \leq t < \infty,$$

$$\sigma_1(t) = \infty, \quad 4 \leq t \leq 16. \tag{60}$$

We look for solutions $\rho^{e\ell}$ in a closed, bounded subset K of C, which consists of all $\rho^{e\ell} \in C$ such that

(i) $||\rho^{e\ell}|| \leq R$,

(ii) $\rho^{e\ell}(s,t) = 0$, $s \leq \frac{16t}{t-4}$, $4 \leq t < \infty$. (61)

Now for R and V sufficiently small, the operator A of (53) is a contractive self-mapping of K. (This is not supposed to be obvious; the proof requires a long series of involved estimates.) Thus, there is a solution of (53), unique in K. Since the solution $\rho^{e\ell}$ and the input satisfy the bound (58), the unsubtracted Mandelstam representation constructed from (13) will converge. Since Eq. (53) guarantees elastic unitarity in the elastic region, we have proved that there exist functions which satisfy crossing symmetry and elastic unitarity, and have an unsubtracted Mandelstam representation.

By means of additional restrictions on v, which are slightly involved and not too informative, one is able to guarantee that the unitarity inequality hold at all energies:

$$\text{Im } A_\ell \geq \rho |A_\ell|^2. \qquad (62)$$

These results are gratifying, in that they show that the Mandelstam program is at least mathematically reasonable. The qualitative properties of the solutions are not very physical, however. The solutions are small, they contain no resonances, and there are no Regge poles in the right-half ℓ-plane. The total cross-section vanishes as $s^{-1-\mu}$ at infinity. The solution $\rho^{e\ell}$ is small compared to the input v, and if v goes to zero, so does $\rho^{e\ell}$. The double spectral function is positive,[30] while physical arguments indicate that there should be oscillations.[31] Certainly, Regge poles in the right plane require oscillations of an extreme sort.

It is easy to show that the solution we have obtained

is analytic in a parameter multiplying the input function
v; i.e., if v(s,t) is replaced by $\lambda v(s,t)$, then $\rho^{e\ell}(s,t;\lambda)$
is analytic in some circle $|\lambda|<\lambda_o$, for fixed (s,t). There
is a unique continuous extension of the solution to larger
λ, reaching as far as the first singularity of the Fréchet
derivative $I-A'(\rho^{e\ell})$. At least initially this extension is
analytic continuation in λ. A singularity of the Fréchet
derivative need not be a singularity of $\rho^{e\ell}$ as a function
of λ.

The singularities of $I-A'(\rho^{e\ell})$ are not yet understood.
One of the main problems of this program will be to understand this operator, which is a linear singular multi-dimensional integral operator of an unfamiliar sort. We can
not say we have made much progress until we at least understand the linearization of the problem at hand! Numerical
studies of the type that have been attempted in the strip
model would be greatly aided if we knew how to handle $A'(\rho^{e\ell})$
for arbitrary $\rho^{e\ell}$. One could then try implicit function theory, Newton's method, etc.

The singularities of the Fréchet derivative undoubtedly
have something to do with CDD poles. Atkinson's method can
be generalized so as to bring CDD poles explicitly into the
equations (cf. Atkinson and Warnock, Ref. 13). One then obtains solutions with narrow resonances, which lie outside the
set K which was considered above. To include CDD poles in
the s-wave, say, we subtract out the s-wave in Eq. (53), and

treat it by the N/D method. The unknowns are now $\rho^{e\ell}$ and $\mathrm{Im}A_o$, say, while $v(s,t)$ and the s-wave elasticity $\eta(s)$ are regarded as given. The left-cut of the partial wave A_o is determined by $\rho^{e\ell}$ and $\mathrm{Im}A_o$. We then get two coupled equations,

$$M(\rho^{e\ell}, \mathrm{Im}A_o) = \rho^{e\ell} \qquad (63)$$

$$N(\rho^{e\ell}, \mathrm{Im}A_o) = \mathrm{Im}A_o. \qquad (64)$$

Eq. (63) is just Eq. (53), but with the contribution of the s-wave absorptive part to $d(s,t)$ isolated and written explicitly. Eq. (64) is the s-wave N/D equation, which may contain an arbitrary finite number of CDD poles. The coupled equations may be analyzed by the contraction mapping principle, but the analysis is more delicate than before, since $\mathrm{Im}A_o$ contains sharp peaks near the CDD poles. By making the CDD residues sufficiently small, these peak regions become so narrow that they do not stand in the way of proving the contractive property of the mapping. We obtain solutions which are similar to Atkinson's, except that they contain arbitrary narrow s-wave resonances. The corresponding Mandelstam representation has single-spectral functions, however, which are necessary for unitarity of the s-wave. The single-spectral function has a peak, a narrow blip, near each CDD pole.

This "CDD ambiguity" in the solution may be ruled out

by requiring ℓ-plane analyticity. The existence of single-spectral terms which dominate the double-spectral terms near the CDD poles implies the existence of Kronecker delta terms $\delta_{\ell 0}$, which spoil the ℓ-plane analyticity. This conclusion, which agrees with early conjectures of the Berkeley school, is a nice example of the qualitative insights to be gained from non-linear analysis in Banach space.

The question remains of how to find physical solutions, since the ones we have found are clearly not very close to physics. If there exist solutions which look more reasonable, how are they related to the iterative solutions we have found? It could happen that the physical solutions are just continuations in λ of the iterative solutions. The continuation would have to pass through one or more singularities of the Fréchet derivative, however, since the Mandelstam crossing-unitarity equation requires subtractions in the physical case. The equation might change discontinuously from unsubtracted to subtracted form at a singularity, just as the form of the N/D equation changes when a CDD pole enters the picture. On the other hand, we cannot rule out the possibility that the iterative solutions are spurious and not directly related to physics.

Kupsch[14] and Atkinson[13] have attempted to study the realistic case by working at the start with subtracted equations. The subtracted equations have solutions, but the unitarity constraint $\mathrm{Im}A_\ell \geq \rho|A_\ell|^2$ cannot be guaranteed, except in the case of one subtraction.[14] The partial-waves are, in fact, out of control at high-energies: the bounds obtained are only $|A_\ell|^2 = 0(s^N)$. Furthermore, ℓ-plane analyticity for small $\mathrm{Re}\ell>0$ is not ensured. The main difficulty is that the methods of estimation used do not take advantage of the oscillations in the double-spectral function which must be present. It seems that the double-spectral function is not the appropriate variable to study. It may be better to formulate the problem in terms of the Froissart-Gribov amplitudes $A^\pm(\ell,s)$ in some appropriate domain of ℓ and s.

Recently, Atkinson[32] has been able to improve the proof outlined above by considering a more subtle norm. He obtains amplitudes with total cross sections which decrease only logarithmically at large s, rather than as a power of s, while still maintaining the inequality of unitarity. This is done without taking account of oscillations, and it probably is close to the limit of what can be done in that way.

The dependence of the solutions on an arbitrary function v(s,t) deserves some comment. From one point of view it is not surprising, since we do not expect that the two-body scattering will be determined independently of many-body scattering. From another point of view, the apparent degree of arbitrariness is rather puzzling. This has been emphasized by Martin,[33] who has obtained some tight inequalities on partial waves below threshold, beginning with assumptions much weaker than those of Atkinson. In particular, if the s-wave of $\pi_0-\pi_0$ scattering is known at a certain energy, the d-wave is narrowly bracketed. It is a quantitative question as to how this can be reconciled with Atkinson's work. It is hoped that numerical solutions of Atkinson's equations, with various v's within his allowed class, will clarify the situation.

In the strip approximation one tries to avoid all input by putting v=0. With v=0, Eq. (53) has the trivial solution $\rho^{e\ell}=0$, and furthermore, the derivative $A'(\rho^{e\ell})$ vanishes at this solution. Thus, $[I-A'(\rho^{e\ell})]_{\rho^{e\ell}=0} = I$ is non-singular, so a non-trivial solution cannot branch off continuously from the trivial solution. We have no idea at present of how to look for a non-trivial solution. The unique small solution, obtained by iteration, is just the trivial one.

The strip model calculations of Collins and Johnson,[7] although much more encouraging than those carried out in the past, have a somewhat ambiguous status. The calculation is not arranged to give the solution of a definite fixed-point problem. Rather, there is an algorithm which seeks agreement between input and output of an N/D equation over a limited energy domain, the input and output being in the form of Regge poles and residue functions. Actually, the input is partially processed by a Mandelstam iteration before it is put in. This mixture of the Mandelstam iteration and the N/D algorithm gives a set-up which is difficult to evaluate in a clear way, since it is not certain at the end that there exists a crossing-symmetric unitary amplitude with the Regge trajectories that are found. This remark is not meant to devalue the effort of Collins and Johnson, which is courageous and valuable, but rather to point out that the mathematical problem underlying the strip model has not yet been clearly formulated, at least in the realistic case with subtractions. If the problem could be formulated in terms of a definite set of equations, the solution of which would give an amplitude both crossing-symmetric and unitary, then we would have a chance of coming to some definite conclusion about the model. We could make a better assessment of ambiguities, by exploring the region of Banach space around a proposed solution. We might refine a rough solution by Newton's iteration, check on local uniqueness and dependence on cut-offs by implicit function theory, etc. As usual, the oscillations in double-spectral functions make the problem difficult, so it may be better to avoid double-spectral functions if we can somehow ensure crossing-symmetry without using them as unknowns.

The preceding survey should make clear that one is still in the foothills, not anywhere close to the peaks of non-linearity. By a lucky choice of route, it should be possible to get closer to the top.

ACKNOWLEDGEMENT

I am very grateful to Professor E. C. G. Sudarshan and Professor Y. Ne'eman for the opportunity to present this review at their symposium.

I have learned recently that B. Webber, in Berkeley, has attempted to reproduce the Collins-Johnson calculation[7], without success. In his closest approach to a solution, the Regge trajectories and residues do not have the satisfactory properties claimed by Collins and Johnson. The situation in the strip approximation is then even less clear than indicated above.

REFERENCES

1. S. Mandelstam, Phys. Rev. <u>112</u>, 1344 (1958); <u>ibid</u> <u>115</u>, 1741 and 1752 (1959).

2. G. F. Chew and S. Mandelstam, Phys. Rev. <u>119</u>, 467 (1960); G. F. Chew, S. Mandelstam, and H. P. Noyes, <u>ibid</u> <u>119</u>, 478 (1960); G. F. Chew and S. Mandelstam, Nuovo Cimento <u>19</u>, 752 (1961); For related developments see V. V. Serebryakov and D. V. Shirkov, Forts. d. Phys. <u>13</u>, 227 (1965).

3. G. F. Chew, <u>The Analytic S Matrix</u>, (W. A. Benjamin, Inc., New York, 1966); see also Ref 5.

4. G. F. Chew, Phys. Rev. Letters <u>9</u>, 233 (1962).

5. S. C. Frautschi, <u>Regge Poles and S-Matrix Theory</u>, (W. A. Benjamin, Inc., New York, 1963).

6. M. Froissart in <u>Seminar on Theoretical Physics</u>, Trieste, 1962, (International Atomic Energy Agency, Vienna, 1963).

7. R. C. Johnson and P. D. B. Collins, Phys. Rev. <u>185</u>, 2020 (1969).

8. S. Mandelstam, Nuovo Cimento <u>30</u>, 1148 (1963); see p. 1162.

9. M. Froissart, Phys. Rev. <u>123</u>, 1053 (1961).

10. A. Martin, Phys. Rev. Letters <u>9</u>, 410 (1962).

11. The papers of Refs. 12-15 form a nearly complete list of work in this program. For a detailed introduction, including references to the mathematical literature, see

R. L. Warnock, in Lectures in Theoretical Physics, Vol. XI-D, edited by K. T. Mahanthappa et al. (Gordon and Breach, New York, 1969.) Similar methods were used earlier in problems of field theory by J. G. Taylor, who reviewed his work in Vol. 9-A of the lectures quoted above.

12. C. Lovelace, Commun. Math. Phys. $\underline{4}$, 261 (1967).
13. D. Atkinson, J. Math. Phys. $\underline{8}$, 2281 (1967); Nucl. Phys. $\underline{B7}$, 375; $\underline{B8}$, 377 (1968); $\underline{B13}$, 415 (1969). D. Atkinson and R. Warnock, Phys. Rev. $\underline{188}$, 2098 (1969).
14. J. Kupsch, Nucl. Phys. $\underline{B12}$, 155, $\underline{B11}$, 573 (1969); Nuovo Cimento $\underline{66A}$, 202 (1970).
15. R. Warnock, Phys. Rev. $\underline{170}$, 1323 (1968); H. McDaniel and R. Warnock, Phys. Rev. $\underline{180}$, 1433 (1969); Nuovo Cimento $\underline{64A}$, 905 (1969).
16. F. Low, Phys. Rev. $\underline{97}$, 1392 (1955); G. F. Chew and F. Low, ibid $\underline{101}$, 1570 (1956); E. M. Henley and W. Thirring, Elementary Quantum Field Theory, (McGraw-Hill, New York, 1962), Chap. 18.
17. See, for instance, R. Roskies, Phys. Lett. $\underline{30B}$, 42 (1969); Nuovo Cimento $\underline{65A}$, 467 (1970); Yale preprints.
18. G. F. Chew and S. C. Frautschi, Phys. Rev. $\underline{123}$, 1478 (1961). See also the fourth paper by Atkinson, Ref. 13.
19. We do not state the equation in quite the most general form. For a precise discussion see R. L. Warnock, in Lectures in Theoretical High Energy Physics, H. Aly (Edi-

tor), Chap. 10 (Wiley-Interscience Publications, New York, 1968).

20. R. L. Warnock, Phys. Rev. 146, 1109 (1966).
21. R. L. Warnock, Nuovo Cimento 50, 894 (1967) and erratum, 52A, 637 (1967).
22. S. Mandelstam, Phys. Rev. 140, B375 (1965).
23. M. A. Krasnosel'skii, "Topological Methods in the Theory of Nonlinear Integral Equations" (MacMillan Co., New York, 1964); J. Dieudonné, Foundations of Modern Analysis (Academic Press, New York, 1960); see also Ref. 11.
24. G. H. Pimbley, Jr., Eigenfunction Branches of Nonlinear Operators, and Their Bifurcations (Lecture Notes in Mathematics, Vol. 104, Springer-Verlag, Berlin, 1969).
25. M. M. Vainberg and V. A. Trenogin, Uspekhi Mat. Nauk 17, No. 2, 13 (1962). Translated in Russian Mathematical Surveys, Vol. 17, No. 2.
26. M. Baker and S. Glashow, Phys. Rev. 128, 2462 (1962); S. Glashow, ibid, 130, 2132 (1963).
27. R. E. Cutkosky and P. Tarjanne, Phys. Rev. 132, 1355 (1963); P. Tarjanne, Ann. Acad. Sci. Fennicae A VI. Physica 187 (1965).
28. L. V. Kantorovich and G. P. Akilov, Functional Analysis in Normed Spaces, (Pergamon Press, Oxford, 1964); R. H. Moore in Nonlinear Integral Equations, edited by P. M. Anselone (University of Wisconsin Press, Madison, 1964).

29. K. Huang and A. H. Mueller, Phys. Rev. **140**, B 365 (1965).

30. It is possible to include some limited departures from positivity, but that does not really help the situation. See the latest paper of Atkinson, Ref. 13.

31. A. Martin, Nuovo Cimento **61A**, 56 (1968); F. J. Yndurian, CERN preprint TH. 1005 (1969).

32. D. Atkinson, private communication. To appear as CERN Report. Th. 1168, submitted to Nuclear Physics.

33. A. Martin, Nuovo Cimento **63A**, 167 (1969); see also F. J. Yndurian, *ibid*, **64A**, 225 (1969), and J. L. Basdevant, G. Cohen-Tannoudji, and A. Morel, *ibid*, **64A**, 585 (1969).

DISCUSSION OF PROF. WARNOCK'S TALK

<u>Moffat (University of Toronto)</u>: The Mandelstam program came to a screeching halt almost ten years ago when it was realized by Mandelstam and Chew that if you had Regge poles then you have to supplement the Mandelstam program with the many-body solution to the problem in order to solve the original unitarity equation. It seems to me that progress on this essential problem has been virtually zero over the last ten years -- this is my impression -- and I want to ask what is being done on this.

<u>Warnock</u>: Yes, I agree that essential progress is almost zero and I don't know of any hope for making essential progress. I still think that some sort of approximate solution of the many-body question will be possible under certain conditions. I should reiterate that there are people who think that one can do the strip approximation and that it might correspond very closely to physics. These people put in a cut-off function, which is arbitrarily chosen, in front of the Mandelstam operator. It is equal to one in the elastic region and then it cuts off in some arbitrary way. As Atkinson has shown, this allows the equation to be well-defined as a reasonable operator in an appropriate Banach space. Atkinson's viewpoint is that we put in the cut-off, but then we compensate for any mistake made in introducing the cut-off by introducing a suitable (but unknown) force function $v(s,t)$. So, in a sense he still has an exact way of talking about the equation. The

work of Collins and Johnson, which has $v = 0$, suggests that one can find a self-consistent scheme with a ρ trajectory with reasonable parameters and a Pomeranchuk trajectory with reasonable parameters. This is a theory in which Regge trajectories eventually turn back at some high energy rather than going on to infinity. Their scheme also gives decent S-waves. So one point of view is that the strip model, with cut-off and including just the inelastic effects which the strip model entails, (that is, two particle states in the t channel produce many particle states in the s channel) gives you a physically realistic picture. Unfortunately, it seems that if this does work out, this will be just a one shot success. Only pion-pion scattering is going to be such a favorable case. It will be very hard to go on to pion nucleon scattering and so on.

<u>Moffat</u>: Could I make this one more quick comment? In regard to your comment about the trajectories, they are contrary to a popular press opinion, but there is no experimental or theoretical justification which justifies the limit to infinity. You can't construct models which turn out to have this very naive behavior. In fact, if you have trajectories rising into infinity, this can cause serious operational problems.

<u>Warnock</u>: Yes, I think that makes things at infinity very wild and maybe that is what we finally have to put up with, but it doesn't seem to me that the Veneziano program has arrived at a sufficient stage of completeness to be forced to take that

point of view. It is a little strange perhaps to have the trajectories turning back at $J = 1000$, but that might very well be what happens.

DEVELOPMENT OF QUANTUM FIELD THEORY*

I. Segal

Mathematics Department,
Massachusetts Institute of Technology
Cambridge, Massachusetts 02139

Abstract: In his talk Prof. Segal reviews recent developments in the foundations of quantum field theory. The topics discussed include the abstract algebraic approach, the mathematical development of the wave-particle duality, hyperbolicity of quantized partial differential equations, the meaning of non-linear equations of motion and the interaction Hamiltonian as a generalized operator in a Hilbert space. In this last context Prof. Segal discusses the problems which arise if one requires the interaction Hamiltonian to be defined at a sharp time. It is emphasized that although this can be done in lower-dimensional space-time, a modification is likely to be required in four-dimensional space-time. Examples of such "distribution" propagators which allow a well-defined interaction representation are given.

---

*An invited talk presented at the Symposium "The Past Decade in Particle Theory" at the University of Texas in Austin on April 14-17, 1970.

I want to talk about physics, without hesitating to use contemporary mathematical tools when they appear relevant. I think it's useful and important to maintain the standards that mathematics has developed--although I'm not so much concerned about the grammar as the architecture of mathematics--basically, whether or not things are clear-cut and effectively organized. On the other hand, the ultimate question in physics is correlation with experimental data. So, towards the end of the talk, I hope to get to something which does make contact with experimental aspects. I'm pleased to be here, and I especially appreciate the ten year interval we are supposed to survey. It seems that some physics colloquia are conducted in rather breathless fashion. The advances of the past year or six months is a little difficult to have much perspective about.

I'd like to begin by summarizing what I hope to talk about, which are some recent developments in the foundations of the quantum field theory and, in order to be somewhat analytical about it, I've been a little artificial and broken up things that go together. I'll begin with developments which go back to the late '20's and early '30's; then I'll work my way to the '50's and '60's; and end with something about possible future developments. Some of these develop-

RECENT DEVELOPMENTS IN THE FOUNDATIONS OF
QUANTUM FIELD THEORY

1. Quantum field phenomenology in terms of nascent operators (C^* algebras)

   Illustrative application: quantization of Klein-Gordon equation with imaginary mass.

2. Invariant representation of a free quantum field in terms of a measure in the space of classical fields (in generalized Schrödinger style).

   (Mathematically explicit treatment of wave-particle duality.)

3. Causal propagation of quantized partial differential equations. (Generalized hyperbolicity of partial differential equations, expressed in terms of rings of operators.)

4. The interaction Hamiltonian as a generalized operator in Hilbert space, -- in the interaction representation.

5. Mathematical analysis of relativistic non-linear equations of motion.

   -- Wick powers relative to the physical rather than free vacuum.

   -- The case of positive-energy scalar fields in two-dimensional space-time.

6. Non-perturbative computation of the S-matrix.

   --Superseding coupling constant renormalization thru the use of 'distribution' (time-smeared) propagators.

Figure 1: Summary of the recent development of Quantum Field Theory.

ments are described in Figure 1.

One subject that has emerged is the treatment of quantum field phenomenology in terms of what I'll call nascent operators. By "nascent operator," I mean just an element of a certain algebra. It becomes concrete when you represent it as a bona fide operator in Hilbert space. But, one of the things that one notices as one reads Professor Dirac's book, which I take as the Bible in these matters since it's the most serious attempt at a logical, deductive treatment of the subject, is that it is carefully left open on what the quantum field operators act. One finds that if one does specify what the operators act on, one gets into trouble. So it's very clever to leave that open. One of the things that has emerged, is that one can solve this problem in a quite clear, simple, and natural way. One can deal with a certain kind of algebra that has all the properties of operators, but they don't operate on anything in any particular. I'll call these operators "nascent"; more conventionally, they are elements of what are known as "abstract C*-algebras."

As a simple application of this development, I'll mention the treatment of the Klein-Gordon equation with imaginary mass in terms of C* algebras. This has become a question of some physical interest, since the work of Sudarshan, Feinberg and others on the so-called tachyons. What one finds is one can quantize this equation in essentially as good a way as the real mass equation, but there is no preferred Hilbert space; there are no stationary states; and there's no vacuum. So these particles would not appear observable as <u>free physical</u> particles. I'll come to that in more detail momentarily.

Secondly, I'd like to talk about the explicit mathematical development of the wave-particle duality. In Professor Dirac's book you'll find a very succinct statement of the basic idea, which is that any linear boson field can be represented as an assembly of harmonic oscillators. In this representation, you're first of all restricting yourself to a discrete spectrum, and secondly you have a representation that is not really invariant; it refers to a particular choice of basis. So, it's quite useful to give a representation of a free quantum field, which we are most familiar with in the particle representation, which originated with Fock in a heuristic way and which was treated in an invariant way and in a mathematically clear way by Cook in terms of tensors on Hilbert space. Besides this particle representation, there's also a representation in which the Hilbert space of the quantum field is the space of all square-integrable functions over something, as in the Schrödinger representation. This is a thing which is fairly definitive in formulating in a clear and precise way the physical idea of wave-particle duality. It is important in particular for work on the construction of quantum fields.

Another interesting subject which has emerged is the treatment of the hyperbolicity of quantized partial differential equations. I think everybody here has been exposed to the classical treatment of the wave equation, or some similar equation, for which the value of a solution at a certain point depends only on the Cauchy (or initial) data in the backward cone, --i.e. not on all the Cauchy data, but only on the data in a certain finite region. This works also for quantized equations. Of course, one has to be careful about what one means by a function of operators, but that's taken care of by the von Neumann-Murray theory of rings of operators. In those terms one has some definite results concerning causal propagation of quantized fields and their hyperbolicity in a generalized sense. That turns out to be important for the purpose of constructing quantum fields; it's what permits one to introduce spatial cut-offs, which make the Hamiltonian halfway finite.

I should say that the problem we're talking about really

originates in Professor Dirac's first paper on quantum theory of radiation, where he gets some beautiful results for the first order perturbations of a quantum field, but one has the drawback that all higher perturbations seem to be infinite. To this day, no one has succeeded in the fundamental problem of formulating this Hamiltonian in a fully clear-cut and effective way. The work I'm talking about is directed largely towards essentially this end. However, one of the partial considerations is this: there's not the slightest doubt that the total Hamiltonian *is* infinite, in a certain sense, and it's only when you make a spatial cut-off that it becomes halfway finite. So, the reason why it appeared to be infinite was that it *was* infinite. That's one of the reasons why there were no apparent vectors in its domain. However it turns out that the spatial cut-off does not affect the finite temporal propagation of the field variables provided you stick to a certain finite region of space which can become arbitrarily large if your spatial cut-off is in a sufficiently large region.

The interaction Hamiltonian as a generalized operator in Hilbert space has had some serious study. Professor Wick systematized and developed a very simple and effective way of normalizing products of free field variables some 20 years ago and this has been developed by a variety of people. In the middle sixties, Gårding and Wightman treated aspects of the time averages of Wick products of free fields. However for dynamical purposes, it's not enough to know the time average. If you want to solve a differential equation of evolution, you have to know what the source term is at a fixed time; otherwise the equation doesn't make sense. A fundamental difficulty of singular non-linear differential equations, and especially of quantized equations, has been that one has to deal with a sharp time. So one has to define and develop the properties of the Wick products at a sharp time. Their appropriate mathematical treatment shows them to be generalized operators in a quite well-defined and effective sense and that, indeed, if one deals with sufficiently nice cases, such as two-dimensional space-time, then the space averages of the Wick powers are even self-adjoint operators. They have a lot of vectors in their domain, in the Hilbert space sense, they're determined by what they do on those vectors, simultaneously diagonalizable, etc. This work has taken place in the past few years.

Finally, I'd like to talk about the meaning of the non-linear equations of motion. When one has a quantized equation such as say $\Box\phi = m^2\phi + \phi^3$, the question has always been, of course, what does the ϕ^3 mean? There's no clear meaning to ϕ^3 a priori. Of course, if ϕ is the free field, one can postulate that the ϕ^3 is the third Wick power. That's a perfectly logical (though not necessarily theoretically compelling) definition. But, in this equation, the ϕ is not a free field; it is the interacting field, which is subject to the physical rather than free vacuum; so there's no a priori conception of what ϕ^3 should mean. In fact, ϕ^3 is divergent, and this is one of the most characteristic features of the quantum field theory that, in spite of its enormous compactness and the compelling simplicity of its formulation, the fundamental differential equation you start with doesn't have a basically clear-cut physical or mathematical meaning.

I want to address myself to this matter. Finally, I'd like to show how some of the results which come out of this give one a natural means of formulating a non-perturbative computation for the S-matrix. In a way, this does something along the lines that Professor Dirac was suggesting last night,* that is, bringing the S-matrix into effective relation with equations of motion. On the other hand, you don't ac-

_____
*At the public lecture held at the University of Texas on April 15, 1970.

tually solve the equations of motion completely, in that you don't actually find the physical vacuum, which at this stage i rather remote. Ten years from now people may have done someth like that. But at this stage of the game you can at least mak rational computations in the interaction representation which should make contact with empirical physics, although the under lying mathematical theory is incomplete.

To proceed to these various points in order (and make some incidental remarks about their history), let me first present a Table about quantum phenomenology (see Table 1) which apart from the notion of vacuum, which is about 10 years old, goes back two decades or more. Here is a comparison of the classical with the now fairly well accepted contemporary formulation. Classically, an observable was an hermitian operator A, a state was a vector in the Hilbert space in which the observables operate, a dynamical transformation was defined by a unitary operator U, and transformed the observable A into UAU^{-1}. The vacuum was the lowest eigen-vector of the energy, where the energy is just the self-adjoint generator of the temporal translation group. During the '20's and '30's, von Neumann initiated a new approach to phenomenology, embodying a formal idea of Jordan and based on abstract algebraic and Hilbert space ideas. This approach became quite complicated, relative to the applicable results obtained. Some powerful and simple ideas were introduced by Gelfand and Naimark in the middle '40's.

COMPARISON OF QUANTUM PHENOMENOLOGIES

Concept Era	Observable	State	Dynamical Transformation	Vacuum
Classical (1925-45 etc.) Main source Dirac's book	Hermitian operator A (Main pro- perty, self- adjoint operator <u>Not</u> Hermitian matrix)	Vector in space on which operators act	Unitary operator U on this space. Transforms $A \rightarrow U^{-1}AU$	Lowest eigen- vector of energy (self- adjoint generator of tempo- ral trans- lation group)
Contempo- rary (Under develop- ment since 1946) Specific applica- tion to quantized fields since 1959	Nascent operator, or element of suitable algebra ("C*") not neces- sarily given as acting on any space	Expecta- tion value form (nor- malized positive linear function- al on this algebra)	Transfor- mation $A \rightarrow A'$ such that $(A+B)'=A'+B'$ $(AB)'=A'B'$ $(cA)'=c'A'$ i.e. n automor- phisms of the algebra	Norma- lized positive linear function of E which is (1) inva- riant under automor- phism group; (2) cor- responding self- adjoint generator in H_E (Hilbert space de- termined by E) is ≥ 0.

Table 1: Comparison of Quantum Phenomenology.

The ideas of this work and of von Neumann's (both of which had a kinship with positivity ideas of M. G. Krein) were brought together in a work of mine in the later '40's which provided a physically conceptual and mathematically effective approach based essentially on the theory of C*-algebras. In the past decade, this standpoint has had considerable acceptance outside of mathematics and has been found useful in field theory, of which more later.

The contemporary point of view would be that an (conceptual) observable is really an element in a C*-algebra. There is assumed to be given a certain Jordan algebra, so-called, having properties which insure that it has representations by concrete operators in Hilbert space, but a priori, there is no distinguished Hilbert space for it to act on. There are a continuous infinity of different Hilbert spaces, and of different action of the given algebra in these spaces. These actions are really inequivalent. This is not the case for a finite number of degrees of freedom, where there is no problem, --the actions are all essentially equivalent, --but only for an infinite degree of freedom, are they really different.

The state is objectively determined only by its expectation values, so you might as well define a state as its expectation value form, or mathematically you say "a normalized positive linear functional" on this algebra. Now what's a dynamical transformation? If it carries A into A', then the

sum A + B should go into A' + B' and AB should go into A'B'--
this is fairly clear. The transformation A → A' is what is
known as an automorphism. If you were dealing with the concrete case, in which you had <u>all</u> operators in a Hilbert space,
every such dynamical transformation would come from some unitary or antiunitary operator. However, in general, it does
not by any means, very far from it. So this one little loophole here, that for an irreducible C*-algebra things like
this exist, makes an enormous difference.

The simplest way of saying that a state is a vacuum goes
like this. When you take this state, let's say E is the expectation value form, there's a certain Hilbert space (depending on the state); a certain self-adjoint operator H_E
which generates temporal development, in this particular representation, and a certain vector V_E. Now what you want is
that H_E should be non-negative, that is, when you take the
system and make it concrete, that the corresponding Hamiltonian should be non-negative. In other words, you have this
abstract algebra and a one parameter group of automorphism,--
of motions. A vacuum then, is one of these positive linear
forms which has this property.

Well, of course, physically it's clear, in general, that
a vacuum won't exist. If you take an indefinite energy situation to which the equations are soluble, you're going to get
a group of dynamical transformations for which you don't expect a vacuum. In general, it just doesn't exist and I men-

tioned an example earlier; the Klein-Gordon equation with imaginary mass has no vacuum. But typically it does, in physically appropriate situations. One can prove for example that when one applies this framework to quantum fields then one gets uniqueness of the vacuum, and recovers the conventional vacuum. (See Figure 2.)

Example. Quantization of $(\frac{\partial^2}{\partial t^2} - \Delta)\phi = m^2\phi$ (*imaginary mass*).

Theorem. *There are nascent operators* $\phi(x,t)$, *unique as such, satisfying this differential equation & CCRs*:

$$[\phi(x,t), \phi(x', t')] = c\text{-number}.$$

In addition, for any Lorentz transformation L, there is a corresponding automorphism α_L *which sends*

$$\phi(x,t) \to \phi(L^{-1}(x,t))$$

But: *there is no "regular" stationary state - a fortiori no vacuum state (For proof, see MIT thesis of M. Weinless)*

Informally: *There is no preferred Hilbert space for those nascent operators to become concrete in.*

Physical interpretation. *There is no apparent way in which particles satisfying the Klein-Gordon equation with imaginary mass might be observed experimentally as* <u>free</u> *particles. The possibility of their being physically relevant as virtual particles in an interacting field is not excluded.*

Figure 2: Structure of imaginary mass field theories.

Let me take a brief look at the example of quantization of the Klein-Gordon equation with imaginary mass. Theorem: there exist nascent self-adjoint operators (which must be smeared over space), and they are unique as such, which in a certain sense satisfy the differential equation, satisfy canonical commutation relations, which are uniquely determined and are Lorentz invariant. The Lorentz group doesn't act via unitary transformations, but rather by automorphisms. However, there exists no regular stationary state. (I should mention that there always is an invariant state, but most of the time this invariant state is not regular, and it gives infinite expectation value for the relevant field variables.) Obviously, if the interesting field variables have infinite expectation values in a state, it isn't a physical state. "Regularity" is a very natural limitation to states which have a remote chance to be physical. (The cited result can be found in the MIT thesis of M. Weinless a few years ago, with a wealth of other stuff of this type.)

There is no apparent way in which imaginary-mass particles like this could be directly observed, but this doesn't exclude the possibility that these particles could be coupled with some other system in interactions in such a way that they will have stable states. You just wouldn't have in and out fields. You might get something like a bound state situation conceivably; mathematically, this is not at all excluded.

I might continue with an indication of how one started

to do quantum fields in terms of C*-algebras. More than a decade ago (cf. the 1957 Lille Conference), when people had gotten thoroughly used to the idea that there were lots of different representations for the canonical variables of quantum field, I found a certain way of making things essentially unique. As you know there is a folk-theorem that says that whenever you had variables, say p_i and q_j, such that the commutator is $[p_i, q_j] = \sqrt{-1}\, \delta_{ij}$, an infinite number of them, and if you made, say a Boguliubov transformation or any kind of transformation which preserved the commutation relations, and got new variables p'_i, q'_j, then it was assumed that the old ones were connected with the new ones by a unitary transformation: $U p_i U^{-1} = p'_i$, $U q_j U^{-1} = q'_j$. That's just plain wrong. It's not true--not for any metaphysical or peculiar reasons, but quite objective, clear-cut reasons. So, this folk theorem is simply wrong; but what is true is this: suppose one takes the point of view that one doesn't observe all the p's or all the q's simultaneously, but only functions of finite sets of them, or limits of such functions in a physically conceptual sense. Then what you can do is this: you take all functions $F(p_1, p_2, \ldots, q_1, q_2 \ldots)$ of finitely many of the p's and q's and to be clear cut I'll take bounded functions. Now you assume that these bounded functions of finitely many of the many canonical variables, at least those should be observable. Then you say anything you can get from these by forming uniform limits should be observable because you can approximate to them in a quite direct way. Now you might think that in this limit you get all bounded operators on this space, but you don't; you get a proper subalgebra, which is a C*-algebra; and so there's a certain new thing which I call a Weyl algebra, consisting of all bounded functions of finitely many p's and q's, plus their limits. Now this Weyl algebra has the property that if you make this transition from the p's and q's to the new p's and q's, there exists a bona fide automorphism of the Weyl algebra effecting it. It doesn't necessarily correspond to any unitary transformation but: any canonical transformation determines a corresponding automorphism. This is the basis of the quantization of this Klein-Gordon equation with imaginary masses and lots of other such equations.

However, when you get to non-linear phenomena, you find some difficulties which are only now being elucidated. Several years later, --initially, at an MIT Conference in 1963, --Haag and Kastler did a variant of the foregoing where you deal with field variables, say $\phi(x,t)$ and what you do is you take all functions of the field variables where x and t range over a bounded set, and use that to form a C*-algebra. Thus, instead of taking a finite set of canonical variables, you're taking a finite region of space-time. Now, this of course gives you a perfectly good C*-algebra, but there is one difficulty with this, which hasn't yet been resolved actually, and that is that there's no corresponding uniqueness theorem. Even different free fields of different masses give apparently different algebras. So, as far as is now known it's an ad hoc construction leading to many different algebras. This leads to complications in the construction of quantum field which I'll come to later.

The particle representation for a free quantum field is a thing I've already explained--the Hilbert-space K is related to the single particle space H in the following way. H^0 is a one-dimensional space containing the vacuum; $K = H^0 \oplus H \oplus H \otimes_s H \oplus \ldots \oplus H^n \oplus \ldots$ where H^n is the space of all symmetric tensors of rank n over H. I am talking of bosons now for simplicity. Now the wave representation represents the Hilbert space K as the space of all square-integrable function over a certain space M. M is just the space of all classical fields. Alternatively, it is the space of all the Cauchy data $\phi(x,t)$ for the classical fields at a fixed time t. It's a big function space, but on this function space there's a natural way to impose a certain measure. The idea is that this makes possible the use of the mathematical techniques developed by Lebesgue, which are quite handy here. This construction represents the vacuum by the function identically one; and the quantized field operators are diagonalized, at the time t -- that's why you call it the wave representation and it is simply an invariant and precise form of what is given in Professor Dirac's book as the representation of linear boson field in terms of harmonic oscillators. An example of what you can do with this is the following: You can use it to get nice domains for Hamiltonians.

Typically what happens is this, if you're trying to construct a relativistic field, you have a free Hamiltonian H and you have a perturbation V. Now, in the nicest cases like a two-dimensional space-time with a non-linear scalar interaction the H and the V when I introduce a spatial cut-off, which as I've explained is permitted, the H and the V are then self-adjoint operators. So it is a relatively nice situation, but they are still rather bad self-adjoint operators relative to each other, --in e.g. the sense, that the commutator of H with V hardly exists; it doesn't have any vectors in its domain worth mentioning. Well, it's a little hard to add such operators. If operators commute you can generally add them easily. If they don't commute, you can add them if their commutator is bounded or something halfway decent. But the H and V that we're concerned with here, are rather singular. The V is something like a non-differentiable function. What one needs is some bunch of vectors, which are nice both from the point of view of the free Hamiltonian (which of course is nice in the particle representation) and the interaction Hamiltonian (which is nice in the wave representation). And this nice domain comes out from this measure space representation. What you do is this: you introduce the space $L_p(M)$ which is the set of all functionals on the measure space M, whose p-th power has a finite integral.

The *particle representation* for a free quantum field:

   (a) Fock (unregularized operator field in space)
   (b) Cook (Thesis, U. of Chicago) invariant construction in terms of tensors over Hilbert space:

$$K = H^0 \oplus H \otimes_s H \oplus H \otimes_s H \otimes_s H \times \ldots$$

The *wave representation*:

$$K = L_2(M)$$

M = Space of all 'classical fields', at a fixed time, $t = t_0$ endowed with a certain Lebesgue measure (total measure of M, 1).

Vacuum: function 1 in $L_2(M)$
Quantized field operators $\phi(x, t_0)$: represented as simple multiplications (after regularization in space)

The *application*: good regularity domain for free and spatially cut off Hamiltonians, certain non-linear fields (e.g. relativistic scalar fields in 2-dimensional space-time).

$L_p(M)$ = all functionals f on M such that

$$\int |f(x)|^P dx < \infty$$

D = common part of all $L_p(M)$, $2 \leq p < \infty$

Theorem: If the single-particle Hamiltonian A is ≥ 1, then for the corresponding field Hamiltonian H,

$$\| e^{-tH} u \|_{pe^{\delta t}} \leq \| u \|_p$$

for all $t > 0$, p ($2 \leq p < \infty$), and a universal constant δ.

Corollary: Both e^{-tH} and e^{-tV} act quite nicely on D and leave it invariant. (V = interaction Hamiltonian such as Wick power of even degree in 2-dimensional space-time).

Figure 3: Free particle representation of field theory.

And then, if you take the common part of all these spaces, that is relatively nice for both the free Hamiltonian and the interacting Hamiltonian (more exactly, the associated semigroups).

I've written down here a theorem. Roughly if H is the free Hamiltonian e^{-tH} exerts a certain smoothing effect. More specifically, if you start out with a state of the quantum field which is in $L_p(M)$ and apply e^{-tH} to it, it's even better than being in $L_p(M)$; it's in $L_q(M)$ where $q=pe^{\delta t}$ and δ is some universal constant. One can also see that if V is the perturbation, then e^{-tV} also acts nicely on $L_p(M)$, and both semigroups leave the common part of all $L_p(M)$, $1 < p < \infty$, invariant. Well it's a technical thing, but you can't get off the ground without some such and it really makes things much simpler--it makes it possible to deal with two-dimensional fields rather easily. So that's a "hard" application of the wave-particle duality, and there are plenty of others, both hard and soft.

Now I want to talk a little about hyperbolicity (See Figure 4). First I will explain what a function of operators is. If you have some operators, a function of them is just any operator in the von Neumann ring which they determine. Alternatively a function of operators is any operator which commutes with all unitary operators which commute with the given operators. (The equivalence of these two concepts follows from a well known theorem of von Neumann.) There's little doubt that's the right notion for present purposes. As an illustration, suppose you wanted to make an S-operator formulation of causality. You might insist that the outfield at a certain point should be a function in this sense of the infield at all earlier points. I tried this many years ago, but it doesn't work well except for mass zero. For non-vanishing mass, there are no normalizable Klein-Gordon wave functions which vanish in the backward cone. A bit later Borchers had a very nice heuristic result. It was not quite a mathematical theorem because it used the von Neumann theory, which dealt with bounded operators, for certain unbounded operators. But what it indicated was that the value of a field at a certain point in space-time was a function of its values at an earlier time restricted to this cone. It's just like the domain of dependence situation for a classical relativistic hyperbolic equation. Borcher's result is very suggestive but it isn't very viable for construction purposes. For this purpose the independence from the spatial cut off is proved by using two things, namely, the Lie formula which is a very natural thing to use; I think everybody here has seen the Lie formula for the exponential of the sum of two operators: $e^{A+B} = \lim_{n \to \infty} (e^{A/n} e^{B/n})^n$. Then you have to use the strict locality of the interaction which is the fact that the Wick products are, when suitably formulated, self-adjoint

A. "*Function of operators*": element of '*ring of operators*' determined by these operators, in Murray – von Neumann sense.

B. <u>Example</u>: possible approach to S-operator formulation of causality (1958)

   $\phi_{out}(x,t)$ should lie in ring operators generated by the $\phi_{in}(x',t')$ at earlier points.
   (Works for mass zero particles; for non-zero mass, difficulty from absence of normalizable wave functions vanishing in backward cone.)

C. Borchers heuristic theorem: in suitable axiomatic fields, the domain of dependence of a point is the same as for a classical relativistic differential equation.

D. Independence of temporal propagation of certain relativistic fields from spatial cut off.

   Source: a) Lie formula: $e^{A+B} = \lim_{n\to\infty} (e^{\frac{A}{n}} e^{\frac{B}{n}})^n$

   b) 'Strict' locality of interaction

   Informally: $e^{it(H_0+V(\delta))} \phi(x,0) e^{-it(H_0+V(\delta))}$, where $V(\delta)$ is independent of the spatial cut-off δ, once it is sufficiently close to 1, as $V(\delta) = \int :\phi(x,0)^{2k}: \delta(x) dx$

Figure 4: Summary of hyperbolicity.

operators that really belong to the von Neumann ring to which they should belong, an entirely local one. Using these, it turns out that the propagation is qualitatively hyperbolic as you intuitively expect.

This leads to the question of what kind of an operator the Wick powers are when you don't insist on temporal smoothing. In other words, at a fixed time, what kind of an object is a Wick power? In general, it just isn't an operator. That is to say, there is no vector in the Hilbert space to which you can apply a Wick power and get another vector in the Hilbert space; however, if you do things right, you get it to be a perfectly well-defined generalized operator, i.e. it simply maps vectors in the Hilbert space into certain other vectors which are in a different space which you can identify. Another point I want to make, of a more formal nature, is that the way to get a viable general theory, for interacting fields, is to modify the definition of the Wick powers by changing over to intrinsic characterization. Certainly the r-th power of the field at the time t should commute with the field itself, at the same time. Further, if you consider the commutator of the r-th power, with the first time derivative of the field then what you want is that the usual differentiation formula for forming commutators should be valid. In other words,

$$[:\phi(x,t)^r:, \dot{\phi}(y,t)] = ir:\phi(x,t)^{r-1}:\delta(x-y)$$

In addition one wants that the expectation value of the r-th power should vanish. There are two points about this definition which make it worth noting here. First of all, in the case of a free field this does give you the same thing as the Wick construction. Secondly, it works for any vacuum, it doesn't need to be the free field vacuum. Nothing in this definition refers to creation or annihilation operators. These statements are summarized in Figure 5.

At first glance the definition might appear a bit circular because it defines the r-th power in terms of the (r-1)-th power; however, eventually the r-1 will be down to 1 where it makes good sense. This means ϕ^2 is well-defined; once ϕ^2 is well defined ϕ^3 is well defined so recursively they are all well-defined. Note also that if the fields are irreducible at a fixed time then these definitions uniquely determine these powers because the only things which commute with the fields are then scalars and the scalars have to be 0 because of vanishing expectation values. So this does give you a unique definition. To make things entirely clear cut, you have to do some regularization which I won't go into except to mention that you have to deal with the unitary operators you get by exponentiating the Wick powers. There is a lot of pathology with unbounded operators and Hermann Weyl introduced many years ago a so-called "Weyl relation" which took care of the Heisenberg relations; and there are nonlinear variations of the Weyl relations which serve to do the

> What kind of operator is $:\phi(x,t)^n:$?
>
> <u>Answer</u>: It is uniquely determined by algebraic conditions $[:\phi(x,t)^n:, \phi(y,t)] = 0$
>
> $[:\phi(x,t)^n:, \phi^*(y,t)] = in:\phi(x,t)^{n-1}:\delta(x-y)$
>
> and $\quad E_{vac}[:\phi(x,t)^n:] = 0.$
>
> N. B. #1   When E_{vac} = free vacuum, this gives a Wick product of the powers of the field operator. But it is applicable to any vacuum. This leads to the next topic.
>
> N. B. #2   For an entirely clean-cut theory, you must regularize spatially and deal with
>
> $$e^{i\int :\phi(x,t)^n: \mathfrak{f}(x)dx}$$
>
> by non-linear variant of 'Weyl relations'

Figure 5: Theorems about generalized Wick powers.

same thing here. You can prove that the generalized Wick powers are uniquely characterized by these purely algebraic conditions.

This leads one to a very simple and natural way of giving meaning to a non-linear quantized equation such as
$$\Box \phi = m^2 \phi + p(\phi).$$
What does this mean actually? (See Figure 6) ϕ is here a conjectural field and we don't know what a nonlinear function of it really means. Well, the definition of a generalized Wick power applies equally well to this type of situation and what you then naturally require is that if you make a kind of spatial smearing on the field at a fixed time that you then get a self-adjoint operator in Hilbert space. In addition you want a physical vacuum, of course, v' let us say, and a self-adjoint operator H' which is the total Hamiltonian made concrete, having the properties earlier indicated. The differential equation itself is the key thing; it says that
$$\Box \phi = m^2 \phi + p(\phi)$$
where $p(\phi)$ is the generalized Wick polynomial of ϕ <u>relative to the physical vacuum</u> v'. (Of course, the free vacuum at a finite time makes no physical sense at all in this connection.)

This is simply a mathematically clear-cut way of describing the Heisenberg field; the ambiguous question, in which infinities reside, of what a polynomial in a (non-linear) quantum field means is here resolved in a natural and invar-

What does quantized equation

$$\Box \phi = m^2 \phi + p(\phi)$$

mean?

Natural answer:

The $\int \phi(x,t) f(x) dx$ are self-adjoint operators in Hilbert space;

There is vector v' (= 'physical vacuum') and a self-adjoint operator H' such that

$$H'v' = 0 \qquad H' \geq 0$$

$$e^{isH'} \phi(x,t) e^{-isH'} = \phi(x, t+s),$$

and:

$$\Box \phi = m^2 \phi + \ :p(\phi): \qquad \text{generalized}$$

Wick polynomial with respect to physical vacuum

Figure 6: Non-linear Field Theory.

iant·way. It is likely that this equation will soon be solved for arbitrary positive-energy scalar fields in 2-dimensional space time; but it is also likely that, without some further modification, it does not have solutions in a four-dimensional space-time. Before taking up the latter, let me mention that certain related fields have been constructed for the cited two-dimensional cases, which some authors have designated the "Heisenberg" field, in the special case of the polynomial $p(\phi) = \lambda\phi^3$ which they treated but there is no apparent reason to believe that the two fields are the same.

More specifically, what has been done so far for such 2-dimensional cases is to solve a simpler but a priori unphysical equation of the form

$$\Box \phi = m^2 + j$$

where the source j is defined in a certain way involving not only the field ϕ but a Hamiltonian dependent both on ϕ and other operator fields. The solution gives a one-parameter group of automorphisms of the space-time version of the Weyl algebra; there is a vacuum corresponding to this motion; generalized Wick polynomials are meaningful relative to this vacuum; the j that occurs is not, in general, the $p(\phi)$ in this sense. For this to be true, the vacuum expectation values of j and of its successive commutators with $\dot\phi$ at the same time would have to vanish (recall that $[p(\phi),\dot\phi] = ip'(\phi)\delta$ for an appropriate δ-function), and this does not appear to be generally the case. Of course, one can <u>define</u> j to be a renormalized version of p(φ), and write j = :p(φ): purely as a matter of notation; but this would hardly be in agreement with normal usage regarding "function" and "differential equation of motion." For a differential equation of the form $\partial\psi/\partial t = F(\Psi,t)$ means normally that when t and $\Psi(t)$ are given, $F(\Psi(t),t)$ is determined according to a definite rule, etc.; and the j is given as a function of values at another time.

Actually, in terms of the generalized Wick product, a differential equation of this type can be obtained, of the form $\Box\phi = m^2\phi + p(\phi) +$ lower order terms; to solve the original equation, one must start with a different p and show that for a suitable one, the lower order terms vanish. Unlike the earlier indicated work which is quite in agreement with contemporary mathematical standards (cf. e.g. "Construction of nonlinear quantum processes," to appear), these latter developments are as yet heuristic. Note in proof: a satisfactory rigorous form has now been given to this idea; but, the question of whether suitable counter-terms exist remains open.

Of course, one can do all this in terms of Hamiltonians, rather than differential equations, if one prefers. In these terms, the (time-independent) Hamiltonian of the true Heisenberg field is

$$H_{phys} = m^2 \int \phi_{phys}^2 dx + \int (\text{grad } \phi_{phys})^2 dx + \int \dot\phi_{phys}^2 dx + \int P(\phi_{phys}) dx$$

$$(P' = p)$$

where ϕ_{phys} denotes the hypothetical Heisenberg field which one is trying to construct; the situation is thus a highly implicit one. The (time-dependent) Hamiltonian which has been studied mathematically is

$$H_{model} = m^2 \int \phi_{free}^2 dx + \int (\text{grad } \phi_{free})^2 dx + \int \dot\phi_{free} dx + \int P(\phi_{free}) dt$$

(materially time-dependent); here ϕ_{free} is a given free field, of some chosen mass. Thus what is being done roughly is to replace the physical field at a finite time (e.g. t_0) by a free field of some mass or other. This is of course quite unphysical; at infinite times, the two fields are presumably equal, but they are presumably very different at finite times. Because of this agreement between the two fields at infinite times, the clearly appropriate usage of free fields at this time for the solution of field-theoretic problems is in the interaction representation; the free-field may then be taken as the physical mass. The work I've described carries implications for the construction of the interaction-representation field but for lack of time I'll have to omit this subject (cf. forthcoming articles in the M. H. Stone and S. Bochner symposiums; in any event, very little is known about the asymptotics of this situation).

I'll now go on to the 4 dimensional case. In the 4-dimensional case, there is the problem that $\int :\phi(x,t)^r: f(x) dx$ is not an operator but becomes an operator only if you smooth it in time. The differential equation of motion can be given a clear-cut mathematical meaning in the interaction representation also, in a certain way. The question then arises as to whether this equation has solutions. Whenever you have a differential equation there are two ways to solve it. One method is by successive approximation which is roughly what people do in physics in perturbation theory. The other method is by step-by-step integration. I want to describe a finite process, --whose convergence, however, is still in question--for a step-by-step integration of the corresponding equation. The S operator as you know is the time-ordered exponential of the interaction Hamiltonian. The trouble is that the interaction Hamiltonian is not a bona fide operator at each finite time t. What you can do is to approximate S by

$$\exp\left[i\int_{t_1}^{t_2} H_I(s)\,ds\right] \exp\left[i\int_{t_2}^{t_3} H_I(s)\,ds\right] \cdots \exp\left[i\int_{t_{n-1}}^{t_n} H_I(s)\,ds\right]$$

and hope that when you integrate from t_{j-1} to t_j that you then get a self-adjoint operator. Sometimes that works but most of the time it doesn't. If I were doing quantum electrodynamics, for example that would not quite suffice. However, if I take

$$\exp\left[i\int H_I(s) f(s)\,ds\right]$$

where f has a small number of derivative, that does exist and what I can do then is this: this formula corresponds to taking a "product integral" over a long interval, let us say from t to t' and chopping it up into small integrals and integrating over each of these small integrals. Now instead of doing that what I might do is take an integral like this:

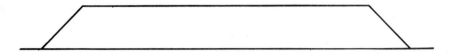

and break it up into functions like this:

which are somewhat smoother than

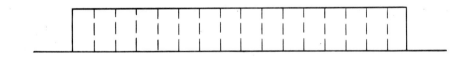

So when all is done I get a means of approximating to an operator which doesn't take ϕ from a sharp time t to a sharp time t' but rather from one fudged time to another. It turns out that you can handle any finite polynomial by using an appropriate regularization. For example for ϕ^4 you would need things which would have one more derivative which would be like

f(s):

(assembled from sine-curves). What's involved here then is
$$e^{i \int H_I(s) f(s) ds}.$$
integrated over such a function f(s). This operator $\int H_I(s) f(s)$ is then bona fide; it has a dense set of vectors in its domain which it maps into the Hilbert space, and a self-adjoint extension; it makes sense to take matrix approximations to it, which are readily computed, and to exponentiate them. There are theorems of the character that: if you have a self-adjoint operator H and a sequence H_n of matrices converging to it, in a certain sense, then it is true that the exponential $e^{iH_n t}$ converges to e^{iHt} under certain conditions. So it makes sense that when you have a bona fide operator to approximate to its exponential by exponentiating matrices. This kind of procedure may well be relatively practical by now on a computer

Thus, step-by-step integration of the differential equation should lead to an effective S operator, while avoiding the more basic but more difficult question for the future of the structure of the physical vacuum if, indeed, this exists. Let me point out that one doesn't have time-dependent unitary operators of the form $S(t,t')$; one has an S depending on a function f which is smooth, and such that $S(f)S(g) = S(f + g)$ if g is "earlier" than f; this replaces the normal propagation identity, $S(t'',t')S(t',t) = S(t'',t)$. What this means is that you've lost the connection between the interaction representation field and the Heisenberg or non-linear field. That connection depends upon having operators at a sharp time and you no longer have these. What this suggests is that there is something more fundamental, perhaps, in the interaction representation which should enable one to circumvent the problems of the existence and properties of the physical vacuum; and that it's probably fruitless to hope to find the interaction-representation field at a sharp time. But it is reasonable to hope that with this one emendation, --the use of the "distribution" propagators $S(f)$ to avoid sharp times,-- four-dimensional fields may be treated in a physically effective way by the approach now established, in substantial part, for two-dimensional fields.

DISCUSSION OF PROF. SEGAL'S TALK

J. M. Jauch, Institut de Physique, Geneva: There is one question which always bothers me with the C*-algebra approach. Presumably, at least people who know this field very well, they think of the C*-algebra as containing among many other things also all the observables.

Segal: I don't think that's correct. That's a naive point of view.

Jauch: This is, of course, not my point of view, so ...

Segal: They contain all the conceptual observables in a certain sense. The much more important observables are of course the S operator and the energy. The energy is not in the C*-algebra but is arrived at by a round-about procedure which involves this algebra.

Jauch: I see.

Segal: There are conceptual observables which are not truly empirically observable. It's just like in elementary quantum mechanics: you may take say cos p + sin q; that's a bounded operator, so it's an observable; but you cannot imagine any experiment which will actually measure a thing like that.

Jauch: I am sorry, you misunderstood my statement. I said, among many other things, which are not observable it should contain the observables.

Segal: It certainly doesn't and it really shouldn't.

Jauch: The bounded observables?

Segal: No, it shouldn't.

Jauch: Oh, I see. Then most of the things that I have read about this in the literature are wrong.

Segal: I think they are oversimplifications. Let me say how you get the main things you want. You want two things. You want the energy and the S operator. Let me say, in a simplified way, how you get the energy. You have the C*-algebra which is the habitat so to speak of your conceptual field observable. On this C*-algebra you construct in some fashion or other (from a given differential equation or its equivalent, usually) a one-parameter group of automorphisms. Having constructed this one-parameter group of automorphisms, it is then a well-defined question whether or not a vacuum exists. If a vacuum exists, you then make the standard construction for this state and this then gives you a certain Hilbert space; certain field operators on the Hilbert space (which are rather useless for empirical correlation); but in addition, it gives you a one-parameter unitary group on the Hilbert space; and that's what you want, really, because that group has as its generator the physical energy operator. This physical energy operator is definitely not in the algebra either in theory or in practice, though in a certain sense it is derived from and affiliated with it.

Jauch: You say that the question then becomes again pertinent by what physical principles do you select your algebra which is the bases in your axiomatic.

Segal: Yes, that's a good question which has again been overdone a bit. First of all, when you are trying to construct a Hilbert space it doesn't matter what bunch of nice functions or operators you start with. If you take the continuous functions on (0,1) or the differentiable functions on (0,1) or the polynomials on (0,1) and complete them relative to the Hilbert metric you end up with exactly the same thing. Now that completion is very analogous to the canonical construction for going from a C*-algebra and a given state to a Hilbert space, so one would anticipate that almost any reasonable C*-algebra, when you go through the standard construction, will end you up with the same Hilbert space. However, there is a very definite problem: that is you don't know a priori any natural C*-algebra to use for constructing the Heisenberg field. Some people have tried to assume that at an initial time t the initial (pre-Heisenberg) field is in the free representation. Physically of course it is absurd to think of the Heisenberg field as being in the free representation but the hope would be that in the end it wouldn't matter; that when you made the canonical construction, the results would be independent of the initial representation. This unicity certainly hasn't been proved in the existing literature, which doesn't explicitly take cognizance of the problem, but it's a serious open question. In other terms, there is no unicity theorem concerning the analogue of the Weyl algebra where you use the bounded functions in the fields in finite regions of space. If you made the same kind of construction, using a different free field to start with, you would get an ostensibly different final answer. It may well require an enormous amount of work to check that the results are essentially the same. That's one reason why I insist on the interaction representation; all that washes out in terms of the interaction representation because it is explicitly in terms of the known physical field, the free incoming field; you know that representation precisely, and all the operators are in terms of it, so there is no ambiguity about the proper representation or C*-algebra.

Jauch: I think Professor Dirac tells us always that one should use the Heisenberg representation.

Segal: Well one should but it is difficult.

Jauch: I see.

Segal: And if you can get the S-operator without doing that I think that's worth doing.

P. A. M. Dirac, Cambridge University: There is some help in using the Heisenberg representation because you have all the vacuum fluctuations.

W. Thirring, CERN: Is this wave representation you mentioned equivalent to the Fock representation?

Segal: It's unitarily equivalent to the Fock representation. That is to say, whenever you have a bunch of operators on Hilbert space you always raise the question of how to diagonalize them, or express them in an otherwise convenient form. (Now concerning this Fock-Cook representation: I think the mathematics, --the self-adjointness and also the invariant form,-- was really done by Cook. It's only formally identifiable with what Fock did. By now people are so familiar with the idea of Schwartz distributions tht it's simple to make a transition. At the time Cook did it, it was by no means trivial, and his formalism has been very useful.) There are three different representations of a boson field, each of which is good for certain things; say, the particle (or Fock-Cook) representation; the wave representation; and one I'll call the holomorphic functional representation. One can express quite explicitly the unitary operator W which carries the particle into the wave representation Hilbert space, and you can write down just what W does to all the relevant operators, etc., to the field operators, the vacuum, the energy, Lorentz group action, etc. Incidentally, W is entirely independent of the nature of space-time. The theory depends only on a given Hilbert space, which is physically the single particle space.

I might say also that most of the work described earlier is done in a way which is independent of detailed assumptions as to the precise structure of space-time; if one were to make space compact, or curve it a bit, etc. it wouldn't change things much. All that's important here is general character of the energy, say, E as a function of the momentum. The energy-momentum dependence, or rather its asymptotic form for high energy, is essentially all that determines the degree of singularity of the theory.

NONLEPTONIC DECAY AND THE CURRENT x CURRENT THEORY*†

S. Pakvasa

University of Hawaii, Honolulu, Hawaii 96822

and

S. P. Rosen

Purdue University, Lafayette, Indiana 47907

In the 1950's the "paradoxical" behavior of nonleptonic decay taught us about strangeness and parity nonconservation, and ever since then we have been trying to understand the decay process itself in terms of this lesson. Complete success may have eluded us, but we can definitely say that some progress has been made in the past ten years. On the experimental side we have measured more and more decay parameters with greater and greater accuracy; and on the theroetical side we have managed to limit ourselves to fewer and fewer theories. At this time we still have one model which seems to

---

*Supported in part by the U.S. Atomic Energy Commission.

†Paper presented to the Symposium "The Past Decade in Particle Theory" University of Texas at Austin, April 14-17, 1970.

account for the gross features of the data, and for some of the fine details too. Important problems remain, but at least we do have a theoretical framework within which to analyze them.

This framework is, of course, the current x current theory according to which all weak interactions are engendered by the interaction of a charged current with itself. Reflecting the developments which stem from parity and strangeness nonconservation, the current is an equal mixture of vector and axial vector parts with a definite phase relationship between them, and its hadronic components are taken to be members of an octet. In fact we assume that the weak hadronic currents combine with the electromagnetic one to generate chiral SU(3) x SU(3) symmetry; when the hypothesis of PCAC is added, we have all the ingredients for the current algebra theory of nonleptonic decay. Given this complex structure, we must be careful to examine the consequences of each element in it, and to separate the predictions of one element from those of another.

We begin with strangeness.[1]-[3] It is an empirical fact that those nonleptonic decays in which the strangeness quantum number changes by one unit are characterized by effective coupling constants of the same order of magnitude as the neutron β-decay constant G_V. Processes in which strangeness changes by two units have either not been detected (e.g. $\Xi \to N + \pi$) or are so small as to be of second order in G_V (e.g. the $K_L - K_S$ mass difference). Thus we can say that nonleptonic decays are engendered as the lowest order effects of a weak interaction Hamiltonian which satisfies the rule $\Delta S = 1$.

From electric charge conservation and the Gell-Mann-Nishijima formula, we know that the $\Delta S = 1$ rule in nonleptonic decays is equivalent to the selection rule $\Delta T_3 = \frac{1}{2}$. This does not necessarily imply that the change in total isotopic spin is also $\frac{1}{2}$, but it does make it an interesting hypothesis: for, if ΔT were equal to $\frac{1}{2}$, the rules for ΔT_3 and ΔS would be automatically satisfied.

The earliest hint that this might indeed be the case appeared in the 2π decay modes of the K-meson.[1]-[3] In 1954, Gell-Mann and Pais observed that, because of Bose statistics and the spinless character of K- and π-mesons, the $\Delta T = \frac{1}{2}$ rule would forbid $K^+ \to \pi^+ \pi^0$ and require the branching ratio of $K_S \to \pi^+ \pi^-$ to $K_S \to \pi^0 \pi^0$ to be 2:1. Empirically the amplitude for $K^+ \to \pi^+ \pi^0$ is approximately 1/20 times that for $K_S \to 2\pi$ and the branching ratio in neutral K decay is about 2.3 (see Table 1). Thus the $\Delta T = \frac{1}{2}$ rule would appear to be a very good

TABLE 1

Test of $\Delta T = \frac{1}{2}$ Rule in $K \to 2\pi$.

	$\Delta T = \frac{1}{2}$	Expt
$\Gamma(+-)/\Gamma(00)$	2.00	2.33 ± 0.05
$\Gamma(+-)/\{\Gamma(+-) + \Gamma(00)\}$	0	$(1.50 \pm 0.07) \times 10^{-3}$

first approximation in the theory of nonleptonic decay.

Since 1954, this conclusion has been tested in numerous decay processes, and it has always been confirmed.[1]-[3] In $K \to 3\pi$, for example, the observed density distribution in the Dalitz plot is, to a high degree of accuracy, given by

$$D(\alpha, \beta, \gamma) = \Gamma(\alpha, \beta, \gamma)[1 - 2M_K T_{max} \sigma(\alpha, \beta, \gamma) y_3] \quad (1)$$

where α, β, γ, are the charges of the final state pions, $\Gamma(\alpha, \beta, \gamma)$ is the decay rate divided by the phase space volume, and $\sigma(\alpha, \beta, \gamma)$ is the slope parameter. In general the $\Delta T = \frac{1}{2}$ rule allows both a $T = 0$ and a $T = 1$ final state; however, because the density of eq^n(1) is linear in the energy variable y_3, the admixture of $T = 0$ is negligible. The consequences of this selection rule and a comparison with the experimental data are shown in Table 2. Roughly speaking $\Delta T = \frac{1}{2}$ appears to be satisfied to within 5-10% in amplitude.

TABLE 2

Test of $\Delta T = \frac{1}{2}$ and absence of $\Delta T = \frac{5}{2}$ in $K \to 3\pi$. The rate and slope parameters are given in eq.^{n}(1) of text. Note that the comparison of different decay modes of same K-meson tests only a $T = 1$ final state, while the comparison of decays of different K-mesons tests $\Delta T = \frac{1}{2}$.

	$T = 1$ final state	Expt (Devlin)
$\Gamma(0\ 0\ 0)/{}^3/_2\Gamma(+\ -\ 0)$	1	1.00 ± 0.06
$\Gamma(+\ +\ -)/4\Gamma(+\ 0\ 0)$	1	1.00 ± 0.03
$\sigma(+\ 0\ 0)/\sigma(+\ +\ -)$	-2	-2.6 ± 0.24
	$\Delta T = \frac{1}{2}$	
$\Gamma(+\ -\ 0)/2(+\ 0\ 0)$	1	0.80 ± 0.044
$\Gamma(0\ 0\ 0)/[\Gamma(+\ +\ -) - \Gamma(+\ 0\ 0)]$	1	0.79 ± 0.04
$\sigma(+\ -\ 0)/\sigma(+\ +\ -)$	-2	-2.0 ± 0.16

In the decay of spin $\frac{1}{2}$ hyperons, the final state is an admixture of S- and P-waves, and the nonrelativistic transition matrix can be written in the rest-frame of the parent particle as

$$Y^a_b = S^a_b + P^a_b\, \underline{\sigma} \cdot \hat{p} \qquad (2)$$

where $\hat{p}$ is a unit vector along the momentum of daughter baryon.

The notation on the left hand side of eq^n (2) means that hyperon Y of charge a decays into a π-meson of charge b plus a baryon whose charge and strangeness are determined by $\Delta S = 1$, $\Delta Q = 0$. If the $\Delta T = \frac{1}{2}$ rule is valid, then the decay amplitudes must satisfy

$$\Delta(\Lambda) \equiv \sqrt{2}\, \Lambda^0_0 + \Lambda^0_- = 0$$

$$\Delta(\Xi) \equiv -\sqrt{2}\, \Xi^0_0 + \Xi^-_- = 0 \qquad (3)$$

$$\Delta(\Sigma) \equiv \Sigma^+_0 + \frac{1}{\sqrt{2}} \Sigma^+_+ - \frac{1}{\sqrt{2}} \Sigma^-_- = 0$$

for both S- and P-waves. In the case of Λ- and Ξ-hyperons, the sum rules of eq^n (3) can be tested by a comparison of the rates and asymmetry parameters for the appropriate decay modes (see Tables 3 and 4); but for the Σ-hyperon we must study the amplitudes directly. When we neglect final state interactions and CP nonconservation and use the fact that the rates for the three Σ-decay modes are very nearly the same, we obtain from eq^n (3) the well-known right angled triangle of Gell-Mann and Rosenfeld (see Fig. (i)).

The data on hyperon decay are summarized in Table 3 and a comparison with the predictions of $\Delta T = \frac{1}{2}$ is made in Table 4 and Fig. (i). We have neglected the effects of CP nonconservation, which are expected to be about 1/10 of 1%, but we have included final state interactions. Again, we see

FIGURE CAPTION

Fig. (i) The Gell-Mann-Rosenfeld and Lee-Sugawara Triangles (Figure taken from O. E. Overseth, International Conference on Particle Symmetries and Quark Models, Wayne State University June 1969).

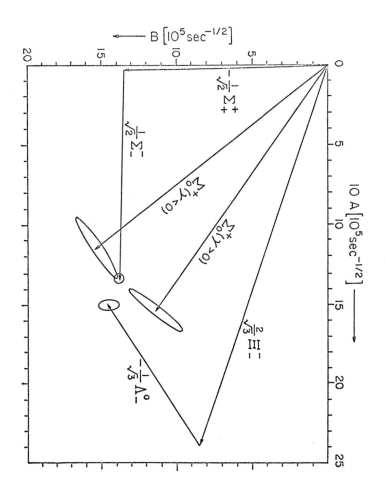

Table 3

Data on Hyperon Decay. The decay distribution is given by:

$$\frac{d\omega}{d\Omega} = \frac{\Gamma}{8\pi}\{[1 + \alpha\underline{\Sigma}\cdot\hat{p}] + \underline{\sigma}[(\alpha + \underline{\Sigma}\cdot\hat{p})\hat{p} + \beta\underline{\Sigma}\times\hat{p} + \gamma(\hat{p}\times(\underline{\Sigma}\times\hat{p}))]\}$$

where $\underline{\Sigma}$ and $\underline{\sigma}$ are the polarization vectors of the parent and daughter baryons respectively, and

$$\Gamma = (2pc/\hbar)[|S|^2 + |P|^2] \qquad \alpha = 2\text{Re}(SP^*)/[|S|^2 + |P|^2]$$

$$\beta = 2\text{Im}(SP^*)/[|S|^2 + |P|^2] \qquad \gamma = [|S|^2 - |P|^2/[|S|^2 + |P|^2]$$

The parameters S and P are given in eq.$\underline{n}$(2) of the text, and p is the magnitude of the momentum of the daughter baryon. Γ is the decay rate, and α, β, γ are the asymmetry parameters. Since $\alpha^2 + \beta^2 + \gamma^2 = 1$ we can write β and γ as

$$\beta = \sqrt{(1 - \alpha^2)}\sin\phi \qquad \gamma = \sqrt{(1 - \alpha^2)}\cos\phi$$

It has become customary to write the decay matrix in the covariant form $i\bar{N}(A + B\gamma_5)\Psi\pi$; the relation between A and S, and between B and P can be obtained by taking the nonrelativistic limit of the covariant expression in the rest-frame of Y.

<u>Λ-Decay</u> $\Gamma_{total} = (0.397 \pm 0.005) \times 10^{10} \text{ sec}^{-1}$

	$\Lambda \to p\pi^-$		$\Lambda \to n\pi^0$
Branching Ratio	0.644 ± 0.008	B.R.	0.35 ± 0.013
α^-	0.65 ± 0.02	α^0	0.645 ± 0.07
β^-	-0.10 ± 0.07		
γ^-	0.75 ± 0.02		
A_Λ^-	$-(1.53 \pm 0.02) \times 10^5 \text{ sec}^{-\frac{1}{2}}$		
B_Λ^-	$-(10.50 \pm 0.33) \times 10^5 \text{ sec}^{-\frac{1}{2}}$		

Table 3 (continued)

Ξ - Decay

	$\Xi^- \to \Lambda\pi$	$\Xi^0 \to \Lambda\pi^0$
Γ^-	$(0.602 \pm 0.013) \times 10^{10} \sec^{-1}$	Γ^0 $(0.330 \pm 0.020) \times 10^{10} \sec^{-1}$
α^-	-0.383 ± 0.038	α^0 -0.36 ± 0.08
ϕ^-	$8° \pm 8°$	ϕ^0 $25° \pm 16°$
A^-	$(2.07 \pm 0.02) \times 10^5 \sec^{-\frac{1}{2}}$	A^0 $(1.56 \pm 0.05) \times 10^5 \sec^{-\frac{1}{2}}$
B^-	$(-6.68 \pm 0.70) \times 10^5 \sec^{-\frac{1}{2}}$	B^0 $(-4.79 \pm 1.10) \times 10^5 \sec^{-\frac{1}{2}}$

Σ - Decay

	$\Sigma^+ \to n\pi^+$	$\Sigma^+ \to p\pi^0$
Γ	$(1.235 \pm 0.02) \times 10^{10} \sec^{-1}$	$(1.235 \pm 0.02) \times 10^{10} \sec^{-1}$
B.R.	0.472 ± 0.015	0.528 ± 0.15
α	0.037 ± 0.049	$-0.98^{+0.05}_{-0.02}$
ϕ	$176° \pm 24°$	$22° \pm 90°$
A	$(0.06 \pm 0.02) \times 10^5 \sec^{-\frac{1}{2}}$	$(1.53 \pm 0.14) \times 10^5 \sec^{-\frac{1}{2}}$
B	$(19.07 \pm 0.34) \times 10^5 \sec^{-\frac{1}{2}}$	$(-11.52 \pm 1.85) \times 10^5 \sec^{-1}$

	$\Sigma^- \to n\pi$
Γ	$(0.621 \pm 0.19) \times 10^{10} \sec^{-1}$
B.R.	1
α	-0.134 ± 0.034
ϕ	$-5° \pm 23°$
A	$(1.89 \pm 0.03) \times 10^5 \sec^{-\frac{1}{2}}$
B	$-0.72 \pm 0.01 \times 10^5 \sec^{-\frac{1}{2}}$

Table 3 (continued)

Ω^- - Decay

$$\Gamma = (0.76^{+0.24}_{-0.37}) \times 10^{10} sec^{-1}$$

$$\Lambda K^- = 13 \text{ events}$$

$$\Xi^0 \pi^- = 8 \text{ events}$$

$$\Xi^- \pi^0 = 3 \text{ events}$$

References for Data

D. Berley et. al.	Phys. Rev. **D1**, 2015 (1970)
F. Harris et. al.	Phys. Rev. Letters **24**, 165 (1970)
S. Olsen et. al.	Phys. Rev. Letters **24**, 843 (1970)
O. E. Overseth	Proceedings of International Conference on Symmetries and Quark Models at Wayne Wayne State Univ. June 1969 (to be published)
P. M. Dauber et. al.	Phys. Rev. **179**, 1262 (1969)
Particle Data Group	Rev. Mod. Phys. (Jan. 1970)

TABLE 4

The $\Delta T = \frac{1}{2}$ Rule in Hyperon Decay.

Λ-Decay	$\Delta T = \frac{1}{2}$	Expt
$\Gamma^-/(\Gamma^- + \Gamma^0)$	0.66	0.644 ± 0.08
α^0/α^-	1.00	1.000 ± 0.068
β^-/α^-	$-\tan(\delta_{11}-\delta_1)$ $= -0.11 \pm 0.04$	-0.16 ± 0.06

Ξ-Decay		
Γ^0/Γ^-	0.5	0.55 ± 0.04
α^0/α^-	1.00	0.94 ± 0.22

Σ-Decay		
$\Delta(\Sigma)$	0	$\Delta_S/S^- = 0.05 \pm 0.07$
(see eq$\underline{^n}$(3) of text)	0	$\Delta_P/P^+ = 0.05 \pm 0.07$
β^0/α^0	$\frac{1}{3}\tan(\delta_{11}-\delta_1) + \frac{2}{3}\tan(\delta_{11}-\delta_3) = 0.03 \pm 0.05$	$0.08^{+0.16}_{-0.02}$
ϕ^+	$176°$	$176° \pm 24°$

Ω-Decay		
$\Gamma(\Xi^0\pi^-/\Gamma(\Xi^-\pi^0)$	2	$8/3$ (?)

Note: The predictions for β/α and ϕ^+ in Σ- and in Λ-decay are taken from O. E. Overseth & S. Pakvasa, Phys. Rev. __184__ 1663(1969) that deviations from the $\Delta T = \frac{1}{2}$ rule are of order 10% or less. We also note that the amplitude for $\Sigma^+ \to n\pi^+$ lies along the P-axis of the Gell-Mann-Rosenfeld diagram; this remarkable result is not dictated by $\Delta T = \frac{1}{2}$, but it does become important in the SU(3) symmetry scheme.

Having established the domain of validity of the $\Delta T = \frac{1}{2}$ rule, we must now consider deviations from it. The very fact that the charged K-meson does decay into two pions implies that the effective weak Hamiltonian must contain a term with $\Delta T = \frac{3}{2}$ or greater. If the amplitude for $K_-^+ \to \pi^{\pm}\pi^0$ is any guide, the strength of this term lies between 5 and 10% of that of the dominant part of the Hamiltonian. Where, we may ask, could such a term originate?

The earliest answer was the electromagnetic interaction which, because it does not conserve isotopic spin, can generate $\Delta T = \frac{3}{2}$ corrections to a "bare" $\Delta T = \frac{1}{2}$ Hamiltonian. These corrections, however, being of order $\alpha = \frac{1}{137}$ relative to the normal weak amplitude, are smaller than the effect we want to explain by a factor of 5 or 6 (see Table 1). Thus if the occurrence of $K^+ \to \pi^+\pi^0$ were an electromagnetic correction to the $\Delta T = \frac{1}{2}$ rule, we would find ourselves in the curious position of needing some dynamical enhancement for the decay amplitude!

In 1964 Cabibbo suggested a way out of this paradox. He showed that in the limit of exact SU(3) for strong interactions, the current x current interaction together with octet dominance forbids $K_S \to 2\pi$. He could then argue that the relatively large value of the $K^\pm \to \pi^\pm \pi^0$ amplitude arises not from an enhancement of charged K decay, but rather from a suppression of neutral K decay. Up to now no accurate calculation of the suppression factor has been made, and so Cabibbo's argument remains a qualitative one.

This leaves us free, if the opportunity presents itself, to set aside the notion that $K^+ \to \pi^+ \pi^0$ occurs as an electromagnetic correction to weak interactions, and to consider alternative sources for deviations from $\Delta T = \frac{1}{2}$. It turns out that the current x current interaction does provide us with such an opportunity.[1] In general the Hamiltonian contains both octet and (27)-plet components, but in the limit of exact SU(3) symmetry for strong interactions it forbids <u>all</u> two-pion decay modes of the K-meson. Therefore we can take the point of view that these decays occur through the intervention of SU(3)-breaking strong interactions, and that this mechanism tends to enhance the octet part of the current relative to the (27)-plet part.

Although there are no convincing calculations of meson decay to support this octet dominance point of view, there are some indications from current algebra, and also from field algebra, that it may be correct. In addition, calculations of S-wave hyperon decay amplitudes tend to substantiate the idea that the weak interactions are dominated by the octet, but contain an intrinsic component belonging to a higher representation. We shall therefore explore this idea in some detail.

The current x current Hamiltonian for nonleptonic decay,

$$H_{NL}(x) = \frac{G}{\sqrt{2}} \{J_\lambda(x), J_\lambda^+(x)\}, \quad G \approx 10^{-5} m_p^{-2} \quad (4)$$

is constructed from a hadronic current $J_\lambda(x)$ which contains an equal mixture of vector and axial vector parts

$$J_\lambda(x) = V_\lambda(x) + A_\lambda(x) \quad (5)$$

The vector and axial vector currents are themselves admixtures of strangeness conserving ($\Delta S = 0$) and strangeness violating ($\Delta S = 1$) terms

$$V_\lambda(x) = \cos\theta_V \, V_\lambda^{(0)}(x) + \sin\theta_V \, V_\lambda^{(1)}(x)$$
$$A_\lambda(x) = \cos\theta_A \, A_\lambda^{(0)}(x) + \sin\theta_A \, A_\lambda^{(1)}(x) \quad (6)$$

where θ_V and θ_A are the vector and axial vector Cabibbo angles. It is interesting that in addition to $\Delta S = 1$ decays, H_{NL} also gives rise to strangeness conserving processes. There is some evidence that these weak effects have been detected in nuclei and we shall discuss them in due course. For the moment we concentrate on the strangeness violating part of H_{NL}.

If all the hadronic currents in eq$\underline{\text{ns}}$ (4) and (6) behave as members of octets and are of the "first class" variety, then H_{NL} conserves CP* and transforms as an admixture of the SU(3) representations 1, 8, and 27. In isospace $H_{NL}^{(1)}$, the $\Delta S = 1$ part of H_{NL}, contains only $\Delta T = \frac{1}{2}$ and $\frac{3}{2}$, and in U-spin space it satisfies $\Delta U = 1$, $\Delta U_3 = \pm 1$. The parity conserving part of $H_{NL}^{(1)}$ is symmetric under the exchange of SU(3) indices $2 \leftrightarrow 3$ (i.e. it behaves like Gell-Mann's matrix λ_6) no matter what the relation between θ_V and θ_A, but the parity violating part possesses this symmetry only when

$$\theta_V = \theta_A = \theta \qquad (7)$$

This is something of an irony because $2 \leftrightarrow 3$ symmetry has a powerful influence upon parity violating decay, but none on parity conserving decay.'

---

*Throughout this talk we shall neglect the effects of CP non-conservation. We refer the reader to Prof. Wolfenstein's talk for a discussion of this important topic.

The equality of Cabibbo angles in eq^n (7) is, of course, no more than the usual statement about the V-A structure of weak currents. It carries with it the tacit assumption that $V_\lambda^{(0)}$ and $V_\lambda^{(1)}$ are members of the same octet, and that the same is true for $A_\lambda^{(0)}$ and $A_\lambda^{(1)}$. In the language of chiral SU(3) x SU(3), the current J_λ belongs to the (8,1) representation when eq^n (7) holds, and H_{NL} belongs to (1+8+27,1). This assignment plays an important role in the predictions of current algebra.

The first noteworthy consequence of a Hamiltonian given by eq^{ns} (4) and (7) is that the amplitudes for $K^+ \to \pi^+\pi^0$ and $K_S \to 2\pi$ vanish in the limit of exact SU(3) for strong interactions. When SU(3)-breaking strong interactions are introduced, the 2 ↔ 3 symmetry of $H_{NL}^{(1)}$ is broken, and the decay amplitudes no longer vanish.[1] However, they still obey $\Delta T \leq \frac{3}{2}$ and hence must satisfy the sum rule:

$$A(K_S \to \pi^+\pi^-) + \sqrt{2}\, A(K_S \to \pi^0\pi^0) = 2A(K^+ \to \pi^+\pi^0) \qquad (8)$$

Present data are consistent with eq^n (8), but they are not sufficiently accurate to provide a definitive test of it.

To gain some insight into the suppression of the $\Delta T = \frac{3}{2}$ part of $H_{NL}^{(1)}$, we employ the apparatus of PCAC and current algebra together with the chiral symmetry properties of H_{NL}. We can then show[2] that when both final state particles are soft, the K → 2π amplitude is proportional to $<K|H'_{nl}|vacuum>$ where H'_{NL} is an isospin-rotated form of H_{NL}.[1] Since the K-meson has isospin ½, the $\Delta T = \frac{3}{2}$ part of the Hamiltonian can make no contribution to the matrix element. Therefore, to the extent that the extrapolation from the mass-shell to the soft-pion limit does not lead to a large variation in the amplitude, the dominant contribution to the decay will come from the $\Delta T = \frac{1}{2}$ part of H_{NL}.[1]

We must now try and estimate the admixture of $\Delta T = \frac{3}{2}$ present. By contracting the K-meson instead of the pions in the current algebra method, we find that the amplitude is proportional, in the soft K-meson limit, to the matrix element of a product of currents between π-meson states:

$$<\pi\pi|H_{NL}^{(1)}|K> \sim <\pi|J_\lambda J'_\lambda|\pi> \qquad (9)$$

We can evaluate the right-hand-side of eq$\underline{^n}$ (9) by inserting a complete set of states between the current operators and then expressing the matrix element in terms of weak form factors.[6]

Three types of intermediate state are likely to enter the calculation: octets, nonets, and singlets. With no detailed assumptions about form factors, we find that when only one of these types is present, the ratio of the $\Delta T = \frac{3}{2}$ amplitude to the $\Delta T = \frac{1}{2}$ one is fixed, and at least of order unity. When two types are present, the ratio is no longer fixed, and it will be small if the contribution to $\Delta T = \frac{3}{2}$ from one type of intermediate state tends to cancel that from the other.

The extent to which the cancellation does occur depends upon the details of the model. In a typical calculation,[6] the result obtained is in good agreement with experiment, but it is very sensitive to the parameters of the weak form factors. Taking this model dependence and also our doubts about the soft-meson extrapolation into account, we cannot regard the $K^+ \to \pi^+ \pi^0$ problem as being solved. Hopefully we may be able to solve it when we know more about weak form factors and low lying meson states.

A more positive statement can be made about a companion process, namely S-wave hyperon decay. By combining the symmetry properties of the current x current Hamiltonian with PCAC and current algebra, we can show that $S(\Sigma^+_+)$, the S-wave amplitude for $\Sigma^+ \to n\pi^+$, depends only upon the (27)-plet component of H_{NL}.[1] The fact that this amplitude is found to be small is then direct experimental evidence for octet dominance.

We also have "theoretical" evidence because, when we calculate $S(\Sigma^+_+)$ by the same general method as was used for the matrix element of eq^n (9), we find that it is much smaller than other S-wave amplitudes.[4]-[12] This result is not very sensitive to the detailed behavior of weak form factors, and it can be explained in terms of rather simple qualitative arguments. Furthermore the extrapolation from the physical region to the soft pion limit is not as serious in hyperon decay as it is for meson decay. Thus we can claim to have some understanding of octet dominance in this case.

It is instructive to follow the argument more closely. As a result of the 2 ↔ 3 symmetry of the parity violating part of $H_{NL}^{(1)}$, the S-wave decay amplitudes satisfy two sum rules[1]

$$\Delta(\Lambda) + \Delta(\Xi) = 0 \qquad (10)$$
$$\sqrt{3}\Delta(\Sigma) - \Delta(\Lambda) = 2\Delta \text{ (L.S.)}$$

which relate the $\Delta T = \frac{3}{2}$ amplitudes in Λ, Ξ, and Σ decay (see eq.[n](3)) to each other and to the violation of the Lee-Sugawara triangle

$$\Delta(\text{L.S.}) \equiv \sqrt{3}\Sigma_0^+ - \Lambda_-^0 - 2\Xi_-^- \qquad (11)$$

If octet dominance is valid, then both the $\Delta T = \frac{1}{2}$ rule and the Lee-Sugawara triangle are satisfied:

$$\Delta(\Lambda) = \Delta(\Xi) = \Delta(\Sigma) = \Delta(\text{L.S.}) = 0 \qquad (12)$$

but if it is not, then eq.[n] (10) is all we have to work with.

The PCAC hypothesis enables us to write the matrix element for $\alpha \to \beta + \pi^{(i)}$ in the soft pion limit as[2]

$$\frac{M_\pi}{f_\pi} <\beta|[Q_5^{(i)}, H_{NL}(0)]|\alpha> + \text{Born Term} \qquad (13)$$

where $Q^{(i)}$ and $Q_5^{(i)}$ are the vector and axial vector charges respectively. The SU(3) x SU(3) properties of H_{NL} imply that

$$[Q_5^{(i)}, H_{NL}(0)] = [Q^{(i)}, H_{NL}(0)] \qquad (14)$$

This, together with 2 ↔ 3 symmetry, tells us that, in the limit of exact SU(3) for strong interactions, the equal time commutator in eq^n (13) contributes only to S-wave decays, and the Born term only to P-waves. Corrections due to strong SU(3) breaking are not expected to distort this result by more than 10%.

Because $Q^{(i)}$ is a generator of isospin for i = 1, 2, 3, the commutator $[Q^{(i)}, H_{NL}]$ contains no isospin states that are not already contained in H_{NL} itself. The only effect of commuting H_{NL} with $Q^{(i)}$ is to rotate each component of a given isospin into another component of the same multiplet:

$$\tfrac{1}{2} \to \tfrac{1}{2}, \quad \tfrac{3}{2} \to \tfrac{3}{2} \qquad (15)$$

This means that, as a result of eq^n (14), the S-wave amplitudes obtained from the first term of eq^n (13) are described by an effective Hamiltonian of the form

$$\lambda[(\overline{B}B)_{\tfrac{1}{2}} \pi]_{\tfrac{1}{2}} + \mu[(\overline{B}B)_{\tfrac{3}{2}} \pi]_{\tfrac{3}{2}} \qquad (16)$$

where the subscripts denote the isospin to which the particles inside the brackets are coupled. By a similar argument, the SU(3) structure of the effective Hamiltonian is

$$f[(\overline{BB})_{8_F}\pi]_8 + d[(\overline{BB})_{8_D}\pi]_8 + g[(\overline{BB})_{27}\pi]_{27} \quad (17)$$

An immediate consequence of eq^n (16) is that, since $\overline{N}\Lambda$ and $\overline{\Lambda}\ \Xi$ both have isospin $\frac{1}{2}$, the S-wave amplitudes for Λ- and Ξ- decay both satisfy the $\Delta T = \frac{1}{2}$ rule (see eq^n (3)). In addition, the three amplitudes for Σ-decay happen to satisfy a "pseudo" $\Delta T = \frac{1}{2}$ rule:

$$\Delta(\Sigma) = \sqrt{2}\ \Sigma_+^+ \quad (18)$$

Combining these results with the consequences of $2 \leftrightarrow 3$ symmetry (see eq^n (10)), we obtain

$$\Delta(L.S.) = \sqrt{\frac{3}{2}}\ \Sigma_+^+ \quad (19)$$

Thus the amplitude Σ_+^+ is a direct measure of the breakdown of the $\Delta T = \frac{1}{2}$ rule (eq^n (18), and of octet dominance (eq^n (19)). This is not surprizing because $\overline{N}\Sigma^+$ has isospin equal to $\frac{3}{2}$ and appears only in the second term of eq^n (16) and the (27)-plet term of eq^n (17).

We know from experiment that the S-wave amplitude for $\Sigma^+ \to n\pi^+$ is small, and we must now find out whether theory agrees. The amplitude we obtain from eq^{ns} (13) and (14) is proportional to the matrix element of a product of currents,

$$S(\alpha \to \beta\pi) \sim \frac{G}{\sqrt{2}} \sin\theta \cos\theta\ <\beta|J_\lambda J_\lambda'|\alpha> \quad (20)$$

and, as in the case of meson decay, it can be evaluated by

inserting a complete set of states between the current operators.[5] When the intermediate states are limited to the baryon octet ($J^P = \tfrac{1}{2}^+$) and decuplet ($\tfrac{3}{2}^+$), and some reasonable assumptions are made about semi-leptonic form factors, this type of calculation leads to octet dominance and the vanishing of $S(\Sigma_+^+)$.

To understand how this comes about, we note first that for the decuplet intermediate state, octet dominance occurs simply because the SU(3) Clebsch-Gordan coefficient for the (27)-plet ($\bar{B}B$) term in

$$\sum_D \langle\beta|J_\lambda|D\rangle \langle D|J_\lambda'|\alpha\rangle = f(\bar{B}B)_{8_F} + d(\bar{B}B)_{8_D} + g(\bar{B}B)_{27} \qquad (21)$$

is small. In the case of the octet intermediate state, Simmons[9] has shown that the contributions from the magnetic and axial vector form factors to the (27)-plet vanish when the corresponding D/F ratios are equal to $\sqrt{3}$, a value quite close to the experimental one; the contribution from the charge form factor contains a large admixture of (27)-plet, but its overall magnitude is much smaller than the magnetic and axial vector terms. Thus we see that each type of intermediate state gives rise to octet dominance on its own; however, we need comparable amounts of both in order to obtain the correct F to D ratio for the effective weak spurion (see

Table 5

S-Wave Spurion from Current Algebra.

$$S_{ij} = 2[dD_6 + fF_6 + tT_{27}]_{ij}$$

Experimental Values (in units of 10^{-5} MeV)

$$d = -1.6, \quad f = 3.6, \quad t \approx 0$$

Theoretical Values as taken from:

(a) Y. Hara, Prog. Theoret. Phys. (Kyoto) 37, 710 (1967)

(b) Y. T. Chiu, J. Schechter, and Y. Ueda, Phys. Rev. 157, 317 (1967).

Each parameter receives contributions from octet and decuplet intermediate states as shown.

Calculation (a)	d	f	t	Calu. (b)	d	f	t
Octet Intermediate State	-0.53	1.58	0.23	.....	0.15	1.9	-0.12
Decuplet-Intermediate State	-2.68	2.23	-0.29	....	-1.9	1.6	-0.28
TOTAL	-3.2	3.8	-0.1	TOTAL	-1.75	3.5	-0.4

Table 5: S-Wave Spurion from Current Algebra.

Table 5).

In the case of $\bar{\Omega}^- \to \Xi^*\pi$, the matrix element for which can also be expressed as a product of currents sandwiched between baryon states, we find a different situation. Neither the octet nor the decuplet intermediate states show octet dominance by themselves, but the non-octet components cancel one another when the two sets of states are taken together. When used in conjunction with the pole model, this result enables us to predict the $\Delta T = \frac{1}{2}$ rule for all modes of $\bar{\Omega}^-$-decay.[12] It also suggests that the wide domain of validity for the selection rule might be due to the happy conjunction of many accidental concellations.

No discussion of the current x current theory would be complete without some mention of divergence problems. In general the matrix elements of a product of two currents diverge quadratically, and so they must be cut off if we are to obtain finite answers in the calculations described above. The most frequent method of introducing a cut-off has been to ascribe dipole form factors to the weak current instead of monopole ones. This procedure emphasizes the low-energy contributions to the matrix elements, and its use has been justified by Nussinov and Preparata[10] in a specific dynamical model based on quarks.

A different approach to the divergence problem has been taken by Nishijima.[13] He begins with two observations: first, that in the original derivation of the Goldberger-Treiman relation, a divergence in the matrix element for pion decay is cancelled by a corresponding divergence in the renormalization of the pion wave function; and second, that this cancellation occurs automatically if the divergence of the axial vector current is allowed to become an interpolating field for the pion-which is one way of stating the PCAC hypothesis. Nishijima therefore suggests that divergences in nonleptonic decay might be avoided if weak currents are treated as interpolating fields for vector and axial vector mesons. Combining this idea with the gauge field algebra of SU(3) x SU(3), he finds that the current x current interaction is equivalent to a K*-pole model, and that the amplitudes for K → 2π and S-wave hyperon decay can be calculated using only the weak coupling constant, the pion decay constant, and the empirical masses of hadrons as input.* His results are in remarkable agreement with experiment.

Another interesting feature of this calculation is that the factor of $\sin^n \theta$ in eq. (20) is needed to obtain the correct magnitude for the parity violating amplitudes. This seems at first sight to be rather surprising because the phenomenological coupling constant for these decays is close in value to G. What

*See note added in proof.

seems to happen is that while the factor of sin θ brings the effective constant down to about G/4, the matrix element $<\beta|J_\lambda J_\lambda'|\alpha>$ is enhanced by a factor $m_{A_1}^2 / (m_{K^*}^2 - m_\rho^2)$ which is large enough (≃4) to overcome the effect of the Cabibbo angle. The need for sin θ also occurs in the other calculations we have described, but the factor which enhances the matrix element is not as easily isolated as in Nishijima's theory.*

Let us now turn our attention to parity conserving processes.[14]-[20] In 1961 Feldman, Matthews, and Salam[14] introduced the pole approximation for nonleptonic hyperon decay, and since then this idea has been the mainstay for all theories of P-wave amplitudes. One reason for this development is that when CP nonconservation is neglected, weak parity conserving poles have the same SU(3) properties as the current x current Hamiltonian; and another reason is that the Born term of eq$\underline{n}$ (13) gives rise to a particularly simple form of the pole model. In other words the pole model seems to represent the effects of both the current x current interaction and current algebra upon parity conserving hyperon decay.

In the most general form of the pole approximation, the P-wave decay amplitude contains baryon poles in the s- and u-channels and meson poles in the t-channel. However, when we examine the Born term of eq$\underline{n}$ (13) in the soft-pion limit, we ob-

*See note added in proof

tain a version of the model in which t-channel poles are absent, and the s- and u-channel poles are restriced to the $J^P = \frac{1}{2}^+$ baryon octet.[2] The effective interaction consists of $(\overline{B}B)$ terms coupled to octet and (27)-plet representations, and it is symmetric under 2 ↔ 3. These symmetry properties imply that the amplitudes satisfy the sum rules of eq$\underline{n}$ (10) when expanded up to first order in the mass-difference parameter[1]

$$\delta = (\Sigma - \Lambda)/(\Xi - N) \qquad (22)$$

Consequently when octet dominance holds, the P-wave amplitudes will satisfy the $\Delta T = \frac{1}{2}$ rule (see eq$\underline{n}$ (3)) and the Lee-Sugawara triangle (eq$\underline{n}$(11)).

In the current algebra calculation symbolized by eq$\underline{n}$ (13), the parity conserving pole in the Born term is proportional to the parity violating amplitude associated with the equal time commutator. It follows that the question of octet dominance in P-wave decay rests upon the properties of S-wave decay. We know from the smallness of $S(\Sigma_+^+)$ that the S-waves are dominated by the octet interaction, and hence we deduce that the same must be true for P-waves. At the present time this is the only "theoretical" evidence we have for the octet and $\Delta T = \frac{1}{2}$ rules in P-wave hyperon decay.

Given the relationship between parity conserving and parity violating amplitudes, we can go beyond the matter of selection rules and actually calculate the P-wave amplitudes in terms of the S-wave ones and the strong D/F ratio. The amplitudes we obtain in this way agree with the empirical ones in sign, but they are too small by roughly a factor of two. The inclusion of a t-channel pole (which vanishes in the soft pion limit) and other symmetry breaking effects improves the calculation to some extent, but a substantial difference between theory and experiment still remains. Thus we have to conclude that we do not understand P-wave decays very well.

This conclusion should not be too surprising in view of the fact that the current algebra analysis of P-wave decay involves some very delicate limiting procedures. It may be that when our techniques for handling these limits are more refined, we will be able to remove the discrepancy between theory and experiment. On the other hand, it may be that by trying to apply current algebra to P-wave decay we are barking up the wrong tree, and that we should really be looking elsewhere for clues to an understanding of the process.

We have known of one such clue for some time, namely the vanishing of the P-wave amplitude for $\Sigma^- \to n + \pi^-$. This result, we must emphasize, is not a consequence of the $\Delta T = \frac{1}{2}$ rule alone: for, in order to deduce $P(\Sigma^-_-) = 0$ from the Gell-Mann-Rosenfeld triangle and the fact that $S(\Sigma_+^+) = 0$, we must make use of the equality of the rates for the three modes of Σ-decay. Conversely, if we assume that $P(\Sigma_-^-)$ vanishes and that two of the three rates are equal, we can deduce the equality of the third rate from the triangle. Therefore, we are free to regard the vanishing of $P(\Sigma_-^-)$ as an independent constraint.

Because the $n\pi^-$ system has isospin $T = 3/2$, it cannot belong to an octet. Consequently the simplest way of ensuring that $P(\Sigma_-^-) = 0$ is to require an octet final state in P-wave hyperon decay.[1] This selection rule automatically gives us the $\Delta T = \frac{1}{2}$ rule for Λ-decay and the Gell-Mann-Rosenfeld triangle for Σ-decay. If we make the additional hypothesis that the effective Hamiltonian be symmetric under the interchange of initial and final baryon states, then we obtain a unique coupling scheme:[1]

$$\mathcal{H}_{NL}(pc) \sim \bar{B}_\lambda^{-3} M_\mu^\lambda B_2^\mu + \bar{B}_2^{-\mu} M_\mu^\lambda B_\lambda^3 + h \cdot c \qquad (23)$$

It has the virtues of being pure octet, of predicting the Lee-Sugawara triangle, and of giving one other relation,

$$3 P(\Lambda_-^0) = P(\Sigma_0^+) = 3 P(\Xi_-^-) \qquad (24)$$

which is in fair agreement with experiment. However, it has the disadvantage of not being itself the outcome of any known dynamical model. The invention of such a model is clearly a major challenge to the theorist.

Although the parity conserving amplitudes for hyperon decay present us with a difficult problem, the corresponding ones for meson decay do not. By means of PCAC and current algebra techniques, we can express the matrix elements for $K \to 3\pi$ in terms of the matrix elements for $K \to 2\pi$. Since the 2π channel satisfies the $\Delta T = \tfrac{1}{2}$ rule to within 5% in amplitude, we expect the 3π channel to do likewise; furthermore from the measured values of the 2π amplitudes we can calculate the rates and the slope parameters for the 3π channel (see eq$^{\underline{n}}$(1)). In both cases we obtain a good agreement with the $K \to 3\pi$ data, and so we can conclude that for meson decay, current algebra does help us to understand the relationship between parity conserving and parity violating channels.

Up to this point we have discussed theories of weak interactions as they relate to the most frequent decay modes of hyperons and K-mesons. There are, in addition to these well-known decays, some rare and exotic phenomena, and the theories can be applied to them too. We do not have much information about these rare processes either because they are suppressed by electromagnetic interactions, or because they are masked by competing interactions of much greater strength; however we do have enough to make a rough comparison between theory and experiment.

The simplest rare process is the strangeness changing radiative decay $A \to B + \gamma$.[21]-[26] Angular momentum conservation prevents it from occurring in pseudoscalar mesons, but it does occur in hyperons and even competes with semi-leptonic decay on fairly favorable terms.[3] In general, weak radiative decay does not conserve parity, and so the photon, like its counterpart the pion in $A \to B + \pi$, can exhibit an asymmetry with respect to the production plane of the parent hyperon. The corresponding asymmetry parameter is one of the two experimental quantities of immediate interest; the other one is the decay rate.

A good example of weak radiative decay is $\Sigma^+ \to p + \gamma$. By means of the pole approximation, we can express its matrix elements in terms of the nonleptonic decay amplitudes and the anomalous magnetic moments of Σ^+ and p. We can then calculate its branching ratio to $\Sigma^+ \to p\pi^0$ and its asymmetry parameter α. Typical results are:

		B.R. ($p\gamma/p\pi^0$)	α
Theory	Graham & Pakvasa	0.28%	0.06
	Ram Mohan[25]	0.44%	-0.04
Expt[26]		0.276 ± 0.051%	$-1.03^{+0.52}_{-0.42}$ [25]

The agreement between theory and experiment is good for the branching ratio, but not for the asymmetry parameter. The experimental value of α is taken from the first and only reported measurement of this parameter, and although it is not very accurate, it does suggest that α may be much larger than the values predicted in eq^n (25). This result, if borne out by future experiments, will raise an interesting problem.

Since the electromagnetic interaction conserves U-spin, the effective Hamiltonian for $\Sigma^+ \to p + \gamma$ must have the same U-spin properties as the current x current interaction. Among these properties is 2 ↔ 3 symmetry, which is sufficient, as long as we ignore CP nonconservation, to ensure that the parity violating part of the amplitude will vanish. Therefore, in the limit of exact SU(3) for strong interactions, we expect α to be zero. When symmetry breaking is taken into account this result will be distorted; but, if the calculated values in eq^n (25) are any guide, this distortion cannot be very substantial. In other words, a large asymmetry in $\Sigma^+ \to p + \gamma$ would be very hard to understand without some anomaly in the breaking of SU(3), or a breakdown in the current x current interaction itself.

Another decay in which the same difficulty might occur

is $\Xi^- \to \Sigma^- + \gamma$. It is the only energetically allowed $\Xi \to \Sigma$ transition, and its asymmetry parameter is expected to be small for the same reasons as in $\Sigma^+ \to p + \gamma$. Unfortunately its branching ratio is predicted to be low ($\sim 0.2 \times 10^{-4}$%) and so, given the general scarcity of Ξ particles, we may have to wait until the '80's before we can study it. Decays with better branching ratios and large asymmetries seem to be predicted only for neutral hyperons.

The second "exotic" phenomenon we wish to consider is that of parity violating nuclear forces.[27]-[32] One of the novel features of the current x current interaction is that, in addition to the usual $\Delta S = 1$ terms, it also contains pure hadronic terms which conserve strangeness. These terms generate a weak nuclear force which, like all weak interactions, does not conserve parity. Since this force has to compete with the normal strong nuclear force, we can detect its presence and study its properties only through parity violating effects.

To detect parity violation we may either search for nuclear transitions which would be absolutely forbidden if parity were conserved, or we can look for pseudoscalar quantities in nuclear radiation. If F measures the strength of the weak force relative to the strong one, then the probability of detecting forbidden transitions is proportional to $|F|^2$, and the anticipated size of pseudoscalar quantities is of order F. Since $F \simeq 10^{-6}$, we see that although the search for pseudoscalars is the more hopeful approach, it is far from being an easy one.

At the present time the best evidence for the existence of parity violating nuclear forces comes from the circular polarization of γ-rays emitted from Ta^{181}. The polarization was first definitely observed by Lobashov, Nazarenko, Saenko, and Smotritskii,[31] and it has subsequently been seen by other groups. They all agree that an effect of order 10^{-5} exists, but they are not unanimous about its exact magnitude.

If we take the standard form of the current x current interaction as given in eqns (4)-(7), we can write the strangeness conserving Hamiltonian as

$$H^{(o)} = \frac{G}{\sqrt{2}} \cos^2\theta \{J_\mu^{(o)}, J_\mu^{(0)+}\} + \frac{G}{\sqrt{2}} \sin^2\theta \{J_\mu^{(1)}, J_\mu^{(1)+}\} \quad (26)$$

where $J_\mu^{(o)}$ and $J_\mu^{(1)}$ are the strangeness conserving and strangeness violating currents respectively. The first term of eqn (26) gives rise to a nucleon-nucleon potential V_ρ which is dominated by ρ-meson exchange, and which contains both isoscalar and isotensor components. The second term generates a one-pion-exchange potential V_π which is pure isovector. Since

$\tan^2\theta \simeq 1/16$, we might expect that V_ρ makes the dominant contribution to $H^{(o)}$; however, the hard core in the strong nuclear force has the effect of enhancing the longer range potential V_π to the point where it is comparable with V_ρ. Thus we must consider both terms in the analysis of parity violating nuclear forces.

In the limit of CP conservation, there are no contributions to V_π from the exchange of neutral pions. As a result, V_π is proportional to n_-^o, the parity violating amplitude for $n \to p\pi^-$. This amplitude can be related by current algebra and SU(3) to the S-wave amplitudes for nonleptonic hyperon decay, and it satisfies the sum rule:$^{(32)}$

$$S(\Xi_-^-) + 2S(\Lambda_-^o) + (\sqrt{3}/2A) S(n_-^o) = 0 \qquad (27)$$

where, from the second term of eq$^{\underline{n}}$ (26), the constant $A = (-1\sqrt{2}) \tan\theta$.

Equation (27) has been deliberately expressed in terms of the parameter A because the $\Delta S = 0$ Hamiltonian of eq$^{\underline{n}}$ (26) is only one among several different possibilities. These possibilities arise in answer to various questions concerning octet dominance and the removal of divergences from weak decay matrix elements. In our discussion of strangeness changing decays we took the point of view that octet dominance comes about through the dynamical suppression of the (27)-plet in the usual current x current theory; alternatively we could have said that octet dominance is built in to the theory from the very beginning, and that the interaction consists of the terms in eq$\underline{\text{ns}}$ (4)-(7) together with others needed to cancel the (27)-plet component. We could also have argued that the current x current interaction is not fundamental, but merely the low-energy limit of an interaction mediated by vector bosons; the interaction between the boson and the weak current can be designed in various ways to yield finite matrix elements and octet dominance. Whatever the model we adopt, the general effect is to add extra terms, usually products of neutral currents, to the Hamiltonian of eq$\underline{\text{n}}$(26). Each model gives rise to its own particular modification, and hence to its own distinctive value of A in eq$\underline{\text{n}}$(27).

It turns out that A changes quite rapidly as we pass from one model to another.[32] In the conventional model of eq$\underline{\text{n}}$(26), A is equal to $(-1/\sqrt{2})\tan\theta$, but when we remove the (27)-plet by adding extra terms, A becomes $(1/\sqrt{2})\cot\theta$ -- an increase by a factor of about 16. One model based upon intermediate bosons predicts that n_-^o should vanish, and another yields A = -4. Since V_π is directly proportional to n_-^o, it follows that the strength of V_π will vary from one model to another, and that by measuring V_π we can determine the correct model of the current x current interaction.

In order to extract V_π from present experimental data, we have to calculate the ρ-exchange potential V_ρ. We can do this either by means of a factorization hypothesis, or from current algebra and the field-current identity. Both methods yield the same result, and they show that V_ρ is much less model-dependent than V_π.[28] Using this result, we find that the conventional model of eq$\underline{\text{n}}$(26) and an intermediate boson model of Segré are consistent with the data while the other models are not. Among other things, this means that octet dominance is likely to be a dynamical effect rather than an intrinsic one. We must be careful, however, not to place too much reliance upon this evidence because the data are all taken from heavy nuclei, and the corresponding theoretical calculations are subject to all the intricacies of nuclear physics.[29]

It may be possible to remove this kind of uncertainty if we can find physical quantities which depend upon only one of the potentials, V_π and V_ρ, and which can be measured in simple nuclear systems. An excellent example of this is the process[29] $n + p \to d + \gamma$: the circular polarization of the γ-ray depends only upon the isoscalar potential V_ρ, and the asymmetry of γ-rays relative to the spin of the incident neutron is determined by the iso-vector potential V_π. A preliminary measurement of the circular polarization has recently been reported, but it is too soon to draw any conclusions other than the effect is of the expected order of magnitude, namely 10^{-6}. Other processes with the desired properties include E1 transitions[30] between certain T = 0 levels in the self-conjugate nuclei ^{10}B and ^{18}F. They depend dominantly on V_π, but unfortunately we have no data on them as yet.

To sum up, we may say that the existence of parity violating nuclear forces in general supports the current x current theory, and the data available right now does not contradict the conventional form of the interaction. This field is still in its early stages, and we look forward to many interesting developments in the next decade.

At the beginning of this survey we decided to use the current x current theory to analyze the phenomenon of non-leptonic decay. Since then we have seen that the theory succeeds in some respects and fails in others; it works well when applied to parity violating hyperon decay, and to parity conserving meson decay, but it has not succeeded so far in explaining the parity conserving hyperon amplitudes or the

parity violating meson ones. Because of the successes of
the theory, it is important to know whether its failures
represent a genuine breakdown of the theory, or whether
they merely reflect our present inability to use it properly.

One way of answering this question is to test the sum
rules of eq$\underline{\text{ns}}$ (8) and (10) which follow from the symmetry
of the current x current interaction, and which correlate
deviations from the $\Delta T = \frac{1}{2}$ and octet rules. Since these
selection rules themselves are satisfied to within 5-10%
in amplitude we shall need amplitudes that are accurate to
about 1% if the test is to be significant. We shall also
need reliable estimates of radiative corrections, final state
interactions, and the effects of SU(3) breaking. Recent experiments[33] indicate that we are on the verge of meeting
at least some of these conditions.

This test of the current x current theory is obviously
a difficult one to carry out, but all the other tests are
just as difficult, if not more so. In the case of parity
violating nuclear forces we are just beginning to overcome
the experimental problems, but the complications of nuclear
physics may well obscure the meaning of the data. In weak
radiative decay, the data should be quite easy to analyze

once we have solved the problem of getting it! And finally, there are both experimental and theoretical difficulties in tests requiring the measurement of neutrino total cross-sections and the polarization of the scattered lepton.[34]

It is not unlikely that, by the time these difficulties have been overcome, we shall have resolved the P-wave hyperon and K→2π decay problems in some other way. Either we shall have learned how to carry out the current algebra calculations correctly, or we shall have developed a new theory -- a possibility not lightly to be dismissed in view of recent hints about the properties of Λ β-decay. Whatever the outcome, the accumulation of precise data and the study of rare and exotic weak phenomena should add new life to the field of non-leptonic decay in the decade of the Seventies.

Note added in proof: *Trofimenkoff (Prog. Theoret. Phys. Kyoto) to be published) has claimed that the model of Nishijima and Sato does not enhance the octet part of the interaction, and does not compensate for the Cabibbo angle suppression. He argues that besides the K*-pole terms considered by Nishijima and Sato, there are other "contact" terms which they neglected. When taken into account these terms remove the attractive results obtained from the pole terms alone. (The authors are endebted to Dr. L. R. Ram Mohan for a detailed discussion of this point.)*

REFERENCES

General

For references up to 1967 and for general purposes see:

(1) S. P. Rosen and S. Pakvasa in Advances in Particle Physics, Vol. 2 edited by R. L. Cool and R. E. Marshak (Interscience, New York 1968) p. 473.

(2) R. E. Marshak, Riazuddin, and C. P. Ryan, Theory of Weak Interactions in Particle Physics (Wiley-Interscience, New York 1969).

(3) Proceedings of the Topical Conference on Weak Interactions at CERN Jan. 1969; CERN Report 69-7.

Saturation of Current Commutators

(4) Y. T. Chiu and J. Schechter, Phys. Rev. Letters $\underline{16}$, 1022 (1966).

(5) Y. T. Chiu, J. Schechter, and Y. Ueda, Phys. Rev. $\underline{150}$, 1201 (1966).

(6) Y. T. Chiu, J. Schecter, and Y. Ueda, Phys. Rev. $\underline{157}$, 1317 (1967).

(7) Y. Hara, Prog. Theoret. Phys. (Kyoto) $\underline{37}$, 710 (1967).

(8) S. Biswas, A. Kumar, and R. Saxena, Phys. Rev. Letters $\underline{17}$, 268 (1967).

(9) W. A. Simmons, Phys. Rev. $\underline{164}$, 1956 (1967).

(10) S. Nussinov and G. Preparata, Phys. Rev. $\underline{175}$, 2180 (1968).

(11) D. S. Carlstone, S. P. Rosen and S. Pakvasa, Phys. Rev. 174, 1877 (1968).

(12) L. R. Ram Mohan, Phys. Rev. D1, 266 (1970).

(13) K. Nishijima and H. Sato, Prog. Theoret. Phys. (Kyoto) 43, 1316 (1970).

P-wave: Pole Models

(14) G. Feldman, P. T. Matthews, and A. Salam, Phys. Rev. 121, 302 (1961).

(15) A. Kumar and J. C. Pati, Phys. Rev. Letters 18, 1230 (1967).

(16) M. Hirano, K. Fujii, and O. Terazawa, Prog. Theoret. Phys. (Kyoto) 40, 114 (1968).

(17) S. Okubo, Ann. Phys. (New York) 47, 351 (1968).

(18) D. S. Loebbaka, Phys. Rev. 169, 1121 (1968).

(19) F. C. P. Chan, Phys. Rev. 171, 1543 (1968).

(20) L. R. Ram Mohan, Phys. Rev. 179, 1561 (1969).

Weak Radiative Decays

(21) Y. Hara, Phys. Rev. Letters 12, 378 (1964).

(22) R. H. Graham and S. Pakvasa, Phys. Rev. 140, B1144 (1965).

(23) J. Pestieau, Phys. Rev. 160, 1555 (1967).

(24) M. Ahmed, Nuovo Cimento 58A, 728 (1968).

(25) L. R. Ram Mohan, Phys. Rev. 179, 1561 (1969).

(26) L. K. Gershwin et al., Phys. Rev. **188**, 2077 (1969).

Parity Violating Nuclear Forces

(27) R. Dashen et al in The Eightfold Way edited by M. Gell-Mann and Y. Ne'eman (W. A. Benjamin, Inc. New York, 1964) p. 254.

(28) See talks by B. H. J. McKellar, by W. D. Hamilton, and by E. Fischbach, D. Tadic, and K. Trabert in Proceedings of International Conference on High Energy Physics and Nuclear Structures, Columbia University, New York, September 1969 (to be published).

(29) R. J. Blin-Stoyle, Proceedings of the Topical Conference on Weak Interactions, CERN, Geneva, 1969 (CERN Report 69-7).

(30) E. M. Henley, Ann. Rev. Nuclear Sci. **19**, 367 (1969).

(31) V. M. Lobashov, V. A. Nazarenko, L. F. Sainko, and L. M. Smotritskii, JETP Letters **5**, 59 (1967).

(32) E. Fischbach and K. Trabert, Phys. Rev. **174**, 1843 (1968).

Miscellaneous

(33) F. Harris et al., Phys. Rev. Letters **24**, 165 (1970); and S. Olsen et al, ibid **24**, 843 (1970).

(34) H. D. I. Abarbanel, Ann. Phys. (N.Y.) **46**, 335 (1968).

TABLE CAPTIONS

TABLE 1: Test of $\Delta T = \frac{1}{2}$ Rule in $K \to 2\pi$.

TABLE 2: Test of $\Delta T = \frac{1}{2}$ and absence of $\Delta T = 5/2$ in $K \to 3\pi$.

TABLE 3: Data on Hyperon Decay.

TABLE 4: The $\Delta T = \frac{1}{2}$ Rule in Hyperon Decay.

TABLE 5: S-Wave Spurion from Current Algebra.

EXPERIMENTAL VERIFICATION OF WEAK INTERACTIONS AND THE MUON PUZZLE[†]

V. Telegdi

Physics Department

University of Chicago

Chicago, Illinois 60637

The muon puzzle, as you know, consists in the fact that we have two electrons, a heavy one and a light one, and that their properties in every circumstance where they could be identical (except for possible kinematic factors) are revealed by experiment to be indeed identical. This situation which was well known in the beginning of the decade has not changed, but has been corroborated by many experiments, and in fact, many of the properties of the muon are now known at the level of accuracy comparable to that of the electron. As Dick Garwin once said, "It is a hell of a lot easier to heat a wire and get electrons that it is to get a beam of muons." Therefore, there has been a great deal of clever experimentation, but the overwhelming majority of my audience today con-

---

[†]Invited talk presented at the symposium "The Past Decade in Particle Theory" at the University of Texas at Austin on April 14-17, 1970.

PROPERTIES OF THE MUON

Property	Value	How Determined
Spin	½	Mesic X-rays
Statistics	Fermi-Dirac	Pairs, Tridents
Mass	$206.76(02)\ m_e$	Energy of 3D-2P transitions in muonic phosphorus[2]
	$206.765\ m_e$	μ_μ/μ_p and (g-2)
Magnetic Moment	$\mu_\mu/\mu_p = 3.18333(4.8)$	from precession in H_2O (13 ppm)
	$= 3.18333(15)$	muonium (5ppm)
Electric Moment	$d_\mu < (5\pm5)\ \text{ecm} \times 10^{-17}$	byproduct of old g-2[3]
Anomaly $a=(g-2)/2$	(1) $116616(31) \times 10^{-7}$ (270 ppm)	Latest CERN experiment[4]
	(2) $1165564 \binom{+14}{-.7}$	Theory to order a^3 is $(a/\pi)^3\ 8.1\binom{+11}{-0.5}$
(1)-(2):	$\binom{+31}{-34} \times 10^{-8}\ 2.8\sigma(7\%)$	
Lifetime:	$t_\mu = 2.198(1)\ \mu\text{sec}$	Direct measurement[5]
Muon neutrino mass	2.5 MeV (90% CL) (!)	Endpoint of Michel Spectrum[6]

Figure 1

2. Muon Electrodynamics

a. $(g-2)$ Expt.[5,8]

b. Muonion hyperfine structure

$\Delta\nu$ = 4.463305 (10)   Chicago double resonance

    = 4.463248 (31)   Yale zero field resonance

c. Muon pairs[7]

QED verified?

d. Muon tridents 2×10^3 km mfp   Tannenbaum
                                 11 BeV/c   BNL

Comment $\alpha = +1$ FD
                                 exp. 0.98 ± 0.23

              $\alpha = -1$ BE

e. μp scattering $\lambda^{-2} = 0.1 \pm 0.03$ BeV^{-2}

                     ($\lambda \sim 3$ BeV)

Figure 1a

sists of theorists. I have nothing to tell them except that this puzzle is still with us as it was at the beginning of the decade, only more so.

My own muon puzzle is why I was invited in the first place. I have thought long and hard about this and I have attributed it to this fact: some of the experiments in which we have been involved have been construed to be particularly strong verifications of the V-A theory of weak interactions[1], and several of the organizers and more prominent advisors of this conference were connected with the theoretical aspects of that topic ... but this is a theory about my invitation which is not yet supported by experiment.

Now in order to give my lecture a little bit of content, I have prepared in Figure 1 a muon wallet card which I will discuss. It contains little new information.

We know that the spins of the muon and the electron are ½ in proper units. We might ask ourselves where we get this information. You just look at mesic X-rays and get the characteristic doublet structure that we see in ordinary atoms. We then move to the next point, the statistics of this spin -½ particle. Obviously, we have a strong prejudice for Fermi-Dirac statistics. We ask is this prejudice founded on any experimental fact? And the answer is that it is indeed. One has to investigate the situation in which more than one muon is present in order to be able to answer the question of statistics at all. There are experimentally two such situations: one is the pair productions of muons by photons, and the other one is the pair production of muons by virtual photons, the incident particle, the carrier of the virtual photon field, being another muon. This configuration (one muon in, three muons out) is known as a trident. The name is also used for the analogous situation for light electrons. Tridents have been for the first time quantitatively examined in a Brookhaven experiment which shall be reported by the Harvard group at the forthcoming Washington meeting. (I shall make a few remarks about this later. The answer is that it quantitatively agrees with the ordinary Fermi-Dirac statistics.)

The mass of the muon is perhaps the best known and least understood quantity in physics. In case you might ask why one whould measure it at all, I'll come to that point. There do exist two independent measurements. The first one is a true measure of the muon mass and simply boils down to measuring the energy of the particular muonic X-ray transition. Then using standard formulae, one gets a measurement to 100 parts in a million. The second measurement is an indirect one and is based on the knowledge of the ratios of the precession frequency of the muon and the proton to each other (roughly speaking, to a ratio fo their magnetic moments). So for this purpose the magnetic moment can be taken to be known with infinite accuracy. This is then combined with a direct measurement of the anomaly of the magnetic moment of the muon, the so-called (g-2) factor. If I now know, so to speak, the muon moment, then I can extract the muon magneton by splitting off the g factor. I effectively have a measurement of the mass. This measurement of the mass is as accurate as is the precession frequency which in turn is a measure of the magnetic moment. Now, the current status of muon physics is one of such sublime and perhaps unnecessary accuracy that there are big fights as to whether this precession frequency ratio is indeed the ratio of magnetic moments or whether they are local chemical shifts and what have you. This need not bother this distinguished audience unduly.

We then proceed to the magnetic moment. The magnetic moment is in fact used as the input to the quantity above. It is currently known in the published literature to 13 parts per million from this precession and there is about a

10 parts per million correction, allowing for the fact that presumably diamagnetic shielding of protons in water and muons in water is not identical, because we measure two precession frequencies. These very refined effects of a few parts per million are of considerable importance if one wishes to study bound muon systems as tests of quantum electrodynamics, and because my contribution to clarifying the muon puzzle is identically zero, I have permitted myself to include a few remarks about these electrodynamic effects.

We then come to a quantity which, if it existed, would merit your attention. But it does not exist; it is the electric dipole of the muon which of course would imply at least time reversal violation. This has been measured as a by-product of the first CERN (g-2) experiment. The muon is moving in a magnetic field and the magnetic field makes a v cross B electric field which, if there were an electric dipole moment, would lead to a spin precession orthogonal to the precession which the magnetic field produces.

Now here comes perhaps the most interesting thing about muons, and that is swallowing the puzzle and using the muon as an admittedly heavy electron for conventional quantum electrodynamics. The merit of the heaviness of the electron is obvious, since when it decays virtually it reaches states of much higher energy or possibly cut-off lengths which are much shorter than is possible with the light electron. Therefore, the measurement of the anomaly of the muon constitutes a rather severe test of quantum electrodynamics. This cut-off distance corresponds to anywhere between 3 to 10 Bev/c. The measurement that I am indicating here is the latest, maybe not the very latest, but the latest printed value that I have of the CERN muon experiment. There is very great accuracy in the anomaly. Here you can compare it to the theoretical prediction. The experiment is sufficiently accurate that one has to worry considerably about contributions of order $(\alpha/2\pi)^3$, and these have never been evaluated in any real fashion. There are bounds on them but they contribute barely at this level. Let me point out that here that there are some uncertainties, +14 and -.7 in the theoretical number. These come from the uncertainty on the hadronic contributions to the anomalous magnetic moment. I will come back to that. If you subtract the theoretical value from the experimental value and compare the difference, you get about 1.8σ or about 7% level of confidence. It is my understanding that the discrepancy is now gone down to 1.5σ, but nobody should be forced to read this in the error

bars. This experiment is the 2nd CERN experiment. The
first one has an accuracy of about 1/2%, and here you see
an accuracy of 2 parts in 10,000. The value of the experiments appears to be not only in the test of quantum electrodynamics but also possibly in the test for these hadronic contributions. The virtual photon which is emitted by a muon can turn into any number of hadronic states, and these corrections can either be considered a terrible bother in checking quantum electrodynamics or can be considered a field of investigation all by themselves.

At the present moment the CERN group is funded for the next generation experiment roughly to the tune of one million dollars, and this experiment will measure (g-2) to the order to 20-30 parts per million in the anomaly. I would like to use this occasion to say that the anomalous moment of the electron is currently the best measured number for which there exists a theoretical prediction, and that this experimental number by the Michigan group is also out of tune with the theory about 1 ½ standard deviations. Its impact on the verifications of quantum electrodynamics in short distances is indeed much less.

The lifetime of the muon is known to be about ½ a part in 1,000 by direct electronic measurement. This number is perhaps not as extravagantly useless as the value of the mass to high accuracy. You may wonder why anybody would like to know the lifetime of any radioactive object to an accuracy of 1 part in 1,000 or better. Let me assure you from personal experience that such measurements are very difficult and expensive. The reason for the importance of this number is that if the V-A theory is valid (and I shall address myself to that question in a moment), then there is a single coupling constant for muon decay. Let's call it G. This number is then related and is almost identical to the coupling constant for the vector part of the semi-leptonic interaction like ordinary beta decay. Another aspect is that muon decay which determines this lifetime because of its structure leads to radiative corrections which are completely calculable, finite and so forth. The analogous situation of course does not hold for the beta decay of the neutron or for any other nuclear decay. Therefore, these comparisons suffer from some uncertainties on the part of the radiative decay constants in beta decay where one wishes to compare them to the muon decay. Nevertheless, the agreement at the present moment is satisfactory as far as one knows about radiative corrections.

Some fraction of this paper will be devoted to the muon puzzle in terms of the µ-ness or muon quantum number. This is a concept which has been introduced to explain an experimental fact and has never had any predictive power of any particular kind. It is almost circular, and the low-brow man on the street wonders what is this quantum number that explains only itself. Maybe there are further low-brow differences between the muon neutrino and the electron neutrino that explains µ-ness. Of course, the most obvious question to ask in that context is, are the masses of the electron neutrino and the muon neutrino equal? The next question is, are they both equal to zero? Mathematical beauty and some amount of experimental evidence have convinced us that the electron neutrino is indeed massless. No such evidence exists for the muon neutrino to date. Of course the mathematical beauty exists, but I'm speaking of experimental evidence. The best measurement of the muon neutrino mass is contained in an unpublished report for a theses by John Peoples from Columbia University. In that report, one measures the end point of a beta decay of the muon, and from that end point one then derives that the mass of the muon neutrino is no larger than 1.5 MEV. This measurement is precise to 2 standard deviations or a 90% confidence level. (It is therefore probably smaller than that.) I think it is very regrettable that one doesn't know any more about the µ neutrino mass, but all experiments to get high accuracy about this number are very difficult, and some are very amusing.

We now go to muon electrodynamics which is to some extent a slight repetition of what I have said already. I described the g-2 experiment. It is good to one and one half standard deviations. Then we have the hyperfine structure of muonium. Muonium is a misnomer for an atom of the structure μ^+e^-. It is copied from positronium, e^+e^-, but it is not a $\mu^+\mu^-$. This atom, of course, has a ground state split by some frequency difference $\Delta\nu$ just like in hydrogen. Incidently, hyperfine structure of ordinary hydrogen is the best known number in physics, known at present to 12 decimal figures, and explained only to 6 figures. There has been a great deal of activity trying to measure this hyperfine structure of muonium. The reason for this activity, aside from its amusement value, is that this case differs from the ordinary hydrogen atom. There one has difficulties in predicting the splitting, not only because as Hofstadter has taught us, the proton is a finite object, but furthermore, it is an object with inner degrees of freedom. It has finite polarizability and, at the exquisite level of accuracy of current experimentation, theories are unable to make a compelling prediction as to the behavior of the proton when squeezed by the electric field of an electron. Muons presumably are unsqueezable and therefore accurate predictions can be made. I go back then to the entry in Figure 1. There are two entries in this table in reasonable agreement with each other. The hyperfine structure of muonium is now known to a couple of parts per million from the most recent experiments measuring the Zeeman effect by double resonance. They are known with somewhat larger error (3 times larger error) from a measurement in zero magnetic field inducing direct transitions between these two levels. These are the experiments done by the orginal discoverer of muonium, Professor Hughes, at Yale. There is now an active contest between Chicago and Yale to get more and more accurate values. The upshot of it is that if you take standard quantum electrodynamics for the muon you get agreement with the Josephson value of α. One more remark which may interest a few people in this audience is that when you calculate the long string of quantum electrodynamic corrections to the hyperfine structure there is one particular term, the so-called logarithmic recoil term, which in the case of the proton is 36 parts per million, but in the case of the muon is 186 parts per million. This is the kind of business where 186 parts per million is enormous. Therefore, you can say that this particular quantum electrodynamics term is now probably verified to 3-4%. This heavy electron has become a household object and in the past few years has become known to perhaps a useless degree of accuracy.

The muon pairs have had a sad history. There was a report when the Harvard people found the violation of quantum electrodynamics in light electron pairs that such violation was also seen in heavy electron pairs[7]. The Harvard results on light electron pairs were then dropped in view of the work at DESY. Ting and his collaborators and more recent work (unpublished at Cornell) on ordinary muon pairs also seems to say that there is nothing particularly wrong[8]. However, muon pairs are not the ideal thing for quantum electrodynamics or statistics investigation. This is because you make a ρ first, then observe the pair decay of the ρ.

Now the muon tridents to which I want to address myself are being investigated by Tannebaum for 11 Bev/c muons and the question is how do you check the statistics? There is an infinite literature on parastatistics and this and that and the other statistics. I know nothing in detail about these things, but the fact of the matter is that if a charged lepton is a bonafide Fermi-Dirac particle, then for every diagram of interest there is an exchange diagram and one has to antisymmetrize the amplitude. One has to add the direct diagram and the exchange diagram with an opposite sign. This is due to the Pauli principle. So we introduce a parameter α which represents the contribution of these interferences between the antisymmetrized diagram and in the desired case, which is Fermi-Dirac statistics, this parameter will assume the value $\alpha=1$. When you fit these experimental measured cross sections you get to a high level of confidence $.98 \pm .23$ for $\alpha=1$. As academic as the subject may appear to be, it is better to settle it experimentally than by prayer.

At this particular point, I am now coming to the weak interactions of the muon, and of course, the central role is played by the muon decay. This is a totally leptonic process for which the theory, at least the V-A theory, is supposed to have absolute predictive power. Thus it is a paradise for the experimentalist. He has a situation where there are many numbers to measure and the theorists just have no way out. They are totally committed: one parameter put in, five to be measured (isn't that great). This rarely happens because as is generally said in the strong interactions physics, "we really don't know how to do these things so we fix it up". Here the situation is beautiful, they are completely tied on the cross, so it would appear.

Figure 2 has to do essentially with the following situation. We work in the university, most of us do, and there is a lot of talk about the noninteraction between scientists and social scientists. In other words, talk about bridging the gap and the third culture and all that sort of thing, and you wonder possibly what you as a physicist could do to bridge this gap. Personally, I have found that experimental physics (theoretical physics is well-known to have predictive power), can be used in the social sciences as a tool to predict major social upheaval, revolution. In Figure 2, we plot a physical parameter which, by the laws of nature, is a constant versus time (units of anno domini), and the parameter in question is the so-called ρ-Michel parameter. When you discover a very sudden discontinuity in the value of this quantity which is supposed to be a unique value and not be time-dependent as seen by the physicist, but not by social scientist, then you know that at that point there was a revolution. In this case it was the revolution of the parity breakdown and of the V-A theory. There is one very amusing feature which, on the other hand, emphasizes the always desirable collaboration between East and West. Go to the last entry...you see that there is a black point

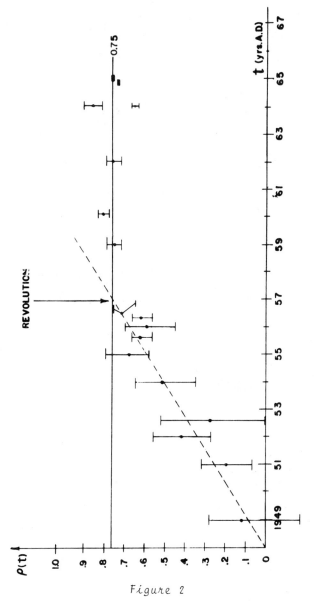

Figure 2

Experimental Value of Michel Parameter

right on the theoretical curve, needless to say, due to
Columbia University.[9]

Slightly to the left of it there are two points which disagreed with the curve. It took great courage in 1965 to pretend that there was no revolution. The bottom point is a measurement performed in England and the top point is the measurement performed in the Soviet Union, and when you take the evitable average, you still agree with V-A.

Let me tell you that everything which man has measured about the muon decay is in perfect agreement with V-A. What is not often emphasized, and may have some semi-philosophical merit is that the converse is not true. That is to say, if you look at the experimental date you are driven to the conclusion that the correct theory is V-A. Now before I do that, let me simply say what the parameters are. When you have a decay, the decay products have a momentum spectrum designated by $S(x)$ where x is the momentum as measured in units of the electron mass. This spectrum has a so-called unpolarized part that is a well-specified function of momentum. This function depends parametrically on a number called the ρ-Michel parameter. So whatever the coupling is, the spectrum has a unique shape where ρ is some function of the coupling constant. If the muon is polarized then there is a correlation between the direction of emission of the electron (i.e., the unit vector of momentum) and a certain parameter δ where δ is in turn a function of the coupling constants. There is also, if you integrate over all this, a total over-all average asymmetry and that is specified in the terms of the parameter ξ. Now here it is always assumed that the momentum (in units of x) is much larger than 1/200 which is the ratio of muon and electron masses. This is the so-called Michel low-energy parameter. So we now know what these two parameters are. We know the energy dependence of the asymmetry. We know there is only one more parameter that we should mention which is the longitudinal polarization of the electron. So we have these quantities to measure and to compare to theory.

The logic of this comparison will be exhibited in Figure 3. At the top there is the general muon decay interaction (local coupling, etc.). This is not the most general form, but since I'm not speaking to a French audience, it makes very little difference. We have all these parameters plus the relation $2\eta^2 < 1-h^2$ which is a general relationship. Now you have a choice: the right-hand way is perhaps the most customary way; the ideal way is to measure everything to be measured, get the parameters, get their relationship to coupling constants, invert the equations and inform the theorist that V=A. This is not possible, primarily because

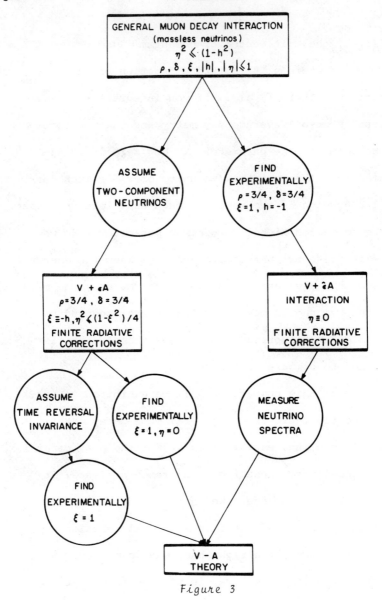

Figure 3

Logic of Experimental Test of V-A Theory

there are not that many practical observables as there are parameters. This is because in the final state there are two neutrinos which are very difficult to handle. So, the problem is fundamentally undetermined in the right-hand approach. In order to get rid of a lot of coupling constants, you must take the left-hand route and assume that you have two component neutrinos. You have then an interaction in the usual ordering to be $V+\varepsilon A$. Everyone in this room knows that when you speak of a Lorentz covariant V or A, you have to specify the order to coupling. I'll come to this fact later.

Thus, if you assume two-component neutrinos, you have only V and A, with one complex number ε, you automatically find the ρ Michel parameter, and δ, the parameter governing the energy dependence of the asymmetry is equal to 3/4. So many of us who have worked very hard for many years to prove that the value of ρ was 3/4 have never proved the V-A theory, just a general class of two component neutrino theories. It is a sad thing to find out after several years of experimentation, but it is a mathematical fact.

You are now in a situation in which the radiative corrections are finite. This is not to be neglected. I think the only really positive thing about muon decay once you swallow V-A is that it is a tremendous test of radiative corrections. None of us would get any agreement with theory if we did not accept the radiative correction. The radiative corrections in turn are computable and finite only for V and A, essentially for V. Thus, while we now have one complex number to measure, this information is contained in the gross asymmetry parameter ξ which in 2 component neutrino theory is equal to the helicity and the low energy parameter η. Once again you can choose to do more or less work. An enormous amount of work was eliminated by assuming 2 component neutrinos. You can eliminate more work by assuming time-reversal invariance. Then of course, the only parameter ε becomes the simple real number and because you have no more real numbers, ξ the asymmetry becomes maximal η, the lower energy parameter becomes 0. You then have concluded that you have the V-A theory with two assumptions, time reversal invariance and two component neutrinos. You can work a little harder. You can show that the decay asymmetry parameter ξ is equal to 1. That proves time-reversal within this framework and then you conclude with V-A.

The right-hand side of this picture is the concoction of a lady by the name of Jarlskog[10] who was a student of the late Gunnar Källén. There you prove things in an extremely pristine Scandinavian spirit. What you do is you do all the experiments you can do, and you report the results that $\rho=3/4$, $\delta=3/4$, $\xi=1$, and $h=-1$. This is compatible with $V=\hat{\epsilon}A$ interaction where $\hat{\epsilon}$ means that V and A are defined in the charge retention order. This is the ordering in which μ and e coupled together and (ν) (ν) coupled together. Of course, even in this picture we have finite radiative corrections. Now you measure the neutrino spectra and in fact if the epsilon is not real, even measure the circular polarization of the resultant neutrinos. After this masterly exercise you finally prove what we all like to see proved. Figure 4 is a summary of the Jarlskog work. It simply consists of charged retention ordering, and then you say, "Well, I won't see any neutrinos so I put them together because I can trace and integrate them out." Thus the electron and muon are coupled together and the γ_5s are put between electron and muon for calculation convenience. Then you assume two component neutrinos, etc. or the general approach and so forth.

We now give you the other side of the medal and that is the compilations of the currently known parameters. In Figure 5 we have a ρ known to be almost 3/4. The great novelties are somewhat insufficiently accurate measurements of η, the first one of these lower energy parameters measured by Derenzo and Hildebrand[11]. Then ξ, the total asymmetry was

OTHER APPROACHES TO DETERMINING DECAY PARAMETERS[10]

Using charge retention ordering, the Hamiltonian is

$$H \sim \sum_i (\bar{e}(\hat{c}_i + \hat{c}_i{}^5 \gamma_5)\Gamma_i \mu)(\bar{\nu}_\mu \Gamma_i \nu_e).$$

a. 1. Assume 2-component neutrinos
    2. Needs semileptonic processes
    3. T good, ε real
    4. $-\xi = k$ depends quadratically on ε.
    5. Need η to 0.1
    6. T complex → need η and ξ

b. General approach.
    1. Need spectrum and polarization of emitted <u>neutrinos</u>!

Figure 4

PRESENT BEST VALUES FOR MUON DECAY PARAMETERS[11]

	V-A	EXP.
ρ	3/4	0.7518(26)
η	0	-0.12(0.31)
ξ	-1	-0.973(14)
δ	3/4	0.7540(85)
R	1	1.00(13)

COUPLING CONSTANTS (CHARGE RETENTION ORDER) IN UNITS OF G_v

	V-A	FIT
S.P.	0	≤ 0.33
A (= x)	1	$0.75 \leq x \leq 1.20$
$\phi(AV)$	π	$\pi \pm 15°$
T	0	≤ 0.28

Figure 5

measured in the Soviet Union. The circles mean that those things have those V-A values even in the 2 component neutrino theory.

We can now quote the latest values and this will give you some room for thought. You know theoreticians believe that certain kinds of things are beautiful at any given time, but later they come back and say to you, "Couldn't you accomodate a little scalar or a little tensor?". For instance, G_a/G_v equal to 1.21 after Adler and Weisberger is beautiful, a postoriori. We ask ourselves, is there a pseudo-scalar admixture? A scalar mixture should be 0 for the favorite theory. Its magnitude in units of G_v is shown only to be less then 1/3. Then G_A, the axial vector coupling, which of course shall have the ratio 1, is contained between 1.2 and 3/4. The phase which in the V-A should be π is correct to within $15°$. The tensor coupling is less than 0.28. The phase in itself is a measure of time-reversal invariance.

I would like now to show you some more things. The scattering that corresponds to this interaction has never been observed. The interactions we should discuss in the weak interaction talk don't really tell us much. µ-capture, neutrino reactions come under the same heading. They are considered today more a study of the form factors than of muon-electron universality. Nowadays V-A is accepted and used as a tool to study matrix elements of certain current operators. Thus the interest in μ capture which agrees to within 10 to 15% with universality is not to test V and A, but in the induced pseudo-scalar coupling. This is the fruit of the PCAC hypotheses and thus the philosophy has shifted. For this reason I shall not talk about the many entries in Figure 6. All of these are interesting, but not relevant to the muon puzzle. Now, the CP violation which Professor Wolfenstein[12] discussed has one effect which involves muons and electrons. That effect is the charge asymmetry. When you go very far away, the K_L decays more into positive particles than into negative ones. This is true both for electrons and muons, and to the present level of accuracy which may be at best 20%, they agree . So this interesting CP effect is also universal.

Now there are a couple more things which are somewhat speculative but nevertheless, I think, should be mentioned here. The most interesting question (also I shall soon say why I don't believe in it) is the question of μ-ness. In other words, one has decided that there exists a private neutrino associated with a heavy electron, and a neutrino associated with an ordinary electron. This is in order, since when you do the experiment $\pi \to \mu\ \nu$, this neutrino in turn does not make any electrons but only muons. One has already introduced electrons and positrons that have lepton numbers 1 and -1. The idea is that the charge of leptons is defined through weak interactions even in the absence of electron-interactions. We can say what's the particle, and not the anti-particle. Thus one has $\ell_i = \begin{array}{cc} e & \bar{e} \\ 1 & -1 \end{array}$ and $\ell_\mu = \begin{array}{cc} \mu & \bar{\mu} \\ 1 & -1 \end{array}$ which is separately conserved. Thus, if you come in with a muon neutrino, you can make only a muon. Incidentally, I would like to say that the general notion of lepton conservation is that you get only μ's.

I'm not speaking of the tests of μ-ness, but ordinary lepton conservation. With mu neutrinos can you make only positive muons? What if they make both positive and negative muons? This has been tested at CERN recently. The precision of the CERN test of lepton conservation for muons is precise to about one part in a thousand level. This is a novel result.

It is well known that attributing leptonic charge for the muon is by no means the only way to guarantee that muon neutrinos will produce only muons and not electrons. This is a revolting tautology in which one invents a thing to explain what you see and you don't predict anything. There exists the competing hypothesis of a multiplicative μ-ness analogous to parity (see Figure 7). Of course you cannot decide whether this quantum number is additive or multiplicative when only a single muon is present. There is the following test which I think was initially suggested by Pontecorvo and Okun. If you take muonium μ^+e^-(let's call the object M), the ordinary muonium, then of course if you have a coupling μe bilinear in muon and electron in the absence of neutrinos, then this muonium could turn spontaneously into μ^-e^+ which is the anti-muonium. If there was a coupling of this particular kind, which is something which we violently dislike since it corresponds to a neutral lepton current. Then when you originally form muonium and, if the conservation law is multiplicative, you will have a matrix element, H_{weak}, between M and $\bar{M}$ and the two states will oscillate one to the other. Now then, this could be experimentally checked. And in fact there has been a recent experiment which I would like to mention. If you put the coupling constant G for the neutral current in H_w, you get for G/G_V a value around 6,000 to a 95% confidence level. In a certain sense we know these days that there's a neutral interaction that has a multiplicative behavior. We know that it is not as strong as electro-magnetism. That's a very refreshing thought. Well anyway, this experiment which in

Muon Weak Interactions

(1) $(\mu\nu)(e\nu)$ - decay
- scattering

(2) $(\mu\nu)$(hadrons) - μ-capture
- ν-reactions
- π decay
- $K_{\mu 2}$, $K_{\mu 3}$
- Hyperon decays
- CP violation

Figure 6

Muness

Additive vs. Multiplicative Quantum Number

Muonium antimuonium transitions $M \leftrightarrow \bar{M}$ indicate $G < 5,800\ G_\nu$ (95% CL)

Violation is $\mu \to e + \gamma$ has a branching ratio of 10^{10}

$\nu_\mu + e$ scattering to 2nd order

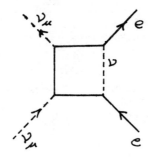

Figure 7

principle is ingenious, is the following. You make this muonium, and the muonium and anti-muonium are of course degenerate of each other except for the weak interaction. Now when you make it in a gas-like argon and when it is oscillating in its anti-muonium phase, it has a negative muon for its nucleus. It can then get attached to an argon atom and make an argon mesic X-ray on being captured. It could never do this if it had not changed its charge. Thus at that stage of the oscillation where it's gone into the anti-muonium it gets quickly captured and makes a mesic X-ray which is extremely characteristic. This value of 6,000 for G/G_V which is not perhaps too informative, is due to the only experiment in this field and was done by Hughes et al, looking for this mesic X-ray. The problem with the experiment, and I'm an experimentalist here to tell you about it, is that the self-same medium which produces this characteristic mesic X-ray, in case that the conversion does occur, also has electric fields which remove the degeneracy between muonium and anti-muonium, thereby effectively cutting down the speed of the conversion by gigantic factors. Many of us are trying to do this muonium anti-muonium conversion in vacuum. It must be said that the interaction is of such nature that it is unappetizing to begin with, but again it's an experimental question.

REFERENCES

1. Sudarshan, E. C. G. and Marshak, R. E., "The Nature of the Four-Fermion Interaction", Proceedings of the Padua Conference on Mesons and Recently Discovered Particles, P.V-14 (1957).
   Feynman, R. P. and Gell-Mann, M., Phys. Rev. 109, 193 (1958).
2. Feinberg, G. and Lederman, L. M., Ann. Rev. Nucl. Sci. 13, 431 (1963).
3. Charpak, G., et al, Phys. Lett. 1, 16 (1962).
4. Bailey, J., et al, Phys. Lett. 28B, 287 (1968).
5. Meyer, S. L., et al, Phys. Rev. 132, 2693 (1963).
6. Booth, J. W., et al, Phys. Lett. 26B, 39 (1967).
7. Blumenthal, R. B., et al, Phys. Rev. 144, 1199 (1966).
   Asbury, J. G., et al, Phys. Rev. Lett. 18, 65 (1967).
8. de Pagter, J. K., et al, Phys. Rev. Lett. 17, 767 (1966).
9. Sachs, A., Phys. Rev. Bull., 10 Sec., J. B. Invited paper;
   Bardon, M., et al, Phys. Rev. Letters 14, 449 (1965).
10. Jarlskog, C., Nuclear Physics 75, 6599 (1966).
11. Derenzo, S., Phys. Rev. 181, 1854 (1969); Derenzo, S., and Hildebrand, R. H., Nucl. Instr. Methods 58, 13 (1968).
12. Wolfenstein, L., "Fields and Quanta" (to be published).

DISCUSSION OF PROF. TELEGDI'S TALK

Nauenberg (CERN):

I seem to recall that one of the experimental facts that led to this idea of the two neutrinos was the lack of decay of the mu into electron and gamma ray. Is there any new experimental information, or is there any point in making more accurate experiments?

Telegdi:

I am very grateful for your asking this question, because if I hadn't made so many jokes I would have had the time to address myself to this point. Clearly, the $\mu \to e^+ \gamma$ is forbidden by muon conservation, by μ-ness, and it was the very origin of the postulate that there was μ-ness. Now you can work this thing out in a theory where it would not be forbidden as some kind of second order process were it not for lepton conservation. The branching ratio experimentally is less than 10^{-8}. You say the violations of lepton conservation, which I like to call the departure of the circular polarization of the neutrino towards ellipticity, is of the order of one part in a thousand. This doesn't conflict with any experiment on lepton conservation. In order to look at lepton violations in $\mu \to e^+ \gamma$ one must redo the experiments to an accuracy which is about

100 times greater than what we have today.

Marshak (University of Rochester):

One has to be careful here about the current-current theory. The number you gave is for the intermediate vector boson theory.

Telegdi:

That's right, for the intermediate vector boson. That's a very important point.

Carron (Rice University):

I would like to ask a general question about the axial vector current. The current commutation relations for the vector current fix the scale and the sign of the vector current, but for the axial current this is not true. The commutation relations are satisfied by A and by -A. I would like to know if there is any principle that fixes the sign of the axial vector current.

Telegdi:

I can't tell you that because I don't know enough about theory. Somebody here can tell you, but in the PCAC there is a prediction for the relative sign of the pseudoscalar coupling constant to that of the axial vector coupling constant. And that appears to be verified by experiment. Now as to the general principle to which you would like to have the answer, I am not competent to give you one. Perhaps Professor Nambu or someone can give you one.

Marshak:

The sign if fixed by experiment.

Nauenberg:

I was hoping that someone would have clarified in this morning's session the situation with the leptonic neutral K decays and the ξ parameter. Could you discuss just what are the measurements in the experimental values of the ξ parameter?

Telegdi:

I can discuss it but I presume that the discussion should ultimately involve numbers which I don't happen to carry in my head -- not even in my wallet card! And the situation is as follows: because of the zero mass of the electron, to a good approximation, K_{e3} has only one form factor. $K_{\mu 3}$ has two form factors and suitable juggling introduces a number which you denote by ξ. This number ξ is measured at least in two completely independent ways, one having to do with what you might call the Curie plot of $K_{\mu 3}$ or simply just the decay spectrum; the other one being a measurement of the muon polarization which evolves in a very singular way as you traverse the Dalitz plot, which nowadays is used to represent the final state population. First of all the experiments have shown considerable disagreement between various spectral measurements. This disagreement is not as striking, perhaps, as it should be because one never uses a decent spectrometer. One could use a bubble chamber, which has very little bias but also little statistics. So there is considerable controversy, but I think it is fair to say that ξ is somewhere of the order of -0.6 to -1.0. I'm told this is the number that theoreticians do not like. But this matter will be cleared up within the next year. Also there are large experiments in progress at the moment, one of which I know about in particular, done at SLAC by a SLAC-Brookhaven collaboration of Sakitt, et al, which perhaps will settle this. I think that this number is of the order of -1 is a fair statement.

Marshak:

I think that the essential point of the experiment is having determined the branching ratio of 0.6 or 0.7.

Telegdi:

I have ommitted the branching ratios.

Marshak:

That is moving ξ in the direction of the polarization measurements.

It is only appropriate to end the weak interaction session by pointing out in the framework of C.P. Snow that the coherence of the theory is also something to be considered.

Telegdi:

Yes, I would also like to say that this great business about these constants of course is not capable of checking the validity of V-A, but V-A provides the universality of the coupling constant in a natural way ... but to counter your remark I didn't want to give a C.P. Snow job!

ACTION - AT - A - DISTANCE*

BY

E. C. G. Sudarshan†
University of Texas, Austin

Abstract: In a talk presented at the symposium "The Past Decade in Particle Theory" Prof. Sudarshan reviews the historical development of "action - at - a - distance" theories. He reviews the formulation of classical action - at - a distance theories and identifies the relevant conserved quantities. This theory is compared with a corresponding classical field theory. The lessons of this classical theory are applied to the formulation of a quantum theory of action - at - a - distance. The problem of unitarity and physical interpretation are described. The relationship with indefinite metric theories of elementary particles is clarified and finally a theory of interactions which includes self interaction but contains only finite computations is formulated.

INTRODUCTION

Physical theory deals with the nature of the elementary entities and their mutual interactions: all fundamental questions pertain to existence

---

*Invited talk presented at the symposium "The Past Decade in Particle Theory" held at the University of Texas at Austin, in April 14-17, 1970.
†Supported in part by the USAEC under contract AEC (40-1) 3992.

entities and their mutual interactions: all fundamental questions pertain to existence and change. In physics they correspond to the existence of "particles" and their "interactions." It has been the practice to assume existence of elementary entities (elements, molecules, atoms, nuclei, etc.) at a certain level and study their mutual interaction (chemical activity, collision characteristics, molecular formation, energy levels of atoms, etc.) The question of the nature and constitution of entities considered elementary at one level is to be tackled at the next level, where new interactions are postulated between the constituents of that object which was previously considered to be elementary. New levels of structure automatically signify new kinds of interactions, but this process of analysis must stop at the point where constituents lose their meaning. In the description of phenomena in particle physics one may well have reached this stage. The interactions between the particles must explain their structure as well. Existence and change became two facts of the same physical system. Not only are forces between particles obtained through exchange of particles, but the very existence of the particles is due to their interactions.

Such a state of affairs leads to very stringent constraints on allowable dynamical laws. If the electron is an elementary classical point charge and if it is to interact with the electromagnetic field in the usual manner deduced from the classical Maxwell equations, then there is a serious inconsistency: the electron should explode. There is no point in using some other force to hold the electron together if it is a point charge unless this force is also infinite. Perhaps the electron is not a point charge even though it is elementary; perhaps it has a finite extension, perhaps the point nature of the electron is only apparent; perhaps the electron does not interact with the electromagnetic field at the same point but interacts nonlocally. In any case the law of electromagnetic interaction when applied to the stability of the elementary effect itself leads inevitably to a contradiciton unless the local law of interaction is generalized to a non-local one.

One may ask whether the introduction of quantum mechanics helps to eliminate the difficulty. It ameliorates it in the case of the Dirac electron, but does not resolve the problem. And in the equally important case of weak interactions it seems to lead to worse troubles as long as one insists on the use of a local relativistic quantum field theory.

ACTION - AT - A - DISTANCE

From the earliest days of dynamics the question of action - at - a - distance versus action - by - contact has been discussed. One gets the impression that the question has been decided in favor of action - by - contact. For example the electrostatic force which is action - at - a - distance is replaced by a local interaction with the electromagnetic field. As we have already noted, Maxwell's theory is not really satisfactory since it contradicts the existence of point particles. Some form of effective non-local interaction is indicated. We can also raise the question: Do the observed phenomena of electromagnetism rule out an action-at - a - distance theory? Could we not reconcile the phenomena by using a suitable physical interpretation of a direct action - at - a - distance? This question was raised and answered in the affirmative by Wheeler and Feynman[1] in two papers (which have been uncritically cited by almost every writer since then). They show that, provided certain entities (collectively acting as the "absorber") are introduced, all the radiative processes can be equally well derived from a theory involving direct action - at - a - distance between charged particles. They confine their attention to electromagnetic interactions only, and their "absorber" is rather poorly defined. They do not solve the riddle of the existence of the electron; they do avoid the infinity due to the self energy of the electron, but at the price of complicating the effective interaction between the charged particles with the field. In other words their theory, even with the introduction of the absorber, is not equivalent to Maxwell's theory. In spite of all this, their work does show that the direct observation of a "propagating influence" such as the electromagnetic field in no way rules out the possibility of action - at - a - distance.

Let us recall that even earlier Whitehead[2] had attempted to construct a relativistic theory of gravitation as action - at - a - distance. He succeeded in showing that the Schwarzchild line element of general relativity describing the exterior solution for a massive gravitating body could be derived from this. This work has been given a satisfactory formulation in terms of an action principle by Schild[3] who derived the conservation laws for such a theory.

The logical questions involved in formulating a direct two-particle action - at - a - distance theory have been analyzed by Van Dam and Wigner.[4] They have shown that by suitably restricting the interaction the ten fundamental conservation laws are satisfied. The individual particle contributions to the four components of the momentum and the six components of the angular momentum are nonvanishing asymptotically. Hence, the total momentum and angular momentum of the collection of particles are conserved in collisions. Van Dam and Wigner conclude that no physical field is involved since energy is neither absorbed nor radiated. As we shall see, this conclusion is not warranted.

Hence we shall be interested in theories which are in accord with the special theory of relativity. If the theory is manifestly covariant the nonlocality in space, implicit in the notion of a potential, automatically implies nonlocality in time. This comes in from the geometry of Lorentz transformations which make simultaneity frame-dependent, in contrast to a nonrelativistic potential. There is, however, a distinct source of nonlocality which came in from the constraint that the theory should explain the "propagation" of the interaction. Such a nonlocality (the "retardation" effect) should be present equally well in a nonrelativistic situation. Conversely, the retardation effect implies a finite speed of propagation. If this is generally true then a (manifestly) relativistic theory must always involve propagated action.

The problem of the divergence of the self-interaction effects implies a need for modification of the local interaction in action - by - contact theories and to a restriction on the type of interaction in action - at - a - distance theroies. We shall see later on in this report that such a theory requires a cancellation of the high wave number (short distance) components of the propagated interaction; this implies in turn that there should be more than one simple coupled "field," and that some of these fields would have the wrong sign of the interaction energy ("negative probability") or be appreciable outside the light cone ("faster than light") or both. An acceptable relativistic action - at - a - distance theory appears to be equivalent to a suitable local interaction theory formulated in terms of an indefinite metric.

The problem of physical interpretation of indefinite metric theories has been studied over the past decade. We presented the status of the subject at that time at the XIV Solvay Congress.[5] Since that time the persuasive exposition (and diligent model construction) of Lee and Wick[6] have drawn despite the failure of their conjecture, much attention to the uses of indefinite metric in field theory. The status of the problem has not changed in any essential aspect. We had stated that "...the proper identification of physical amplitudes is part of the dynamical problem in an indefinite metric quantum field theory." In an action - at - a - distance theory this problem is even more acute. We shall see that the methods of indefinite metric quantum field theory can be borrowed for developing the physical interpretation of relativistic action - at - a - distance theories.

In this report let us first deal with a classical theory of particles interacting through direct action - at - a - distance and show in what sense it is equivalent to a locally coupled field theory. We note that in classical theory the mediation is by a field rather than by exchange of particles! In a quantum theory, the quantized field is also a quantized particle assembly. Many of the essential questions can be raised and answered within the classical framework, particularly the question of the degree to which the mediating field is to be taken to be physical and the important role played by the choice of the boundary conditions. Attention is then turned to the quantum theory of the system and the structure of the corresponding quantum field theory and the choice of physical amplitudes. In addition to the important question of the conservation of probability ("unitarity") we discuss also the questions of analyticity, the connection between spin and statistics and the TCP theorem.

CLASSICAL THEORY OF INTERACTING PARTICLES

Let M_a denote the proper masses of a collection of interacting particles and x_a^μ, their coordinates. If

$$ds_a^2 = (dx_a^0)^2 - (dx_a^1)^2 - (dx_a^2)^2 - (dx_a^3)^2$$

are the proper times of the various particles, an action function for the particles with two-body interactions is:

$$A = \frac{1}{2}\int m_a \dot{x}_a^2 \, ds_a + \frac{1}{2}\iint K_{ab} \, ds_a \, ds_b ,$$

where K_{ab} are invariant functions depending on the coordinates of the particles and their relative four-velocities. For a purely scalar interaction we may put

$$K_{ab} = K([x_a - x_b]^2),$$

while for a vector interaction we may put

$$K_{ab} = \dot{x}_a^\mu K([x_a - x_b]^2) \dot{x}_{b\mu}$$

and so on. For the electromagnetic direct interaction

$$K_{ab} = e^2 \dot{x}_a^\mu \dot{x}_{b\mu} \, \delta([x_a - x_b]^2),$$

while in the Whitehead theory of gravitation

$$K_{ab} = -G^2 \dot{x}_a^\mu \dot{x}_a^\nu \, \delta([x_a - x_b]^2) \dot{x}_{b\mu} \dot{x}_{b\nu} ,$$

the dots denoting differentiation with respect to the appropriate proper time labels. Without loss of generality we can assume that K_{ab} is symmetric in a and b. Manifest covariance of the theory demands that it be a Lorentz invariant quantity.

The variation of the action is

$$\delta A = ([\dot{x}_a \delta x_a] - \int \ddot{x}_a \delta x_a ds_a) + [\int \frac{\partial K_{ab}}{\partial \dot{x}} \delta x_a ds_b]$$

$$+ \iint \left\{ \frac{\partial K_{ab}}{\partial x_a} - \frac{\partial}{\partial s_a} \left(\frac{\partial K_{ab}}{\partial \dot{x}_a} \right) \right\} \delta x_a ds_a ds_b .$$

The equations of motion are obtained by considering variations which vanish at the end points. We obtain

$$m_a \ddot{x}_a = \int \left\{ \frac{\partial K_{ab}}{\partial x_a} - \frac{\partial}{\partial s_a} \left(\frac{\partial K_{ab}}{\partial \dot{x}_a} \right) \right\} ds_b .$$

The coefficients of the variation of the coordinates at the end points yield the expression for the conjugate physical quantities. By writing

$$\delta x_a = \varepsilon \quad \text{(independent of a)},$$

we obtain the expression for the total (independent of a) momentum of the system:

$$P_\mu = m_a \dot{x}_{a\mu} + \frac{1}{2}\int \left\{\frac{\partial K_{ab}}{\partial \dot{x}_a^\mu}\right\} ds_b + \frac{1}{4}\left(\int_{s_a}^\infty \int_{-\infty}^{s_b} - \int_{-\infty}^{s_a} \int_{s_b}^\infty\right) \left\{\frac{\partial K_{ab}}{\partial x_a^\mu}\right\} ds_a\, ds_b,$$

which is conserved by virtue of the equations of motion, and could have been obtained by direct integration of them. By choosing

$$\delta x_a^\mu = \epsilon_\nu^\mu x_a^\nu \qquad (\epsilon_\nu^\mu \text{ independent of a}),$$

we get the expression for the angular momentum

$$M_{\mu\nu} = m_a(\dot{x}_{a\mu} x_{a\nu} - x_{a\mu}\dot{x}_{a\nu}) - \frac{1}{2}\int_{-\infty}^\infty \left(x_{a\mu}\frac{\partial K_{ab}}{\partial \dot{x}_{a\nu}} - x_{a\nu}\frac{\partial K_{ab}}{\partial \dot{x}_a^\mu}\right) ds_b$$

$$-\frac{1}{4}\left(\int_{s_a}^\infty \int_{-\infty}^{s_b} - \int_{-\infty}^{s_a}\int_{s_b}^\infty\right)\left(x_{a\mu}\frac{\partial K_{ab}}{\partial x_a^\nu} - x_{a\nu}\frac{\partial K_{ab}}{\partial x_a^\mu} + \dot{x}_{a\mu}\frac{\partial K_{ab}}{\partial \dot{x}_a^\nu}\right.$$

$$\left. - \dot{x}_{a\nu}\frac{\partial K_{ab}}{\partial \dot{x}_a^\mu}\right) ds_a\, ds_b,$$

which is again conserved.

These equations simplify for the special interactions we have considered. For the scalar interaction

$$\frac{\partial K_{ab}}{\partial \dot{x}_a} = 0$$

$$P = m_a \dot{x}_{a\mu} + \frac{1}{4}\left(\int_{s_a}^\infty \int_{-\infty}^{s_b} - \int_{-\infty}^{s_a}\int_{s_b}^\infty\right) \frac{\partial K_{ab}}{\partial x_a^\mu} ds_a\, ds_b$$

$$M_{\mu\nu} = m_a (\dot{x}_{a\mu} x_{a\nu} - x_{a\mu} \dot{x}_{a\nu})$$
$$+ \frac{1}{4}\left(\int_{s_a}^{\infty}\int_{-\infty}^{s_b}\int_{-\infty}^{s_a}\int_{s_b}^{\infty}\right)\left(\frac{\partial K_{ab}}{\partial x_a^\mu} x_{a\nu} - \frac{\partial K_{ab}}{\partial x_a^\nu} x_{a\mu}\right) ds_a\, ds_b$$

These expressions have a very simple interpretation. The first terms in P_μ and $M_{\mu\nu}$ are the free particle contributions. Remembering that K_{ab} is a function of $(x_a - x_b)^2$, it follows that the two double integrals really are the same terms except for the interchange of the particle labels. The double integral

$$\int_{s_a}^{\infty}\int_{-\infty}^{s_b}\left(\frac{\partial K_{ab}}{\partial x_a^\mu}\right) ds_a\, ds_b$$

refers to the total impulse that has "left" b by proper time s_b, but not yet "reached" a by proper time s_a (but will reach later!). It is therefore the impulse in transit. (Counting indiscriminately over all pairs accounts for the factor 2.) Similarly, in the expression for the angular momentum it is the moment of this "impulse in transit" which accounts for the second term. We shall assume that these mutual contributions to momenta and angular momenta vanish when the particles are spatially separated far enough so that we may neglect them asymptotically. We have already noted that by virtue of the summation over all the particles, the kernel K_{ab} is symmetric. Because the various expressions have to be real, K should be real and symmetric. There is a "retardation," but it contains both retardation and advancement. Since action and reaction are equal, if we have retarded action, we must have (as viewed from the second particle) an advanced reaction. The only natural choice is to have both retarded and advanced impulse transmission in a time-symmetric fashion.

We note that the only dynamical entities are the particles themselves and the momentum and angular momentum reside either in the motion of the particles or in their mutual ("potential") contributions. There is no other vehicle for the dynamical quantities. In particular, there is no possibility of "radiation" leaving the system. If we want at all to consider energy leaving a collection of particles, there should be some other particles (to which it has been lost!) which act as absorbers or suppliers of energy. Isolated particle or particles do not, in the very nature of things, lose energy to empty space!!

Having paid attention to the essential question of the "absorber" to which the particles transmit energy and momentum etc., we can now raise the question of "radiation," i.e. the energy momentum and angular momentum transmitted by the particle or particles of interest to the "absorber." We have assumed that the law of mutual interaction is such that asymptotically these mutual contributions may be neglected between any pair of particles. To be able to talk about radiation we must therefore make the following provisions: (1) There should be an arbitrarily large number of absorber particles which are at all times sufficiently far away from the first set of particles. (2) We should compute only the energy, momentum and angular momentum that are emitted from the particles of interest to the absorber particles, and the calculation of the radiated quantities of interest takes into account only these. (3) The reaction of the absorber particles on the particles of interest is to be totally ignored. The absorber is therefore not just any collection of particles, but of those that are in such a relation to each other as to produce no advanced field at any of the particles. Under these circumstances the reduction in the energy of the particles is, in fact, being transmitted to the absorber and the loss in energy is independent of any first hand specification of the composition of the "absorber."

INTRODUCTION OF THE FIELD

For many purposes it is advantageous to introduce a field as a suitable functional of the particle trajectories. We choose a specially simple case (to illustrate the principles) where the interaction is purely scalar and given by the real symmetric invariant function

$$K_{ab} = g_a g_b \, \bar{\Delta}(x_a - x_b)$$

where $\bar{\Delta}$ satisfies the equation

$$\left[\frac{\partial^2}{\partial t^2} - \underline{\nabla}^2 + m^2 \right] \bar{\Delta}(x) = -\delta(x).$$

The action function is now

$$A = \int m \, ds_a + \tfrac{1}{2} g_a g_b \iint \bar{\Delta}(x_a - x_b) \, ds_a \, ds_b.$$

Let us <u>define</u> the function

$$\phi(y) = g_b \int \bar{\Delta}(Y - x_b) \, ds_b.$$

Then we have the two equations

$$m\ddot{x}_a = g_a \frac{\partial \phi(x_a)}{\partial x_a}$$

$$(\Box^2 + m^2) \phi(y) = -g_b \int \delta(y - x_b) \, ds_b,$$

provided the original interaction involved summation over <u>all</u> <u>pairs including self-interactions</u>. (We already see an inconsistency since $\bar{\Delta}(0)$ is undefined!) This is exactly the same form as the equations of motion of the particles coupled to a scalar field.

The Action for the system involving the coupled system of particles and fields which lead to the above equations is the following:

$$A = \frac{1}{2}\int m_a \dot{x}_a^2 ds_a + \frac{1}{2}\int [\partial_\mu \phi(y) \partial^\mu \phi(y) - m^2 \phi^2(y)] d^4y + g_a \int \phi(x_a) ds_a$$

In one sense there is a great difference between this system and the system of particles under direct interaction. In the present case the field carries its own energy and can act as the repository of energy, momentum and angular momentum. Despite this difference, which we will discuss presently, we note that the similarity between the two systems is more than appears in the equations of motion. The expression for the conserved quantities can be transformed from the action - at - a - distance form to the action - by - contact form. We shall do this for the conservation of momentum. In the field theory form, the total momenta are:

$$P_\mu = m_a \dot{x}_{a\mu} + \int\int \left\{ \partial_\nu \phi \cdot \partial_\mu \phi - g_{\mu\nu}(\partial_\sigma \phi \cdot \partial^\sigma \phi - m^2 \phi^2) \right\} d\sigma^\nu$$

where $d\sigma^\nu$ is the 3-volume element in the y variable orthogonal to the ν direction. But

$$\int \left\{ \partial_\nu \phi \partial_\mu \phi - \frac{1}{2}g_{\mu\nu}(\partial_\sigma \phi \partial^\sigma \phi - m^2 \phi^2) \right\} d\sigma^\nu$$

$$= \int \left\{ \partial^\nu \partial_\nu \phi \cdot \partial_\mu \phi - \frac{1}{2}\partial_\mu (\partial^\sigma \phi \cdot \partial_\sigma \phi - m^2 \phi^2) + \frac{1}{2}\partial_\mu (\partial^\nu \phi \cdot \partial_\nu \phi) \right\} d^4y =$$

$$\int (\Box + m^2) \phi \partial_\mu \phi d^4 y = -g_b \int \partial_\mu \phi(x_b) \, ds_b.$$

Hence

$$P_\mu = m_a \dot{x}_{a\mu} - g_b \int_{s_b}^{\infty} \partial_\mu \phi(x_b) \, ds_b =$$

$$= m_a \dot{x}_a - \frac{1}{2} g_a g_b \int_{-\infty}^{s_a} \int_{s_b}^{\infty} \frac{\partial \bar{\Delta}(x_a - x_b)}{\partial x_a} \, ds_a \, ds_b,$$

where the factor $\frac{1}{2}$ is inserted to avoid double counting. If we so choose we can further explicitly symmetrize the expression in a and b and obtain the form we had written before. Thus the two systems give the same result. But how are we to reconcile this with the fact that the field seems to contain its own autonomous vibrations independent of other particles?

The resolution of this dilemma lies in the observation that as we <u>defined</u> the field, the field had no autonomy. It was completely defined in terms of the particle variables. Hence the question of whether $\phi(y)$ vanishes at infinity cannot be set by us independently of the configuration of particles. If we have an "absorber," the contribution to ϕ from them must vanish at the position of the particles of interest. In other words, the apparent degrees of freedom of the field are really the degrees of freedom of the myriads of particles constituting the absorber, and the boundary conditions imposed on the field are really disguised assumptions about the configurations of particles.

Let us then ask: Why bother with the fields? First, the introduction of the field enables us to restore, albeit formally, a local interaction structure. Second, in radiative processes we need not explicitly talk about the "absorber" but instead replace it by a boundary condition on the field. The conserved quantities can be seen simply as the sum of the free particle energies of the particles and the "free" field energy. Despite these advantages one must be cautious in the use of the field since the number of degrees of freedom in a <u>finite</u> universe for such a theory is <u>not</u> <u>arbitrarily large</u>; it is probably no more than the number of degrees of freedom of the collection of particles.

We may make this last remark more explicit by recalling the definition

$$\phi(y) = q_b \int \bar{\Delta}(y - x_b) \, ds_b.$$

Suppose we separated the particles into two groups α which are in the absorber and β in the set of particles under study. We can now rewrite $\phi(y)$ in the form

$$\phi(y) = \phi_\alpha(y) + g_\beta \int \bar{\Delta}(y - x_\beta) \, ds_\beta.$$

Remembering that

$$\bar{\Delta} = \frac{1}{2}(\Delta_R + \Delta_A) \; ; \; \Delta = \frac{1}{2}(\Delta_A - \Delta_R) ,$$

we could write

$$\phi(y) = -\frac{1}{2}g_\alpha \int \Delta(y - x_\alpha) ds_\alpha + g_\alpha \int \Delta_A(y - x_d) ds_a$$

$$+ \frac{1}{2} g_\beta \int \Delta(y - x_\beta) ds_\beta + g_\beta \int \Delta_R(y - x_\beta) ds_\beta ,$$

with the first and second terms satisfying the force field equation although they are completely determined by the particles. By definition, the second term does not contribute since an absorber does not produce an advanced field. We have therefore

$$\phi(y) = \phi_{in}(y) + g_\beta \int \Delta_R(y - x_\beta) \, ds_\beta .$$

which is the standard form for a general solution to the equation of motion satisfied by ϕ. But we find that the "in" field ϕ is not entirely arbitrary since it does depend upon the configuration and disposition of the particles, both in the absorber as well as elsewhere. If we genuinely had freedom to choose the "in" field, we would have had new degrees of freedom.

However, such a field manifests degrees of freedom and participates in statistical mechanical phenomena as if it had its own degrees of freedom. Of course it "borrows" these degrees of freedom from the "absorber" and the other particles; if the total number of particles involved is finite, the number of degrees of freedom of this field will also be finite, reminiscent of the phonon modes of a Debye solid.

Since the Green's function for the differential equation is not uniquely determined, it follows that the decomposition of a given solution into a particular integral (proportional to a specific Green's function) and a complementary function (which is interpreted as the incoming free field) is not unique. In other words the correct boundary conditions to choose depend upon what is chosen to be the "in" field and the Green's function. In an action - at - a - distance theory the natural Green's function is the time-symmetric half-advanced, half-retarded kernel. For classical theory one likes to choose the retarded function as a rule. And finally quantum theory works with the causal Stückelberg-Feynman propogators which are retarded for positive energy particles but advanced for those of negative energy.

GENERAL INTERACTIONS IN CLASSICAL THEORY

The considerations given above used a very simple interaction structure. Most of these considerations can be effected if m^2 or g^2 were replaced by negative quantities. These cases correspond respectively to waves which travel faster - than - light and waves which carry negative physical variables. But despite these apparently strange properties the calculations of the previous sections can be carried out.

The restriction to purely scalar interactions was made for the simplification of the algebraic steps. In fact the case $m^2 = 0$ has been treated by several authors both for electromagnetism and gravitation. Apart from a difference in appearance there is no significant difference. It is reassuring that the Maxwell-Lorentz equations of motion are recovered from the action - at - a - distance theory with the kernel

$$K_{ab} = e^2 \dot{x}_{a\mu} \dot{x}_b \, \delta([x_a - x_b]^2).$$

Not only are the successes reproduced but also the failures. Using an "absorber" we can get "radiation" which can equally well be obtained with the retarded boundary conditions and vanishing "in" fields.

The more significant generalization, one which is dictated by the requirements of consistency, is to consider a general two-body kernel. We shall again take up a scalar interaction model and simplify the calculation. Accordingly we start from a kernel K_{ab} which is symmetric, real and Lorentz invariant. If it is not factorizable in the form

$$K_{ab} = g_a \, g_b \, K([x - y]^2),$$

then there exists a collection of factorizable kernels of which K_{ab} is the resultant:

$$K_{ab} = g_a(\lambda) \, g_b(\lambda) \, K_\lambda([x - y]^2).$$

For every invariant function K_λ we introduce the spectral decomposition (valid under relatively mild constraints);

$$K_\lambda([x-y]^2) = \int_{-\infty}^{\infty} \rho_\lambda(k) \bar{\Delta}(x-y; \sqrt{k})\, dk.$$

where the weight function ρ_λ is real but not necessarily positive and the (K) integration goes from $-\infty$ to ∞. We have thus finally

$$K_{ab} = g_a(\lambda) g_b(\lambda) \int_{-\infty}^{\infty} \rho_\lambda(k) \bar{\Delta}(x-y; \sqrt{k})\, dk.$$

We must now introduce a family of fields $\phi(x,\lambda, k)$ which are coupled to the particles by

$$g_a = g_a(\lambda) \sqrt{\rho_\lambda(k)}.$$

This general form means that the direct action - at - a - distance can be replaced by a local coupling of the particles and the fields. Both faster - than - light and negative weight fields are in general included.

Not only may negative weight fields (i.e. those for which the spectral weight is negative) appear, but if the self-interactions are to be finite and thus are not to lead to an inconsistency, then they *must* be included. If we consider the static limit this is satisfied if the self-field at the origin vanishes; this is simply the integral over the total spectral weight. But if one allows for the propagation of the field before it is absorbed, it is clear that the detailed manner in which cancellations occur do depend on the velocity dependence (i.e. tensorial character) of the interaction.

We finally wish to note that an arbitrary velocity dependence could, at least formally, be expanded as a series of tensors in the velocities and would therefore correspond to a mixture of interactions. No essentially new point of principle seems to emerge from a consideration of these mixed forms

The problem of coupled systems can now be solved at least in a perturbative scheme. The complete trajectory can be obtained in terms of a power series expansion in the coupling constants. With suitable admixture of negative fields, the various orders of perturbations make sense. These are finite self-interactions effects; and in suitable cases such as the Whitehead-Schild model for gravitation, the inertia of the moving body may be changed ("curved space") by the interaction.

The conclusions we draw from the classical theory are the following:

(1) We can transform an action - at - a - distance theory of particles into an action - by - contact theory of particles coupled to fields.

(2) Self-interaction is a necessary adjunct to this demonstration of equivalence.

(3) There is no possibility of radiation emission from, or absorption by, a system of particles unless there are other particles present in the theory and they are arranged to have

no advanced fields at the location of the original set of particles.

(4) The degrees of freedom of these classical fields are really the degrees of freedom of the "absorber" particles which are of no interest to us otherwise.

(5) To make the general theory we may have to consider the coupling to a multitude of fields with all possible real values for the square of the mass form $-\infty$ to $+\infty$.

(6) Not only may some of the fields so introduced have negative weights, but the requirement of a meaningful self-interaction demands that there are many of them.

(7) Self-interaction effects are expected even within simple models of classical action - at - a - distance theories, but they are finite and calculable in perturbation theory.

QUANTUM MECHANICAL SYSTEMS WITH RELATIVISTIC ACTION - AT - A - DISTANCE

Our main interest is in relativistic quantum theory. In relativistic field theory local couplings have been used traditionally, but then we also have invariably the diseases of covariant local Lagrangian field theories. Consider the simple case of a spin $\frac{1}{2}$ (fermion) field ψ of (bare) mass M and a spin 0 (boson) field ϕ of (bare) mass m with local interaction; the action density is

$$\mathcal{L} = \bar{\psi}(i\partial\!\!\!/ - M)\psi + \frac{1}{2}(\partial^\nu \phi)(\partial_\nu \phi) - \frac{1}{2} m^2 \phi^2 + \mathcal{H}, \quad \mathcal{H} = g\bar{\psi}\psi\phi.$$

In the interaction picture, the fields satisfy the free field equations and commutation relations:

$$(i\partial\!\!\!/ - M)\psi = 0, \quad \delta(x^0 - y^0)\{\psi_r^+(x), \psi_s(y)\} = \delta_{rs}\delta(\underset{\sim}{x} - \underset{\sim}{y})$$

$$(\partial^\nu \partial_\nu + m^2)\phi = 0, \quad \delta(x^0 - y^0)\left[\phi(x), \partial^0\phi(y)\right] = i\delta(\underset{\sim}{x} - \underset{\sim}{y}).$$

The transition amplitudes are defined in terms of the S-matrix

$$S = 1 + \sum_{n=0}^{\infty} \frac{i^n}{n!} \int d^4 x_1 \cdots \int d^4 x_n T(\mathcal{H}(x_1) \cdots \mathcal{H}(x_n)).$$

Using Wick's algebraic identities the time ordered products (denoted by T(....) above) are converted into a linear combination of normal products and these in turn lead to Feynman's graphical rules for computing scattering amplitudes. The amplitudes so calculated, at least in the lower orders of perturbation theory, can be obtained more directly by simply iterative substitution in the Heisenberg equations of motion obtained from the above action density.

The amplitudes so derived have two principal defects which must be resolved in some manner before physically meaningful conclusions can be drawn. The first concerns the fact that the interaction modifies the particle masses. The mass shifts themselves can, in principle, be computed as power series in the coupling constant. We must now view the action density in the form

$$\mathcal{L} = \mathcal{L}_0 + \mathcal{H}_0$$
$$\mathcal{L}_0 = \bar{\psi}(i\partial\!\!\!/ - M_0)\psi + \frac{1}{2}(\partial^\nu \phi)(\partial_\nu \phi) - \frac{1}{2}m_0\phi^2$$
$$\mathcal{H}_0 = g\bar{\psi}\psi\phi + (M_0 - M)\bar{\psi}\psi + \frac{1}{2}(m_0^2 - m^2)\phi^2 ,$$

and the S-matrix should be given with $\mathcal{H}_0$ considered as the interaction density. This mass renormalization is an <u>essential</u> step in the proper physical interpretation of the theory irrespective of the magnitude of the mass shifts.

The second and more serious difficulty is that except for some of the lowest order contributions, the contributions of the S-matrix are infinite in each order; and this comment also applies to the mass shifts. The source of this difficulty can be traced to the divergence of the integrals over the contraction function due to the confluence of the arguments, which in turn is directly to be traced to the local coupling postulated. And this appears in even more pronounced form for other interactions like the chiral four-fermion interaction in the theory of weak interactions.

There is, of course, a sense in which some physical predictions can be made in the theory. One shows that in a class of theories we can absorb all the divergence into renormalizations (albeit by an infinite amount) of the masses and the coupling constants. Quantum electrodynamics of electrons has been witness to the success of this program. But the method leaves too many questions unanswered. The mass shifts and effective coupling strengths are themselves physical quantities and should not be infinite: For example, it would be tempting to identify the mass difference between the charged and neutral pion to be due to the electromagnetic mass shifts. We must, accordingly, arrange to have a theory in which the renormalizations of the masses and coupling constants are themselves finite. This would involve a modification of the theory so that the propogation functions for the fields fall off faster for larger values of the momenta.

Similar conclusions would be arrived at if we started out with an action - at - a - distance theory. In this version we start from an action:

$$A = \int \bar{\psi}(x)(i\not{\partial} - M)\psi(x)d^4x + \iint \bar{\psi}(x)\psi(x)K(x-y)\bar{\psi}(y)\psi(y)d^4x\,d^4y,$$

involving the coupling between the fermion fields at the points x and y through the real symmetric kernel $K(x - y)$. For simplicity let us first consider

$$K(x - y) = g^2 \bar{\Delta}(x - y).$$

In this case we can define the (auxiliary) field

$$\phi(x) = g \int \bar{\Delta}(x - y)\bar{\psi}(y)\psi(y) \, d^4y.$$

Then the equations of motion can be rewritten

$$(i\not{\partial} - M)\psi(x) = g\phi(x)$$

$$(\partial^\nu \partial_\nu + m^2)\phi(x) = g\bar{\psi}(x)\psi(x)$$

In other words, the symmetric interaction of the scalar densities at the points x and y is equivalent to the local coupling to the meson field $\phi(x)$.

We find, as in the classical theory, a Green's function for the propogation of the ϕ field which is different from the standard contraction function. We have already remarked that the identification of the suitable Green's function is equivalent to the specification of the boundary conditions. The standard Green's function for propogating meson fields is the causal Stückelberg-Feynman contraction function

$$\langle T\{\phi(x)\phi(y)\}\rangle_0 = \Delta_F(x - y) = \bar{\Delta}(x - y) + i\pi\Delta^{(1)}(x - y).$$

This Green's function can be obtained by starting with the $\bar{\Delta}$-function and choosing the boundary condition such that the "in" field is the free field plus a term

$$\phi_2(x) = ig\pi \int \Delta^{(1)}(x - y)\bar{\psi}(y)\psi(y)d^4y.$$

In other words, the S-operator

$$S = 1 + \Sigma_{n=0}^{\infty} \frac{i^n}{n!} \, d^4x_1 \ldots d^4x_n T\mathcal{H}(x_1)\ldots\mathcal{H}(x_n)$$

yields the scattering amplitudes provided the matrix elements are taken not for the field $\phi(x)$ but

$$\phi_{in}(x) = \phi(x) - i\pi g \int \Delta^{(1)}(x - y) \bar{\psi}(y) \psi(y) d^4y,$$

and the quantization is taken with respect to $\phi_{in}(x)$ rather $\phi(x)$.

We are now in a position to discuss the general scalar kernel.

$$K(x - y) = \int_{-\infty}^{\infty} \rho(k) \bar{\Delta}(x - y; k) dk.$$

As long as $\rho(k)$ is positive, a field of mass $\sqrt{k}$ may be introduced along with it the coupling constant

$$g(k) = \sqrt{\rho(k)}$$

If $\rho(K)$ is negative we have to associate it with a field quantized according to the indefinite metric:

$$\delta(x^0 - y^0)[\phi(x), \partial^0 \phi(y)] = -i\delta(\vec{x} - \vec{y})$$

For such a quantization scheme, the states with an even number of quanta of the field will have positive norm and the states with an odd number of quanta will have negative norm. The theory of the quantization according to the indefinite metric is now too well known to be reviewed here. It suffices to say that the corresponding contraction functions will have opposite signs. An immediate consequence of this is that if we have a linear superposition of fields of various masses making up a linear field

$$\phi(x) = \Sigma C_r \phi_r(x)$$

the effective contraction function is

$$<T(\phi(x)\phi(y))>_0 = \Sigma_r |C_r|^2 \eta_r \Delta_F(x - y; m_r),$$

where η_r is the "metric" of the field ϕ_r which has the value $+1$ for ordinary fields and -1 for fields with indefinite metric. By choosing

$$C_r = |\rho(k)|^{1/2}$$
$$\eta_r = \rho(k)/|\rho(k)|,$$

we can therefore, reproduce the scalar kernel $K(x - y)$ that we started out with, at least as far as the real part is concerned. (We have already remarked that the imaginary part of Δ_F can be obtained by a suitable choice of boundary conditions).

For a fermion field of spin 1/2, apart from inessential numerical factors, the propogator is

$$S(p,m) = (p,m) = (\not{p} - m + i\varepsilon)^{-1}.$$

For a linear combination

$$\psi(x) = C_r \psi_r(x)$$

the effective propogator is

$$S(p) = |C_r|^2 \eta_r (\not{p} - m + i\varepsilon)^{-1}.$$

If we take

$$\eta_2 = \eta_3 = -1, \qquad \eta_1 = +1; \qquad (C_2/C_1)^2 = \frac{m_1^2 - m_3^2}{m_2^2 - m_3^2};$$

$$(c_3/c_1)^2 = \frac{m_1^2 - m_2^2}{m_3^2 - m_2^2},$$

$$S(p) = |c_1|^2 (m_1^2 - m_2^2)(m_1^2 - m_3^2)(\not{p} - m_1 + i\varepsilon)^{-1}(\not{p} - m_2 + i\varepsilon)^{-1} \times$$

$$\times (\not{p} - m_3 + i\varepsilon)^{-1}$$

so that $S(p) \sim p^{-3}$ for large values of p. This softening of the high momentum components leads to a softening of the small distance behavior of the kernel $K(x - y)$ and to convergence in perturbation theory of processes involving the fermion fields.

For boson fields (of spin 0) the situation is even simpler. Apart from inessential factors

$$\Delta(p,m) = (p^2 - m^2 + i\varepsilon)^{-1}.$$

Hence, if we take

$$\phi(x) = \phi_1(x) + \phi_2(x); \quad \eta_1 = \eta_2 = +1,$$

we get the effective propogator

$$S(p) = (m_1^2 - m_2^2)(p^2 - m_1^2 + i\varepsilon)^{-1}(p^2 - m_2^2 + i\varepsilon)^{-1}.$$

Since this propogator also goes as p^{-4} asymptotically, convergence in each order for any interaction is guaranteed.

At this point, we may profitably compare our findings with the classical theory and verify that the situation is quite similar. In quantum theory we must pay particular attention to the question of the conservation of probability in all physical processes.

UNITARITY QUESTIONS AND THE THEORY OF SHADOW STATES

If the quanta of the abnormal fields were not present either in the initial or final state of any physical process there should be no difficulties with the probability interpretation. Even if they are neither present initially nor finally in a particular process the unitarity condition would demand in general that any intermediate state in which there are abnormal quanta should contribute its share to the imaginary part of the amplitude. We should therefore require that the imaginary part of any physical amplitude gets contributions only from states containing no abnormal quanta. Once this is done all questions of probability conservation ("unitarity") would be automatically taken care of, since the states containing abnormal quanta would now no longer be counted amongst physical states.

To illustrate the notion of shadow states and the role that they play in a quantum theory, let us take a simple quantum mechanical system with the Hamiltonian $\mathcal{H} = \mathcal{H}_o + V$ where V is a suitable interaction. If we were to study the interaction representation, the state vector $|\phi(t)\rangle$ in the interaction representation has the time dependence

$$i \frac{\partial}{\partial t} \phi(t)\rangle = V_I(t) |\phi(t)\rangle$$

where

$$V_I(t) = \exp(i\mathcal{H}_o t) \, V \exp(-i\mathcal{H}_o t)$$

(We shall assume that the division between $\mathcal{H}_o$ and V is so arranged that $\mathcal{H}$ and $\mathcal{H}_o$ have the same spectra.) The formal solution to this equation is

$$|\phi(t)\rangle = T(\exp[i \int_{-\infty}^{t} V_I(t^1) dt^1]) |\phi(-\infty)\rangle$$

$$= \{1 + \sum_{n=1}^{\infty} \lambda^n \frac{i^n}{n!} \int_{-\infty}^{t} dt_1 \int_{-\infty}^{t_1} dt_2 \ldots \int_{-\infty}^{t_{n-1}} dt_n \, V_I(t_1) \ldots V_I(t_n)\} |\phi(-\infty)\rangle$$

$$\equiv \{1 + \sum_{n=1}^{\infty} \lambda^n \frac{i^n}{n!} \int_0^t dt_1 \ldots \int_0^t dt_n \, T(V_I(t_1) \ldots V_I(t_n))\} |\phi(-\infty)\rangle$$

If $|\phi(t)\rangle$ is the corresponding Schrödinger state vector then

$$|\Phi(t)\rangle = e^{-i\mathcal{H}_0 t}|\Phi(t)\rangle,$$

then its solution is

$$|\Phi(t)\rangle = \{e^{-i\mathcal{H}_0 t} \sum_{n=0}^{\infty} i^n \int_{-\infty}^{t} dt_1 \, e^{i\mathcal{H}_0 t_1} V e^{-i\mathcal{H}_0 t_1} dt_2 e^{i\mathcal{H}_0 t_2} V e^{-i\mathcal{H}_0 t_2} \times$$

$$\cdots \int_{-\infty}^{t_{n-1}} dt_n \, e^{i\mathcal{H}_0 t_n} V e^{-i\mathcal{H}_0 t_n}\} |\Phi(-\infty)\rangle.$$

Let us now take the exact eigenstates of the total Hamiltonian

$$|E\rangle = \frac{1}{2\pi} \int_{-\infty}^{\infty} e^{iEt} |\Psi(t)\rangle \, dt$$

$$= \sum_n i^n \int_{-\infty}^{0} e^{i(E-\mathcal{H}_0)\tau_1} V d\tau_1 \int_{-\infty}^{0} e^{i(E-\mathcal{H}_0)\tau_2} V d\tau_2$$

$$\cdots \int_{-\infty}^{0} e^{i(E-\mathcal{H}_0)\tau_n} V d\tau_n \frac{1}{2\pi} \int_{-\infty}^{\infty} e^{i(E-\mathcal{H}_0)\tau} V d\tau |\Phi(-\infty)\rangle.$$

This could be rewritten in the form

$$|E\rangle = \sum_n \frac{1}{E-\mathcal{H}_0+i\epsilon} V \frac{1}{E-\mathcal{H}_0+i\epsilon} V \cdots V \frac{1}{E-\mathcal{H}_0+i\epsilon} V \Pi_E \times |\Phi(-\infty)\rangle$$

where

$$\Pi_E = \frac{1}{2\pi} \int_{-\infty}^{\infty} e^{i(E-\mathcal{H}_0)\tau} d\tau,$$

is the projection operator to energy E of the unperturbed Hamiltonian. If we note

$$\Pi_E \mid \Phi(-\infty) > \; = \; \mid \Phi_E >,$$

we have the final expression

$$\mid E> \; = \; \{1 + \sum_{n=1}^{\infty} \frac{1}{E-\mathcal{H}_0+i\epsilon} \; V \; ^n\} \mid \Phi_E>.$$

This expression, which looks typically "nonrelativistic" will be completely equivalent to the covariant Feynman rules if we choose V and $\mathcal{H}_0$ to be appropriate integrals of the action densities of a local field theory. The scattering amplitude is obtained by taking the energy conserving matrix elements of the summand in the expression for $\mid E>$, so that we get

$$F_{\beta\alpha}(E) = <\beta \mid VGV \mid \alpha> \; = \; <\beta \mid V \frac{1}{E-\mathcal{H}_0+i\epsilon} V \frac{1}{E-\mathcal{H}_0+i\epsilon} V..V \mid \alpha>$$

We note that the scattering amplitude so obtained satisfies the unitarity relation which may be written in the symbolic form

$$F_{\beta\alpha} - F^*_{\beta\alpha} = <\beta \mid V \frac{1}{E-\mathcal{H}_0+i\epsilon} V...V \mid \alpha> - <\beta \mid V \frac{1}{E-\mathcal{H}_0-i\epsilon} V..V \mid \alpha>$$

$$= <\beta \mid V \frac{1}{E-\mathcal{H}_0+i\epsilon} V...V \mid \nu> \delta(E-E_\nu) <\nu \mid V \frac{1}{E-\mathcal{H}_0-i\epsilon} V..V \mid \alpha>$$

$$= \sum_\nu F_{\beta\nu} F^*_{\nu\alpha} \; 2\pi i \delta(E-E_\nu),$$

where, to avoid inessential elaboration, we have considered time-reversal invariance to be valid so that V is real; and anomalous thresholds have been ignored. The imaginary part of the amplitude corresponds to physical intermediate states.

We observe that if we were interested only in constructing a solution to the fully interacting problem we could write

$$|E\rangle = \{1 + \sum_{n=1}^{\infty} (G_o(E)V)^n\}|\Phi_E\rangle,$$

where $G_o(E)$ is <u>any</u> Green's function for the Hamiltonian $\mathcal{H}_o$:

$$(E - \mathcal{H}_o)G_o(E) = 1.$$

The solution we deduced above corresponds to the retarded Green's function

$$G_o(E) = (E - \mathcal{H}_o + i\epsilon)^{-1}$$

If we chose instead the time-symmetric Green's function

$$G_o(E) = \frac{1}{2}(E-\mathcal{H}_o+i\epsilon)^{-1} + \frac{1}{2}(E-\mathcal{H}_o-i\epsilon)^{-1} = P\frac{1}{E-\mathcal{H}_o},$$

the resultant "scattering amplitude" would be entirely real:

$$F_{\beta\alpha}(E) - F^*_{\beta\alpha}(E) = 0.$$

This would not satisfy unitarity if any physical scattering were to take place.

Suppose we partition the set of all states of the quantum mechanical system into two classes: those of the first class would be called the "physical states" and those of the second the "shadow states;" we must make sure that this partitioning is compatible with the Hamiltonian $\mathcal{H}_0$. Hence the Green's function $G_0(E)$ is diagonal with respect to this partition:

$$G_0(E) = \begin{pmatrix} G_{op}(E) & 0 \\ \hline 0 & G_{os}(E) \end{pmatrix}$$

The interaction V would in general have both diagonal and non-diagonal elements:

$$V = \begin{pmatrix} V_{pp} & V_{ps} \\ \hline V_{sp} & V_{ss} \end{pmatrix}$$

Let us now choose the Green's function which is retarded with respect to the "physical states" and time-symmetric with respect to the "shadow states." We write:

$$G_0(E) = \begin{pmatrix} \left(\dfrac{1}{E - \mathcal{H}_0 + i\varepsilon}\right)_{pp} & 0 \\ 0 & \left(P\dfrac{1}{E - \mathcal{H}_0}\right) \end{pmatrix}$$

With this choice of Green's function we get an expression for the scattering amplitude which obtains contributions both from the "physical states" and the "shadow states." However, if we compute the imaginary part of the amplitude in the physical domain we get the result

$$F_{\alpha\beta} - F^*_{\alpha\beta} = 2\pi i \sum_{\nu}{}' F_{\alpha\nu} F^*_{\alpha\nu}$$

with the primed summation going only over the physical states. In other words, the scattering involves only the physical states, even though the dynamics involve both the physical states and the shadow states.

Having arrived at this juncture we can now deal with the physical interpretation of indefinite metric theories. We use the usual diagrammatic approach to perturbation theory but any intermediate state which contains one or more abnormal particles should be treated as a shadow state. This implies that we subtract out the imaginary part coming in from any such intermediate state. This is a consistent procedure and may be carried out in any order of perturbation theory. The shadow states would automatically exclude themselves from the unitarity sum.

We may now remove the restriction that anomalous thresholds do not appear. Even in such a case, we must recognize that the scattering amplitude defined by the perturbation series is defined only for physical energies. In this domain the imaginary parts are given by the unitarity relation. The analytic continuation of the scattering amplitude may contain imaginary parts elsewhere but they will not be in the physical domain. It follows that anomalous thresholds may exist even after the shadow states are eliminated.

It is remarkable to note that despite the radically new approach to the theory of the scattering amplitude in the low energy domain (before the opening of the thresholds for the shadow states) the local analytic properties of the scattering amplitude in this theory are exactly the same as in the usual (inconsistent!) local field theory. It follows that the usual results of quantum field theory like the connection between spin and statistics, the relative parity of particles and antiparticles, the TCP theorem as applied to masses and lifetime of particles, crossing symmetry in the small etc. are all equally true in the present theory. The essential difference lies in the fact that in the present theory the scattering amplitude is not a single analytic function, but is only piecewise <u>analytic</u>, it is continuous for physical values (since the shadow state constirubitons have the usual threshold behavior.)

Since the scattering amplitudes are only piecewise analytic (even though we have the usual unitarity relationships including the one between total cross section and the imaginary part of the forward scattering amplitude), it is not easy to see to what extent dispersion relations should hold exactly. To the extent that the forward dispersion relations do hold, we have to assume that the onset of the shadow states is only at sufficiently high energies.

We are now in a position to deal with the quantum theory of relativistic action - at - a - distance! The auxiliary fields that we have introduced (whether they be of positive weight or of negative weight) are all associated with the time-symmetric (one-particle) propagators. They are therefore not candidates for being considered as physical quanta; we might consider any state in which at least one of these quanta is included as a shadow state. This is as it should be since we are interested in considering the theory of action - at - a - distance! We proceed as before and exclude the imaginary part coming from any intermediate state involving any of these auxiliary quanta. Since the contribution (in the physical domain) came only from states which can be realized, without any violation of the action - at - a - distance framework we can use the usual rules of Feynman diagrams, provided it be understood that the imaginary parts coming from shadow intermediate states is to be subtracted out.

We have now come to a new conclusion! <u>We can construct finite relativistic quantum theories, the effects of interaction including self-interaction and obtain meaningful finite answers</u>. Action - at - a - distance and indefinite metric theories turn out to have basically the same theoretical framework so long as it is realized that the identification of the physical amplitudes is part of the dynamical problem. We now have the solution to the basic problem posed centuries ago: Is it action - by - contact or action - at - a - distance? The answer is: they are only two ways of looking at the <u>same</u> theory. We have now moreover the complete framework for the theory. While the earlier work on finite quantum electrodynamics is a contribution in this general context much more detailed work is necessary to implement the programme in particle theory. Perhaps this will be accomplished during the next decade.

REFERENCES

1. J. A. Wheeler and R. P. Feynman, Rev. Mod. Phys. **17**, 157 (1945); **21**, 425 (1949).

   R. P. Feynman Phys. Rev. **74**, 1430 (1948).

2. A. N. Whitehead: "The Principle of Relativity" (Cambridge, 1922).

3. A. Schild, Phys. Rev. **131**, 2262 (1963); A. Schild and I. A. Schlosser, J. Math. Phys. **6**, 1299 (1965).

4. Van Dam and E. P. Wigner, Phys. Rev. **138**, B1576 (1965).

5. Fundamental Problems in Elementary Particle Physics. XIV. Solvay Congress, Interscience Publishers, New York 1968. More references are given here.

6. T. D. Lee and G. C. Wick, Nucl. Physics B9, 209 (1969).

7. W. Pauli and F. Villars, Rev. Mod. Phys. **21**, 434 (1949).

HEADLIGHTS OF THE LAST DECADE IN FIELD THEORY[*]

John R. Klauder

Bell Telephone Laboratories, Murray Hill, N.J.

Abstract: Professor Klauder's talk is divided into three main parts: one dealing with the connection of the axiomatic approach to scattering theory, a second dealing with the advances in algebraic quantum theory and a third dealing with constructive field theory. In the first part Wightman's reconstruction theorem and its consequences are emphasized. The relation between the asymptotic conditions and the test function spaces in the Haag-Ruelle and LSZ-theories respectively is treated in some detail. Other topics commented upon in this part are conditions on the Lehmann spectral function which reduce the theory to a theory of generalized free fields, and Borchers' results on the equivalence classes of local fields. In the second

---

[*]Presented at the Symposium "The Past Decade in Particle Theory," at University of Texas in Austin, April 14-17, 1970.

part the Gelfand-Naimark-Segal construction is described. The consequences of the weaker, physical, criterion for equivalence between representations of an algebra of measurable operators is discussed, together with the question of implementing automorphisms of an algebra by unitary transformations. In the third part the main attention is given to the limiting problems involved in the removal of spatial cut-offs. The recent results concerning explicit models, mainly in two-dimensional space-time, are reviewed. The talk concludes with some remarks on a theorem on the anti-commutation relation.

Axiomatic field theory means different things to different people. On the one hand, it can be viewed as a precise set of rules to follow and the consequences which can be derived therefrom. Actually, this is not a bad definition. At least if one understands what is meant by the definition one has a framework in which to work. However, the spirit of the game is much broader. One is really concerned with obtaining the relationship of certain outputs to certain inputs by the application of controlled techniques. From this point of view I believe that it can become appealing even to an ex-electrical engineer like myself.

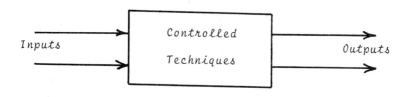

Fig. 1

Engineer's View of Axiomatic Field Theory

It is, of course, difficult to review all of the progress and research made during the last ten years. Therefore, I must be selective. For example, I will not mention the application of related mathematical techniques to the advances in statistical mechanics. I shall confine my attention to field theory questions and will only comment in passing on two advances made in perturbation analysis. On the one hand, there is the clarification of the Bogoliubov-type prescription for renormalization. In this work, Hepp and Speer clarified the distribution nature of the products making up the Feynman graphs. This is a term by term clarification of the perturbation series. On the other hand, we have the proof in certain models, such as those studied by Jaffe, that the entire perturbation series is not convergent anyway; so it is at best an asymptotic series. This result could have been expected. Thirring, in the past decade, had shown that this was the case for the $(\phi^3)_4$ model, the ϕ^3 theory in four dimensions. Jaffe studied the $(\phi^4)_2$ model, (and higher powers) in two dimensions, i.e. 1 space and 1 time. In this model, the particular Feynman graphs can be controlled.

Let me divide my talk roughly into three parts: one dealing with the connection of the axiomatic approach to scattering theory; a second dealing with the advances in algebraic quantum theory; and a third, showing that not only formal problems have been studied but that concrete models have been analyzed as well. I should like to comment on the so-called constructive field theory in which genuine relativistic fields are investigated. Again the most simple ones, from the point of view of mathematical difficulties, are the $(\phi^4)_2$ theory, the Yukawa theory in two dimensions, etc. The $(\phi^4)_2$ theory is the one in which most progress has been made.

At the end, I will make a few remarks on the canonical anti-commutations relations in which an important theorem has been proved.

Let me begin by setting the framework with the remark that the decade began with one set of proposals for quantum field theory. These axioms are shown in Fig. 2 and are commonly associated with the name of Wightman. Requirement 0 assumes that there is a Hilbert space which carries a unitary representation of the Poincare group $U(a,\Lambda)$. This representation fulfills the traditional spectrum requirements, and $|0\rangle$ is a unique, invariant vacuum state. Condition I requires that the fields $A(x)$ be smeared out over space and time to become operators, just in the fashion advocated so long ago by Bohr and Rosenfeld. II asks that the operator $A(x)$ transform in the standard fashion under the action of the unitary representation. III states that if points x and y are space-like, then the boson fields in question will commute. Armed with the fact that the vacuum is assumed to be a member of the invariant domain D, one can construct the matrix elements $W_n(x_1,\ldots x_n)$, the so-called Wightman functions, and study them.

One of the important results of the previous decade was the reconstruction theorem based on the Gelfand-Naimark-Segal

Wightman Framework

0. Hilbert Space, $U(a,\Lambda)$, Spectrum, $|0\rangle$

I. Operator-Valued Distribution $A(x)$, Dense Invariant Domain D

II. Covariance of $A(x)$

III. Locality: $[A(x), A(y)] = 0$, $x \sim y$

$$W_N(x_1, \ldots, x_n) \equiv \langle 0|A(x_1) \ldots A(x_n)|0\rangle$$

Conclusion

    i) Reconstruction Theorem (GNS)

        $\{W_N\}$ determines Theory

    ii) Spin and Statistics, Weak Locality = CPT Invariance, Global Nature of Local Commutivity, Equivalence Classes of Local Fields, Haag's Theorem

Fig. 2

Synopsis of Wightman Framework

construction. I shall mention the GNS construction several times. This important mathematical technique was used over and over in this decade; and one of its consequences is that the Wightman functions or distributions W_n completely determine the theory. Every physical question you need to know about the theory is contained within the W_n's themselves. By the complete exploitation of the Lorentz covariance one proves the many wonderful results that have come out of this work, some of which have also been obtained from previous techniques as well: the connection between spin and statistics; the equivalence of weak locality and CPT invariance. Weak locality is a consequence of the stronger form III. Thus if locality holds we have CPT invariance. Other examples are the global nature of local commutivity and the interesting results of Borchers on the equivalence classes of local fields. For example, if we have an irreducible field A that satisfies III, and if we bring up another field B which satisfies a similar relation with A, then it follows that B must fulfill the same relation with itself, i.e. B is local. If moreover, A and B have the same in-fields they will have the same out-field. Thus these two fields describe the same scattering theory. In other words, there is a collection of operators, all of which describe the same physical situation. We also have Haag's theorem, suitably paraphrased by saying "Thou shalt not use the free field representation for interacting theories having translational invariance." To discuss interesting theories other representations of the operators must be utilized.

Cyclic local algebras are algebras on a space spanned by polynomials in the field operator A(x) acting on the vacuum where the point x lies in a bounded region of space and time. For example, x's lying in this room during the period of this talk. As a final example, let me mention the generalized free field characterized by the most general c-number commutation consistent with the Lehmann spectral representation. This field has been a good testing ground for many of the questions or attempts at constructing a model theory. For instance, if one takes the Lehmann spectral representation of the two-point function and requires that the weight $\rho(m^2)$ that appears there vanishes above some particular mass, then the field in question can be nothing but a generalized free field having a c-number commutation relation. Therefore, to get realistic theories, one has to respect such properties.

The decade also began with two propositions about the connection of field theory with scattering theory. One put

forward by Haag and another more familiar one put forward
by Lehmann, Symanzik and Zimmermann. During this decade
this conflicting situation was clarified, particularly by
Ruelle and Hepp. It was also shown how the two formulations are related. In order to complete the connection,
one has to make additional assumptions. Let me emphasize
that you don't get something for nothing; to get something
out, you have to put something in. You can't have a scattering theory unless there are asymptotic fields. There must
be a Hilbert space corresponding to the in-particles and we
require it to coincide with the physical Hilbert space.
This implies that it has a more or less simple spectrum.
This is condition IV below (Fig. 3). In simple cases, this
spectrum is composed of a vacuum, a one-particle hyperbola,
and a two-particle continuum. The first problem to be analyzed is the so-called one body problem. You have to find
an operator B^+ within the polynomials of the basic field,
which when acting on the vacuum has an output vector partially lying in the one particle subspace. Products of such
operators acting on the vacuum form time-dependent vectors,
$\phi(t)$ in line B of Fig. 3. B^+ is the operator, and $\tilde{f}(p)$ is
the space-time smearing function expressed in momentum space.
The $\phi(t)$ states are studied as time goes to either plus or

Scattering Theory

IV. $H^{in} = H$, Spectrum

A) One Body Problem

$B^{+}\,\Omega = \phi_{\ell} + \psi$

B) $B(f,t)^{+} = \int e^{i(p_0-\omega)t}\,\tilde{f}(p)\tilde{B}(-p)\,d^4p$

$\phi(t) = \pi B(f_n,t)^{+}\,\Omega$

a) $\tilde{f}_n(p) \in S(ms) \subset S(R^4)$

$||\phi(t) - \phi^{ex}|| \leq \alpha|t|^{-1/2}$

a') $supp\ \tilde{f}_n \cap supp\ f_m = 0$

$||\phi(t) - \phi^{ex}|| \leq \alpha_N|t|^{-N}, \forall N$

b) $\tilde{f}_n(p) \in S(R^4)$

$|(\psi^{ex}, \phi(t) - \phi^{ex})| \to 0$

b') $supp\ \tilde{f}_n(p) \cap supp\ \tilde{f}_m(p) = 0$

$|(\psi^{ex}, \phi(t) - \phi^{ex})| \to 0$

C) $\lim \pi B(f_n,t)^{+}\,\Omega = \pi a^{ex}\,(f_n)^{+}\,\Omega$

D) Covariant definition of ϕ^{ex}

E) Smooth Reduction Formulae

F) Dispersion Relations

Fig. 3

Synopsis of Scattering Theory

minus infinity. Here there is a divergence of viewpoints and the two different viewpoints relate to the Haag-Ruelle theory and the LSZ formulation. Suppose we restrict the test functions to lie in momentum space so that they do not see the continuum that lies above the one-particle hyperbola. This restriction we indicated by S(ms), and we can say that all-states are restricted to the mass shell. In such a case, then the vectors in question converge as t goes to $+\infty$ or $-\infty$ with a characteristic rate of $t^{-1/2}$. Thus as t goes toward infinity, the vector would tend to the vector ϕ^{ex}. Since the Hilbert space is complete there will be a vector ϕ^{ex} waiting for it. This is the asymptotic state characterizing this particular interpolating state. If one makes the stronger assumption that the test functions have non-overlapping momentum or velocity components for each of its constituent B^+'s, then stronger degrees of convergence can be proved. In this case the rate is faster than any inverse power of time, as indicated by formula (a') in Fig. 3. These are the propositions of Haag. When one restricts the test functions involved to be outside the continuum, then one has this so-called strong convergence. However, if one expands the allowed test functions so as to see the continuum as well as the one-particle spectrum, then one does not have strong convergence but only the weak form of convergence such as that used by LSZ. Moreover, convergence need not be as fast as above thanks to more slowly decaying contributions from the continuum. Basically, distinctions between the strong and the weak asymptotic condition lie in this seemingly subtle property, that is where you permit the test functions to lie. The limit state space that one obtains can be shown to have all the properties of the free collection of operators operating on the vacuum. These operators fulfill the standard kind of commutation relations and are just, as would be expected, the in-or the out-particle creating operators. Although we have chosen a particular t-plane in which to operate, one can make a boost to other frames and demonstrate that the theory is covariant. In these cases there are smooth forms of reduction formulae, dispersion relations and so forth. So that finally there is a connection between the Wightman Framework and the Haag-Ruelle form of scattering theory, and ultimately with dispersion formalism. At this level scattering theory can be compared with experimental results.

Another abstract development of great intrinsic import, which does not yet enjoy contact with experiment, has

taken place in this decade. It is also of great utility
for the constructive quantum field theory that I shall mention toward the end. This development is algebraic quantum
theory. Here one postulates an algebra of observables and
at the initial state of the game does not say what are the
entry requirements for an operator to be an observable. By
putting a norm on the elements of the algebra each observable becomes associated with a real number. We can think of
these as operators in Hilbert space and then the norm is the
operator norm. At the insistence that the particular norm
that is imposed fulfills the additional relation of line iii
in Fig. 4 the algebra becomes essentially a C* algebra. This
is the algebra that has become currently fashionable. C* algebras have the interesting property that for all faithful
representations, they are representation independent. It
doesn't really matter what choice of representation you start
with if you assign them a norm fulfilling line iii, you will
get the same kinds of operators in the algebra at the end.
To get back to physics from the algebra one introduces states.
The states are very analagous to forming the trace with a density matrix to yield the expectation value of an operator in
that state. Technically a state would be a positive continuous linear function; the meaning of linearity is clear while

ALGEBRAIC QUANTUM MECHANICS

i) A = Algebra of Observables

ii) Norm on A, $||A||$

iii) (C^*) $||A^*A|| = ||A||^2$

   (representation independent)

iv) States: members of dual space

   $\omega(A)$, positive, continuous, linear functional

   $\omega_o(A) = Tr(\rho A)$, & more

v) Weakly compact: $\omega_\alpha \to \omega$, i.e.

   $|\omega_\alpha(A_n) - \omega(A_n)| < \varepsilon_n$, $n = 1,...N$

   $(\alpha > \alpha_o)$

vi) GNS Construction

   $\omega(C) = <v_1, \pi(C) v_1>$

   $v_1(A) = \omega(A^*)$

   $(\pi(C) v_1)(A) = \omega(A^*C) = v_1(C^*A)$

   $v(A) = \sum c_n \omega(A^*B_n)$

   $<v,v> = \sum \bar{c}_n c_m \omega(B^*_n B_m)$

   cyclic, possibly reducible, possibly non-faithful

vii) Physical Equivalence

   $\pi(C) \longleftrightarrow \pi'(C)$

   $|\omega(A_n) - \omega'(V^*A_m V)| < \varepsilon_n$, $n = 1,...N$

   $\pi(K) = 0 \longleftrightarrow \pi'(K) = 0$

   All faithful representations are "equivalent"

Fig. 4

Synopsis of Algebraic Quantum Mechanics

positivity asks that an arbitrary positive operator has positive expectation values. The states are members of the so-called dual space which are the linear functionals on the algebra. This approach is doubly interesting because not only does it contain the density matrices familiar to us from the usual quantum mechanics but it contains more objects as well and these latter have been quite useful in statistical mechanical applications. Now one property of linear functionals on Banach algebras (which is what we happen to be dealing with) is that they are weakly compact. This means that any generalized sequence of states has a convergent generalized sub-sequence which will converge to some particular state. This is a very important property. Namely one knows that if one constructs, for example, a sequence in a butchered theory (i.e. in all sorts of cut offs in momentum space and real space), that the states constructed in this fashion will have a convergent generalized subsequence.

Now let me talk about the Gelfand-Naimark-Segal construction (step vi in Fig. 4) which as I mentioned earlier was useful in the Wightman reconstruction theorem. If there is given any state, (i.e. a positive continuous linear functional on the algebra) one can form a Hilbert space

in which the operator C is represented by some operator
$\pi(C)$ and the expectation is constructed by taking the
inner product with a vector v_1. In one particular functional realization of that Hilbert space, the arguments
of the vectors are the elements of the algebra itself.
If $v_1(A)$ is chosen as the function $\omega(A^*)$, the action of
the operator C upon v_1 is defined in the third line of
step vi of Fig. 4. The most general vector is a linear
sum of these kinds of quantities as shown in the next
line. The inner product of these vectors is then defined
in the last line. The theory has vectors, a definition
for inner product and a representation of the operators.
This particular construction is cyclic in the sense that
all the vectors in the Hilbert space can be obtained by
application of the operators of the theory to the particular state v_1. It may not be an irreducible representation of the algebra and it even may not be faithful. Several of the operators may be mapped into the zero element.
These concepts become physically more interesting when one
talks about the physical equivalence of theories. We could
have two representations $\pi(C)$ and $\pi'(C)$ constructed from
two different states, ω and ω'. One could ask if they are
unitarily equivalent. In other words, is there a matrix V
which connects them for all C. However, experimentalists

do not measure all quantities. They only measure a finite number of things, thus a different kind of definition of equivalence must be introduced. Suppose we have operators A_n that we wish to measure, and when we measure we impose only certain accuracies ε_n on our measurements. There is only a finite number of these operators. Then we ask, can we find some V which connects these measured values in our space with the measured values in the other space to the stated accuracies (Fig. 4). If this is the case, it is a much weaker requirement than unitary equivalence. The criterion for physical equivalence is not full unitary equivalence but simply the fact that the kernels (that is those operators K mapped into 0) are the same for the two theories. If we restrict ourselves to faithful representations (for which nothing but the 0 operator goes into 0) they are all physically equivalent representations. From this point of view the various representations for operators should not be given any great physical significance and all of them belong to a common physical framework.

How does all this fit in with quantum theory? We have been talking of quantum mechanics, a rather general thing. Let's talk about quantum theory. By this I mean the new frame of axioms introduced by Haag and Kasler and shown in

Fig. 5. (I should emphasize that the C* algebraic approach to quantum mechanics was studied already in the past decade by Segal). In this framework, one associates with each bounded space-time region (again, this particular room and the duration of this talk) an algebra of observables and

ALGEBRAIC QUANTUM THEORY

I. $0 \to \mathcal{U}(0), \; \mathcal{U} = \overline{\bigcup \mathcal{U}(0)}$

II. $0_1 \subset 0_2 \to \mathcal{U}(0_1) \subset \mathcal{U}(0_2)$

III. $0_1 \sim 0_2 \to [\mathcal{U}(0_1), \mathcal{U}(0_2)] = 0$

IV. $A \stackrel{L}{\to} A^L, \; \tau_L(\mathcal{U}(0)) = \mathcal{U}(L0)$

V.   Spectrum, Continuity in L of states

Automorphisms: $\mathcal{U} \stackrel{\tau_\lambda}{\to} \mathcal{U}$

i) Local: $\tau_\lambda(\mathcal{U}(0)) = V_\lambda(0) \, \mathcal{U}(0) \, V_\lambda^{-1}(0)$

$V_\lambda(0) \in \mathcal{U}; \; V_\lambda(0) \notin \mathcal{U}; \; \underline{No} \; V_\lambda(0)$

ii) Global: $\tau_\lambda(\mathcal{U}) = V_\lambda \, \mathcal{U} \, V_\lambda^{-1}$

$V_\lambda \in \mathcal{U}; \; V_\lambda \notin \mathcal{U}; \; \underline{No} \; V_\lambda$

$(V_\lambda \notin \mathcal{U} \to V_\lambda \in \mathcal{U}'')$

Fig. 5

one requires that the full **algebra** should be that obtained by taking the union over all possible local algebras suitably closed in the norm. The rest of the postulates are evident, physically desirable quantities. If one space-time region is space-like related with a second space-time region, then it is plausible that the associated algebras should commute (Line III). There is a Lorentz postulate: namely that for every observable A there should be a Lorentz transformed observable A^L. This postulates a kind of a map of the A's onto themselves which we can implement by some automorphism. This should be a rule for assigning the new operator which should be given alternatively by mapping the space-time region itself. This concept of automorphism, somewhat similar to the concept of unitary transformation in traditional quantum mechanics, takes on several different guises.

Let's consider a general map of an algebra into itself by a means of an automorphism. Suppose we look at this problem from both a local and a global point of view. The local point of view asks the question, can we realize the particular automorphism of this particular space-time region by means of a unitary transformation? There are three possible answers:

1. The unitary transformation we have is itself a member of the algebra. It is an observable.
2. The V is not a member of the algebra. If the V is not a member of the algebra, I will assume it is a member of an expanded family of operators. This expanded algebra is obtained by relaxing the conditions on the norm, and is a larger algebra than the one with which we began.
3. A third possibility is that there is no V at all. You cannot realize this map by a unitary transformation.

The same story arises in the global case. When there is a global automorphism, can we, or can we not realize it by a unitary transformation? The unitary transformation may be within the algebra, it may be within its expanded form, or there may be no unitary transformation at all. Well known examples fall into these categories. The Lorentz transformation for example can be realized by a unitary transformation in the extended algebra. This is true for both the local and global requirements. A case of no unitary connection between two local algebras is two different mass fields in a 5 dimensional space-time. The situation in which the local V exists but not the global V corresponds to broken symmetry. We heard in Professor Nambu's talk that broken symmetry leads to massless particles, but this is not an absolute rule and there are exceptions. So it is not surprising that in the axiomatic framework the most positive result is sort of an anti-Goldstone theorem shown in Fig. 6. If the current does not connect to a massless one particle state then the automorphism on the algebra is indeed representable by a unitary transformation and the generator of this transformation is the formal integral of the current over the space-like surface. In other words if you haven't broken the symmetry, then the generator survives as an operator.

Anti-Goldstone Theorem

$$\text{If } \langle \underset{\sim}{k} | j_\mu | 0 \rangle = 0$$

with $|\underset{\sim}{k}\rangle$ a massless one particle state,

$$\text{then } \tau_\lambda(\theta) = V_\lambda \theta V_\lambda^{-1}$$

$$\text{with } V_\lambda = e^{i \int j^\mu d\sigma_\mu}$$

Fig. 6

Let us go on to the very interesting results that have been obtained in actually constructing model field theories. These results are summarized in Fig. 7. There is, as mentioned earlier, the neutral scalar field with a quartic interaction in a world with 2 space-time dimensions. This is a superrenormelizable theory. The boson self energy is a finite graph in two dimensions because there are two loops or d^4p effectively upstairs and p^6 downstairs. The only divergent term is the vacuum energy term due to a volume divergence. Therefore one is dealing with a rather well behaved theory. The basic idea of the approach is as old as quantum field theory itself. One takes the original Hamiltonian and introduces the necessary cut-offs to make it completely well defined. For example one introduces a momentum space cut-off that is like a convolution of the field. You smear out the field and only work with this smeared field in the interaction. Also you have a space cutoff for the interaction so that the interaction is effective over only a finite distance. This double cut-off makes the potential a well behaved quantity. When added to the free Hamiltonian it can be shown that the combined terms do yield a well-defined theory. It would seem obvious that two

CONSTRUCTIVE FIELD THEORY

i) Cut-Offs: $H_\kappa(g) = H_0 + V_\kappa(g)$

Space Cut-Off
$$V_\kappa(g) = \int g(x) : \phi_\kappa^4(x) : dx$$

Momentum Space Cut-Off
$$\phi_\kappa(x) = \int \kappa(x-y) \phi(y) dy$$

Butchered Theory Defined:
$$: (\phi^4)_2 :$$

ii) $\lim_\kappa \exp(-itV_\kappa(g)) = \exp(-it V(g))$

iii) $\lim_\kappa (H_\kappa(g) - z)^{-1} = (H(g) - z)^{-1}$

$$H(g) \geq -E(g)$$

iv) Local Field:
$$\phi(x,t) = e^{itH(g)} \phi(x) e^{-itH(g)}$$

$$g = \underline{const}. \text{ over } |x| + |t|$$

$$e^{-itH(g)} = \lim_n \left(e^{-itH_0/n} e^{-itV(g)/n} \right)^n$$

v) Space Cut-Off Removal:
$$\omega_g(A(t)B) = (\Omega_g, A e^{-tH(g)} B\Omega_g)$$

$$\to \omega(A(t)B)$$

vi) GNS Construction:

$H(g) \to H$, $\Omega_g \to \Omega$

$H \geq 0$, $H\Omega = 0$

$[H, P] = 0$, $K(?)$

Fig. 7

operators A and B should have a well-defined sum, but this is not always the case and one must study that. In any case, in the butchered theory, with the double cut-offs, H is well defined. This point we had essentially reached years ago. What's new is an honest attempt to remove the cut-off and to show that the resultant quantities can now be well defined and satisfy the expected Lorentz properties.

We would like first to remove the momentum space cut-off. This process is summarized in line ii), iii) and iv) of Fig. 7. Consider the limit as κ goes to a delta function. Since V is an unbounded operator, we would not like to work with it directly, but let us analyze its unitarily associated operator by putting V up in an exponent. With the aid of this elementary transformation, one can show that one can in fact remove the momentum space cut-off and one is left with a bona fide local potential. The operator V(g) that one has at the end of this process is unbounded above. That's obvious because it is like ϕ^4. It is also unbounded below because of the subtraction terms that come from the Wick ordering.

In the next case one wants to know whether the total H operator can converge as the momentum space cut-off is removed. Here I have indicated that the resolvant is a quantity which is useful to study. Take the operator in question, subtract a complex number z with Im z ≠ 0, and invert the quantity. $(z-H_K)^{-1}$ is a well-defined, bounded operator. Then study the limits of this family of operators as the cut-off is removed. If I may steal from the Texan Neal Armstrong this may seem like one small step to a physicist but I assure you it is one giant leap for a mathematical physicist. I will briefly indicate what is involved. The resolvant $(z-H_K)^{-1}$ is dependent on the momentum space cut-off. One studied the difference between two resolvents. It satisfies a familiar resolvent-type equation:

$$(z-H_K)^{-1} - (z-H_{K'})^{-1} = (z-H_{K'})^{-1} (V_K - V_{K'}) (z-H_K)^{-1}.$$

This difference is expressed in terms of the product of one resolvent as each of the outside terms and the difference of the potentials inside. What one has to show is that the resolvent difference converges using the Cauchy kind of criterion. Then one is happy that the limit does exist. Other technical ingredients are necessary which I'm not spelling out. It would be direct if the V_K were a family of uniformly bounded operators. Then these bounded operators would already converge by themselves. If one can show that $V_K(z-H_K)^{-1}$ were a uniformally bounded operator, then things would also go through. However, this is not the case in the ϕ^4 theory and more sophisticated methods are necessary. Let me be very brief and indicate what is going on. We categorize functions depending on their ability to be integrated over some finite measure. That is, certain functions can appear once or twice or three times and still provide a finite integration. In other words, $\int |\psi|^r d\mu < \infty$. The larger r the "better" is ψ. The resolvent is a good operator since it takes good functions to better functions. This is due to the H in the denominator. The difference in the potentials takes better functions to poor functions. The final resolvent takes the poor functions and makes them good again. Crudely speaking, these are the kind of estimates involved in showing that such a quantity actually does converge.

$$||(z-H_K)^{-1} - (z-H_{K'})^{-1}|| = ||(z-H_K)^{-1}(V_K-V_{K'})(z-H_{K'})^{-1}||$$

Good ← Poor ← Better ← Good

As a consequence of this one finds that the Hamiltonian exists as a self-adjoint operator as proved by Glimm and

Jaffe. It will define a unitary transformation and it is bounded below in its energy. There is a lowest energy to the problem and this can be reincorporated on the left-hand side to readjust the lowest point of the Hamiltonian. This function $E(g)$ is the only divergent quantity. As g, the spatial cut-off, is removed this vacuum energy would tend toward infinity.

Armed with a bona fide local Hamiltonian, but one still dependent on the space cut-off, we can construct a local field. To begin with take the field $\phi(x)$ that one started with and map it in the usual fashion to define the field $\phi(x,t)$. Suppose g has a uniform value over a large enough region and that region is sufficiently large so that from the point x expanded into +t and -t it was still uniform in the causal-influence region of x, then one would expect that $\phi(x,t)$ would not even care that g dropped off outside of the causal region. A plausible argument, but it needs to be supported. Here again, following suggestions of Segal and employing the property of the self-adjointness, one can argue that it is in fact the case. Here the idea is to use a so-called product representation where we form the exponentional of $H(g)$ as the iterated relation shown in line iv of Fig. 7. H_o and V appear separately. Ultimately we will take the limit as n goes to infinity. But for now forget the limit, we will go to the limit at the end. What happens is that those terms in H_o and V act independently. First H_o operates then V, then H_o etc.. H_o has the effect of propagating the operator in a causal way because that's the way H_o is constructed. In V we recognize that, since the field will commute with itself at equal times and since V is a construct of the field, that it can do no more than spread out the influence by an arbitrarily small amount. Once we've done that the next H_o comes in and propagates causally. We can choose the ε arbitrarily small, in fact we could link it with the n in such a way (i.e., $\varepsilon = n^{-2}$) that when the limit $n \to \infty$ is taken, the region of propagation comes back to a perfect cone. Thus, we have constructed a local field which has its time evolution locally implemented by a unitary transformation.

The space cut-off g is still there. We get rid of that with a space cut-off removal, and here comes the import of the C* algebra techniques which were discussed earlier. With the aid of the local field lets build a C* algebra. For example, we could take spectral resolutions based on $\phi(x,t)$ smeared over space and time but confined to some particular bounded region. Let A and B be operators of these local algebras. We form expectations in the vector Ω_g, the vacuum vector which is present when one still has the space cut-off. It's a lengthy argument to show that not only is there a vacuum vector when the space cut-off is still present but that it is unique. However, armed with this information one is assured that $\omega_g(A(t)B)$ in line v of Fig. 7 exists. Then the weak compactness property that I talked about before (i.e. every generalized sequence will have a convergent generalized subsequence) insures that these states will converge to something as the space cut off is removed. Regularity properties of that convergence, which are under control, insures that every good thing we want also follows.

In particular, the state ω to which we would apply the Gelfand-Naimark-Segal construction (recall that every state could be associated with a Hilbert space construction) has all the right properties. Moreover, H(g) will converge to the physical Hamiltonian through this artifice of constructing the C* algebra, and then taking the limit of states and reconstructing the physical Hilbert space, the vacuum vectors Ω_g will converge in the same sense. The physical Hamiltonian obeys good spectrum properties, annihilates the vacuum, and no longer has a space cut-off so that it has a fighting chance to commute with the space translation generator. There is a physical field operator $\phi(x,t)$. There is a physical momentum operator, and the physical ϕ and π satisfying the canonical commutation relations. All of this is summarized in Fig. 7. There are things that are still unknown about this model. The last dot on the i's has not been put on the boost generator, the only generator in this two-dimensional Lorentz group. In addition one does not know whether the vacuum is unique or whether the field fulfills the cluster property. This is an important ingredient for the independence of experiments done at CERN and at Brookhaven. In other words, expectation values of two far-separated measurements should factorize (cluster property). One does not know whether the representation of the physical ϕ and π is irreducible or whether it is reducible. Suppose some of those questions could be answered. Suppose we knew, for example, that the ϕ and π that we got in the physical space was irreducible. That would imply that every other operator could be constructed from it, in particular, the physical Hamiltonian. It would be so nice to be able to <u>look</u> at that Hamiltonian constructed in terms of the physical irreducible ϕ and π and to learn where we went wrong when we originally wrote H as a function of ϕ and π. Why didn't we know enough to construct the Hamiltonian originally in terms of ϕ and π? How is this construction different? Can we learn from the construction of the physical Hamiltonian in terms of these quantities how to by-pass some of this material and go directly to the physical theories? If that is possible then I think these kinds of techniques will be able to be extended to more interesting theories.

I think I should also comment on other models that have been studied. Besides the $(\phi^4)_2$ theory, there is the Yukawa theory in two dimensions (i.e. $(\bar{\psi}\psi\phi)_2$). Here already there is an infinite mass renormalization as soon as one goes to the momentum space cut-off limit. Thus one has a bona fide infinity. There is a theory discussed by Hepp $(\bar{\psi}\psi\phi^2)_2$ in which one has an additional charge renormalization although none of these theories are yet confronted with wave function renormalization. The $(\phi^4)_3$ again is a mass renormalization problem but no wave function renormalization problem. In a sense this is the golden era of the leading logarithmic divergence. One has not yet been able to attack a theory with quadratic divergence of its boson self energy, which would entail a logarithmic divergence for the wave function renormalization. Hopefully this will be rectified in the near future.

Let me use the remaining minutes to comment on a very nice theorem regarding the anti-commutation relations. Its import can be seen from the last remarks that I have made. One way to construct theories is to butcher the original formulation so that the Fock representation, i.e. the free field representation, actually applies to the problem. Once it is butchered you can put in the familiar representation and the "only" problem left is to remove the cut-offs. It would be very nice if we could start with the physical representation of the problem and construct the Hamiltonian that belongs with it. There are analogous statements for the anti-commutation relations and there are certain boundary rules that one must respect. In Fig. 8 we summarize this situation where # denotes either dagger or no dagger. Suppose one takes spinor fields which fulfill the canonical anti-commutation relations and smears them with smooth functions over the spatial variables, where the number of spatial dimensions s will play a role in this particular calculation. These now become bounded operators. This is a simple property of the fact that the

CANONICAL ANTICOMMUTATION RELATIONS

i) $\{\psi(\vec{x}), \psi^+(\vec{y})\} = \delta(\vec{x}-\vec{y})$

$\psi(f) = \int f(\vec{x}) \psi(\vec{x}) d^3x, \quad ||\psi|| = ||f||$

ii) Locality

$\left|\left| \left[\psi^{\#}(h), \{\psi^{\#}(g), \psi^{\#}(f,t)\} \right] \right|\right| \leq ct^s \, ||h|| \, \sup|g| \, \sup|f|$

iii) $s \geq 2$

$\lim_{t \to 0} \left|\left| \left[\psi^{\#}(h), \{\psi^{\#}(g), \dot{\psi}(f)\} \right] \right|\right| = 0$

iv)

$\lim_{t \to 0} \left|\left| \left(\frac{\psi^{\#}(f,t) - \psi^{\#}(f)}{t} \right) \left| \begin{array}{c} vac, \text{ or} \\ 1 \text{ particle} \end{array} \right> \right|\right| < \infty$

v) Irreducible

$i \dot{\psi}(x) = L\psi(x) + M\psi^+(x)$

$\psi(x,t) = L'\psi(x,0) + M'\psi^+(x,0)$

where $\psi^{\#}$ denotes ψ^+ or ψ.

Fig. 8

number operator for fermions has only the eigenvalue 0 or 1. The bound for these smeared fermion operators is given by the L^2 norm of the smearing function involved (shown on line i). Suppose one throws into this system of the canonical anti-commutation relations the properties of locality. Locality says spinor fields anti-commute for a space-like separation. One builds the object in line ii. There is an anti-commutator inside a commutator. One studies the norm of this collection of three bounded operators purely by geometry, purely by that fact that influence cannot propagate outside the light cone. One can learn that this quantity is bounded by t^s times some other quantities. For the next ingredient, throw in the fact that s will be greater than or equal to two, $s \geq 2$. Let's take the expression we have and subtract from it its value at $t = 0$. That is a harmless step because the equal time anti-commutator which is a c number drops out in the next commutator. Take this quantity and divide it by t so on the right side we are left with $(t)^{s-1}$. If s is two or greater we can let t go to 0 and learn that the right hand side is 0. So the limit of this expression as t goes to 0 is zero. Since $[\psi*(f,t)-\psi*(f)]/t$ becomes $\dot{\psi}$, it

would be nice to bring this limit inside. As I say you
never get anything for free in this game. There is a
price to be paid to bring the limit inside and make ψ.
If the vacuum and the one particle state are allowed
to see this limiting operation and still remain valid
vectors in Hilbert space then you can do this. The next
ingredient we could throw in would be irreducibility.
Let's suppose that the fields at fixed time form a complete family (i.e. everything can be constructed out of
them). Anything that commutes with all these fields must
be a multiple of unity. If $\dot\psi$ anti-commutator with ψ is
a multiple of unity it has no choice but to be expressible as a linear combination of ψ and ψ^+. So one learns
from these inputs that the field equation for ψ is trivial.
Thus to find $\psi(x,t)$ solve the equation of motion in line
v to find some <u>linear</u> combination of its value at time 0.
Hence, interaction cannot be obtained for pure spinor fields
which fulfill anti-commutation relations and locality in
space dimensions greater than two if these field are irreducible. While this elegant theorem of Powers has a rigorous foundation it is in accord with the heuristic predictions of perturbation theory which, under the condition
$s \geq 2$, always has at least a logarithmically divergent wave
function renormalization.

DISCUSSION AFTER KLAUDER'S TALK

JAUCH (GENEVA): Now my first question refers to your section 1 on scattering theory. You described the two different asymptotic conditions, the one with the strong convergence and the other with the weak convergence. Presumably a good asymptotic condition should enable us to prove certain things like unitarity for the scattering operator. Now I have never been able to prove unitarity for the scattering operator if you postulate weak convergence only. I have tried very hard until I learned from Kato that the proof is impossible because the theorem is wrong and, therefore, I would be very much interested to know what kind of additional assumptions you need in order to get unitarity in scattering theory in the case of the weak asymptotic condition that you formulated.

KLAUDER: Well, I think weak convergence is generally just not enough to secure the unitarity of the S-matrix and it probably may be model dependent, problem dependent, or whatever else has to be thrown in. I would hate to have to say that one throws in enough to make a strong convergence because that's not what you want--you want something in between.

JAUCH: Yes, that would be very nice to see. My second question concerns your section two on the algebraic approach to axiomatic quantum mechanics. Now, I think it would be very useful to emphasize the distinction between mathematical axiomatics and physical axiomatics. In the mathematical axioms, the only requirement which one has to fulfill is that the axioms are independent and sufficiently complete so that one can draw conclusions. In the physical axiomatics you should have an additional requirement satisfied, namely that the axioms have, if possible, a physical interpretation and physical motivation. Now this has always been my difficulty with the algebraic approach which you have described--it satisfies the first conditions very well, but I have difficulties with the second requirement; namely you have operations in your axioms which are extremely difficult to interpret physically. For instance, you have an algebra constructed over the complex numbers. You have an algebra, that means you have additivity of your operators. These are questions which for me are a handicap in axiomatic theory. I have always tried to axiomatize quantum mechanics in such a way that every axiom that one introduces has at least one physical interpretation. This is my second question. Why do you do it that way?

KLAUDER: There are, according to my electrical engineering point of view, many roads to approach these problems. It is not operationally clear to me that this is a minimum set of assumptions nor is it clear that each of them is transparently physical, although I think their motivation is quite clear. The complex numbers are a hang over from the Hilbert space--other people are investigating other schemes.

THIRRING (CERN): I was wondering how is the situation, is the S matrix non-trivial at the ϕ^4 theory?

KLAUDER: That is a wonderful question and it goes into the list of untreated things. One does not know if the one particle problem can be solved prior to going to the asymptotic limits. One does not know the spectrum of the energy. One does not have control already over where that one particle will sit.

GRAVITATION AND PARTICLE PHYSICS[*][†]

S. Deser

Brandeis University, Waltham, Mass.

Since this audience is primarily composed of particle physicists rather than general relativists, I will couch my discussion of the significance and status of general relativity in terms of the concepts of particle physics. Since this will keep me away from results in the more geometrical end of the theory, such as collapse and singularity theorems or the null infinity analysis of general relativity, I particularly want to stress that this "geometrical" approach has been a most fruitful one, in which much understanding has been gained in the last decade, and where most relativists have been working. However, the concepts are quite different from, and there is as yet little direct relevance to particle theory. At the other end of the spectrum we have just heard from Professor Salam about the possibility that the quantum gravitational field will cure all the ills of field theory.

---

[*]Invited talk presented at the symposium "The Past Decade in Particle Theory" at the University of Texas at Austin on April 14-17, 1970.

[†]Work supported by USAF OAR Grant 70-1864.

With his resounding propaganda still ringing in our ears, and since my own work[1] here fortunately falls just outside the past decade, I will not discuss this further.

The main experimental piece of news which has occurred in general relativistic spectroscopy is, of course, the observation by the Weber group[2] of a 2^{++} excitation[3] at roughly 0 GeV. (Very preliminary results also seem to indicate that there is no 0^{++} companion at this energy). I will not go into the experimental uncertainities, though this excitation is certainly a striking confirmation of the symmetry scheme proposed by A. Einstein some time ago[4]. Unlike most other experiments in particle spectroscopy, we don't know the location or nature of the accelerator responsible for production, although there have been speculations in this connection. As far as off-shell experiments are concerned, the predictions of the quasi-state approximations of Einstein's model seem to be very much holding their own in agreeing with the "four tests", so general relativity still seems to be the best model of gravitation. It appears as if the next year or so will give us really reliable data in both these areas, incidentally.

General Relativity as Self-Coupling

I will start my discussion by reviewing the physical significance of the gravitational field equations. To do this, I will sketch a very direct "derivation", which involves neither geometry nor fancy gauge invariance (that is, the general coordinate group), but rather consistency and universality arguments within a flat space, Lorentz invariant framework appropriate to our elementary particle viewpoint.

First of all, I recall the standard reasons for excluding lower spin fields: A massless spin zero field alone excludes light bending, as the spin zero field could couple only to the trace of the energy-momentum tensor, which vanishes for the Maxwell field. Spin one leads to repulsion between like masses, which is also ruled out empirically. The spin two field is the next candidate, and it can accomodate coupling to the energy density, that is, the energy-momentum or stress tensor of the matter field. Empirically, the range of this field seems to be at least of the order of the size of say a galaxy and no smaller than the solar system. So, zero is a reasonable approximation for the mass.[3a] Thus, we postulate an infinite range, spin two field. [There is as yet no compelling empirical reason to discuss tensor-scalar mixtures, in any case, their tensor part is the essential one].

Suppose we start with an action which describes a free massless pure spin two field satisfying a wave equation (for simplicity, I will omit details such as indices from my discussion here[7]).

$$I \sim \int (\partial \phi_{\mu\nu})^2 d^4x \rightarrow \; "\square"\phi_{\mu\nu} = 0. \tag{1}$$

The d'Alembertian in quotes is just the spin 2 analogue of the gauge invariant operator ($\square \eta_{\mu\nu} - \partial^2_{\mu\nu}$) in Maxwell theory. This theory possesses a very simple Abelian gauge invariance (which is the exact analogue of the gauge invariance of the free Maxwell field); an immediate consequence is an identically satisfied conservation law. That is, the vanishing of the divergence of the left side of the field equations is an identity:

$$\partial_\nu \; ("\square"\phi_{\mu\nu}) \equiv 0. \tag{2}$$

it corresponds, in Maxwell theory, to the identical vanishing of $\partial_\nu (\partial_\nu F^{\mu\nu})$.

The above identity implies that a linear local coupling of dynamical matter systems to this gravitation-carrying tensor field will be inconsistent. In other words, we can't have a linear theory of gravitation in any normal sense, and we must part from simple $j^\mu A_\mu$-like interactions. The inconsistency arises in the following way: A linear coupling of matter to the gravitational field results in equations of the form

$$"\Box"\phi_{\mu\nu} = T_{\mu\nu}^{\text{matter}}. \tag{3}$$

However, Eq. (2) requires that $T_{\mu\nu}^m$ be conserved. But the energy-momentum tensor of matter is no longer conserved by itself, because of its coupling to gravitation, which implies exchange of energy between the matter field and the gravitational field:

$$\partial_\nu T_{\mu\nu}^m = -\partial_\nu [T_{\mu\nu}(\phi) + T_{\mu\nu}^{\text{INT.}}] \neq 0. \tag{4}$$

The very insertion of the coupling $\sim \phi_{\mu\nu} T_m^{\mu\nu}$ leading to Eq. (3) has thus led to the contradiction of Eq. (4). The only local[5] way out of this dilemma is to postulate universal coupling of all energy to the gravitational field, including gravitational energy itself. This will mean a non-linear generalization of Eq. (3), which we now have to obtain. [We will see later that all of this is paralleled in the Yang-Mills theory, and for the same reasons]. While the above ideas were certainly familiar to Einstein, their explicit use dates from the fifties and sixties[6]. The derivation I am about to present will lead us uniquely to the Einstein equations in one single closed form step, using only the initial abelian gauge and Lorentz invariances of the long range free spin two field as input.

We write the action of a free massless spin two field in first order form in terms of a field $h^{\mu\nu}$ and a conjugate momentum $\Gamma^{\alpha}_{\mu\nu}$ which are to be varied independently:

$$I_o \sim \int [h\partial\Gamma + \eta^{\mu\nu}(\Gamma\Gamma)_{\mu\nu}]d^4x. \tag{5}$$

The first term in the integrand is the kinetic energy density term. This simple quadratic action leads to linear field equations: When we vary the field h, we find that a divergence of the field momentum vanishes,

$$\delta h \to \partial\Gamma = 0. \tag{6}$$

by varying the momentum Γ, we find the relationship between it and the field amplitude

$$\delta\Gamma \to \Gamma = \partial h. \tag{7}$$

These two equations together imply our old friend the free field wave equation, i.e., the "d'Alembertian" equation of Eq. (1),

$$"\square"h = 0, \tag{8}$$

which satisfies the identity Eq. (2).

If this is not to remain a trivial, uncoupled field with no physical consequences, we are, by our earlier argument, forced to put in on the right side of the field equation (8) at least the energy-momentum tensor corresponding to the initial action (5):

$$I_o \to T^{\mu\nu}_{(o)} \sim \Gamma\Gamma + \partial(h\Gamma); \tag{9a}$$

We thus want an improved set of equations,

$$"\Box" \, h^{\mu\nu} = T^{\mu\nu}_{(o)}. \tag{9b}$$

[To compute the stress-tensor $T^{\mu\nu}_{(o)}$ is a reasonably simple task. As we see in Eq. (9), $T^{\mu\nu}_{(o)}$ has two parts. One is the standard "$p^\mu p^\nu$" kinetic energy. The other is the derivative of the $h\Gamma$ term in I_o, which is typical of a higher spin field. This term doesn't occur for spins up to one, but is one of the things that messes up higher spin theories.]

At this first stage, we write down an action which results in Eq. (9), that is, an action which incorporates the energy density of the original field. Since the latter is quadratic, we add to I_o the cubic action,

$$I_c \sim \int h\Gamma\Gamma \, d^4x, \tag{10}$$

being prepared for further iteraction if necessary. At this stage the total action is

$$I_o + I_c = \int [h\partial\Gamma + (\eta+h)\Gamma\Gamma] \, d^4x, \tag{11}$$

which results in the field equations

$$\delta h \to \quad \partial\Gamma + \Gamma\Gamma = 0 \equiv R_{\mu\nu}(\Gamma);$$

$$\delta\Gamma \to \quad \partial h = (\eta+h)\Gamma. \tag{12}$$

In Einstein language, which we are carefully not using, the first of these equations would express the vanishing of the Ricci tensor. The other one is the usual relation between the affinity and the metric. This relation looks innocuous until you invert it:

$$\Gamma = (\eta+h)^{-1} \partial h. \qquad (13)$$

In this inversion are hidden all of the non-linearities of the Einstein theory through the matrix inverse $(\eta+h)^{-1}_{\mu\nu}$ of $(\eta+h)_{\mu\nu}$.

Now let's differentiate the second equation in (12). (You have to trust that all signs and indices work out as I go along). The result is immediately

$$"\Box" h = \partial \Gamma + \partial(h\Gamma). \qquad (14)$$

Using the other field equation in (12) we now recover the energy-momentum tensor $T^{\mu\nu}_{(o)}$ as the right side of (14)

$$\partial \Gamma + \partial (h\Gamma) = \Gamma\Gamma + \partial (h\Gamma) = T^{\mu\nu}_{(o)}. \qquad (15)$$

This theory as it stands clearly satisfies our initial demand that the free part of the energy-momentum tensor is included as a source.

What about the energy-momentum tensor which is generated by the cubic action term I_c we had to add in? Well, the miracle is (but it is not a miracle, it is achieved by judicious choice of variables) that this cubic action makes an identically vanishing contribution to the stress tensor:

$$T^{\mu\nu}_c \equiv 0 \rightarrow T^{\mu\nu}_{tot} = T^{\mu\nu}_{(o)} + T^{\mu\nu}_{(c)} \equiv T^{\mu\nu}_{(o)} \qquad (16)$$

[For the technically minded, I calculate all stress-tensors by the usual prescription -- which has nothing to do with geometry, but rather with Lorentz invariance and the conserved "Noether current" generating this invariance -- that they are the variations of the corresponding action with respect to an external metric $g_{\mu\nu}$, evaluated at $g_{\mu\nu} = \eta_{\mu\nu}$. This is just a simple way of arriving at $T^{\mu\nu}$ and does not endow $g_{\mu\nu}$ with any physical significance. In inserting the $g_{\mu\nu}$ dependence, I have the choice of specifying the tensor character of my variables (h,Γ) with respect to $g_{\mu\nu}$ in any (consistent) fashion. I have taken h to be a contravariant density and Γ a tensor, which makes it clear that the integrand of I_c, namely hΓΓ, is already a scalar density without having to insert <u>any</u> explicit $g_{\mu\nu}$ dependence. Thus, we have $\delta I_c/\delta g_{\mu\nu} \equiv 0$.]. Therefore, the iteration process stops right here, and I can replace $T^{\mu\nu}_{(o)}$ simply by the $T^{\mu\nu}$ of the whole field. At this stage, then, the theory already satisfies our demand of self-coupling, i.e., ("□"h) = $T^{\mu\nu}_{tot}$. It will not surprise you that it also is identical to general relativity.

That is, by simple identification of variables, calling the combination of the Minkowski metric and our spin two field a contravariant metric density our action (11) is precisely the full Einstein action:

$$I_E = I_o + I_c = \int g(\partial \Gamma + \Gamma\Gamma) d^4x, \equiv \int R(x) d^4x$$

$$g^{\mu\nu} \equiv \eta^{\mu\nu} + h^{\mu\nu}. \qquad (17)$$

In all this we have not looked for geometry. Everything has been dictated by the fact that if the field is to be coupled to anything, it must be also coupled to itself in the same well defined way.

So much for the derivation of the free field equation with self interaction, that is to say, of the Einstein equation without sources. When matter is present, we are inescapably led to the standard Einstein minimal coupling prescription. Namely, I claim that the matter action simply is to be rewritten by replacing the Minkowski metric by $\eta + h$:

$$I_m(\eta) \to I_m(\eta+h). \qquad (18)$$

The reason this substitution will do the trick is the following: on the one hand, what enters the gravitational field equation (14) is the variational derivative of the action I_m with respect to h. That is just the Euler-Lagrange nature of the equation. On the other hand, what we want in (14) is the matter stress tensor (in the presence of this coupling) and this is precisely the same $\delta I_m/\delta h$ derivative if and only if the coupling is introduced according to (18). For you will recall that the stress tensor of a system is just given by the prescription

$$T_m^{\mu\nu} = \delta I_m(\eta \to g)/\delta g_{\mu\nu}\big|_{g=\eta}. \qquad (19)$$

Thus, since $\delta I_m(\eta+h)/\delta h \equiv \delta I_m(g+h)/\delta g\big|_{g=\eta}$, the coupling (18) is correct. But the coupling prescription (18) just says that, although everything is taking place in flat space, every material systems feels "as if" it is living in a curved space with the metric $\eta + h$, so we drop the words "as if", and say indeed that we have been forced to a curved-space geometry. In this language the theory is invariant under the wider group which characterizes the Einstein theory. However, it stresses that the driving principle in the derivation was not the group invariance; it was rather the forced play of self-interaction. There were two invariances involved in the universal coupling (including self-coupling) result. One, the linear massless spin 2 gauge invariance implying the conservation identity, the other ordinary Lorentz invariance, whose conserved Noether current was $T^{\mu\nu}$.

Amusingly, one can do exactly the same thing to arrive at another geometrical theory which was originally discovered by by Nordström[8] a little bit before 1960. Nordström's theory was rederived in a modern form by Freund and Nambu[9] with a different motivation. Their theory demands a scalar field which is coupled to the only suitable quantity, the trace of the energy-momentum tensor. It is instructive that, unlike general relativity, this is not a forced play for a spinless system. However, if one does demand coupling to the trace of the energy-momentum tensor, then the above derivation can be carried through[10]. Simply start the whole process with the field variable $h^{\mu\nu}$ taken to be a priori a scalar:

$$h^{\mu\nu} = h\, \eta^{\mu\nu} \qquad (20)$$

The field equation which is then derived is equivalent to the vanishing of the scalar curvature for this conformally flat metric:

$$\Box h = T^\alpha_\alpha(h) \leftrightarrow R(h\eta) = 0 \qquad (21)$$

This Nordström field equation is identical in form to the Freund-Nambu field equation. Both derivations proceeded from the same demand but one was, so to speak, from a higher spin point of view while the other was directly at a scalar level.

Let me clarify the gravitational result by showing that the same procedure[7] can and must be followed to arrive at a Yang-Mills theory. We start with a triplet of free massless spin one fields which posesses constant rotational invariance in the internal space. This invariance leads to a conserved current according to the Noether theorem. If we wish this field to couple to a dynamical source, then because of the identical vanishing of the divergence of the left side of the field equation, we must incorporate the Noether current on the right. We are thus led to a cubic action of the same general type as in the Einstein case. We start from a primary action in first-order form:

$$I_o = \tfrac{1}{2} \int (\underset{\sim}{F} \cdot (\partial \underset{\sim}{A} - \partial \underset{\sim}{A}) + \tfrac{1}{2} \underset{\sim}{F}^2) d^4 x. \tag{22}$$

The invariance of the action under the spacetime-independent transformation

$$\delta \underset{\sim}{A}(x) = \underset{\sim}{A}(x) \times \underset{\sim}{\omega} \tag{23}$$

leads to the conserved Noether current associated with I_o,

$$\underset{\sim}{j}_{(o)} = \underset{\sim}{F} \times \underset{\sim}{A}. \tag{24}$$

To get self-interaction, we write a new field equation

$$\partial \underset{\sim}{F} = \underset{\sim}{F} \times \underset{\sim}{A} = \underset{\sim}{j}_{(o)} \tag{25}$$

the right side of which arises by adding the cubic action

$$I_c = \int \underset{\sim}{F} \cdot \underset{\sim}{A} \times \underset{\sim}{A} \, d^4 x. \tag{26}$$

Again, $\underline{j}_{(c)} \equiv 0$, so that $\underline{j}_{tot} = \underline{j}_{(o)}$. The total action is precisely the Yang-Mills action:

$$I_{YM} = I_o + I_c = \tfrac{1}{2} \int [\underline{F} \cdot (\partial \underline{A} - \partial \underline{A} + \underline{A} \times \underline{A}) + \tfrac{1}{2} F^2] d^4x. \qquad (27)$$

If you vary $\underline{F}$ in this theory you get the usual Yang-Mills relation between field strength and potential. Namely, we have $\underline{F} \sim \partial \underline{A} - \underline{A} + \underline{A} \times \underline{A}$. Note that this is far less non-linear than the Γ-h relation (13), because I_c is like $q^2 p$ rather than $p^2 q$ in gravitation. The actions, however, are both formally cubic, and the procedures are identical in both cases. If you vary with respect to $\underline{A}$, you get the field equations

$$\partial \underline{F} = \underline{F} \times \underline{A}. \qquad (28)$$

I also mention the fact that the Sugarwara model, which is the model that attempts to take the $T^{\mu\nu}$ route to dynamics without Euler-Lagrange equations, bears the same relation to the Yang-Mills theory as the Nordström-Freund-Nambu theory has to General Relativity Theory. That is to say, because it is really a scalar, not a vector theory, one may, but need not, introduce self interaction. If one starts with Lagrangian in which the F^2 term of Yang-Mills is replaced by A^2, and simply imposes a principle of self-interaction, one gets [11] the Sommerfield-Sugaware model of dynamical theory of currents in SU(2) or SU(2) X SU(2). The same principles are operative in all these cases.

DYNAMICS OF THE EINSTEIN FIELD

Let me now discuss the physical content of the self-coupled Einstein field. The past decade was the key period as far as understanding the nature of gravitation as a dynamical field is concerned. Remember that we atart out with a free massless spin two field which, therefore, has two helicity 2 degrees of freedom. The question is, has this been spoiled by the self-coupling in any way? Normally, we don't think of changes in the number of degrees of freedom to be induced by coupling. However, we must remember that here is a theory where the coupling fundamentally involves the kinetic energy and in fact the very definitions of what the dynamical variables are. Therefore, it takes some work to show that the number of degrees of freedom remains unchanged. Further, a naive calculation of the energy-momentum implies that it vanishes, not surprisingly, since translation generators can hardly function when all coordinate transformations leave the theory identically unchanged! So one also has to understand these physical quantities in the absence of any a priori preferred coordinate frames. The work was started by Professor Dirac,[12] and independently by Arnowitt, Misner, and myself[13] in the late fifties. This is the canonical Hamiltonian approach, which may be carried out despite the very complicated form of the

Einstein equations. As we have seen, these equations look simple in the cubic form I gave above, but they turn complicated when the affinity - metric relating equations are analyzed. As for the coordinate-invariance responsible for losing our initial energy-momentum, the solution is very instructive. Basically, general relativity is like a parameterized theory in dynamics, in which the independent invariable is an arbitrary parameter, and the associated "hamiltonian" therefore vanishes. Only here, there is no "real" time and one has to "deparametrize" into some particular coordinate frame to discuss dynamics and time evolution. Of course, one doesn't lose coordinate invariance by picking a frame (you couldn't if you tried). However, it is then important to understand the physical, invariant, significance of the energy, momentum, gravitational radiation, etc. Here the role of asymptotic boundary conditions is essential, and Cartesian frames in an asymptotically flat space *are* preferred, just as in special relativity, with respect to these concepts, which we have taken over from it.

One convenient choice of coordinates is the equivalent of the radiation gauge in electrodynamics. One then takes into account the cons-raint equations similar to the Gauss equation of electrodynamics. The net effect of this **is** that one still has two degrees of freedom. These can be thought of as the transverse-traceless (in the flat-space sense) spatial components of the metric, g_{ij}^{TT}, with conjugate variables π_{ij}^{TT} which are basically the time derivatives of g:

$$(p,q) \leftrightarrow (\pi^{ijTT}, g_{ij}^{TT}) \qquad (29)$$

[The transverse-traceless conditions simply remove four of the six variables in a symmetric rank two tensor]. The reason we come down to two pairs of variables is that four of the twelve (g,π) are gauge variables (like $\underset{\sim}{A}^L$ in electrodynamics), four others are constraints (like $\underset{\sim}{E}^L$) while the $g_{0\mu}$ are Lagrange multipliers (like A_o) determined by the gauge choice. The two degrees of freedom which are left have the obvious transversality and tracelessness properties that you would expect from examination of the linear theory. [In this language the electrodynamical variables are $\underset{\sim}{A}^T$ (or $\underset{\sim}{B}$) and $\underset{\sim}{E}^T$].

The full action intterms of these reduced variables (that is, the effective Lagrangian in radiation gauge, for example) can be written entirely in the usual $p\dot{q} - H$ form:

$$I \sim \int [\pi^{TT} \partial_o g^{TT} - H(TT)] d^4x \qquad (30)$$

In this form the (perfectly non-vanishing) energy density H is a function only of these transverse traceless variables.

Classically the conjugate variables π^{TT}, g^{TT} obey the normal Poisson bracket relations:

$$[\pi^{TT}(\vec{r}), g^{TT}(\vec{o})]_{PB} = \delta^{TT}(\vec{r}). \tag{31}$$

The right hand side of this relation is a delta function in the space of transverse-traceless tensors. So, the fundamental "non-geometrical" aspects of classical general relativity closely resemble the way we handle calssical field theory (and presumably also quantum field theory), and there are no surprises at this level. For example, one can show that when the boundary conditions of asymptotically flat space are fulfilled, there exist Poincaré generators which satisfy the (Poisson bracket) Poincaré algebra. In this way one shows that the theory which in the radiation gauge is of course not manifestly Lorentz invariant, nevertheless is Lorentz invariant. One may also handle gravitational radiation, etc. in a clearcut way, separating it from "coordinate waves" and in general, proceed to develop gravidynamics in parallel with electrodynamics. One interesting sidelight here is that the total energy of a gravitational system, including, of course, the gravitational self-energy, has the characteristics we have come to expect from our derivation of the field equations: Instead of the Newtonian expression, $m = m_o - (\gamma m_o^2/2\varepsilon)$ where ε is the characteristic size of the system, we get, in the Einstein expression, the full energy m rather than m_o appearing in the self-interaction:

$$m = m_o - \gamma(m^2/2\varepsilon)$$

This has amusing consequences for classical point particle models - both purely mechanical ones where m_o is constant, and "classical electrons" where $m_o \sim m_o^{mech} + (e^2/8\pi\varepsilon)$. I leave it to the reader to solve the above quadratic equation and solve the $\varepsilon \to 0$ limits, and to draw his conclusion as to the significance of this (purely classical) result. The above form also suggests that while the attractive gravitational self-interaction decreases the total energy, it cannot bring an initially positive m_o below zero. When applied to a purely gravitational (source-free) configuration, it leads to a positive energy for the Einstein field[27,28] (this is expanded on in the discussion).

Quantization and Relations to Particle Physics

Let me now briefly comment on the question of quantization. This is an open ended subject in itself, and in a way I feel that Professor Salam's talk has relieved me of the necessity of saying anything about it, since according to him, the gravitational field seems to help all other fields to get quantized correctly. However, I mention that the problems involved occur at various levels. Aside from attempts to quantize in terms of the Hamiltonian method that I discussed here, there have been quite different attempts by DeWitt, Faddeev, Feynman, Mandelstam, Schwinger, and other people.[14-19] It is not altogether clear that everyone is talking about the same quantum relativity, or that there is one and only one such consistent system. Are there different ones corresponding to different formulations? Is there one which follows the standard Hamiltonian form? I don't know. I do know that one can quantize canonically in a non-rigorous way by pressing the usual buttons, which is about all we know for Yang-Mills or even coupled electrodynamics. Then there are some problems specifically due to the gauge invariance which arise here, but are already present at the much simpler level of Yang-Mills theory.[19] If the quantization proceeds by Feynman methods or by the path integral methods of Mandelstam, certain spurious auxiliary fields are brought in, as Professor Salam mentioned. Nonetheless these methods do seem to lead to a consistent scheme. However, rather than worry about the morass of possible tricky problems, let us go on and ask, why does one quantize gravitation? Professor Rosenfeld has said that it is only because it is there, rather than on grounds of logical necessity. I would disagree with the latter, but agree with the former. Let us say, anyway, that quantization certainly is a tempting thing to consider once one has a dynamical system. A particularly provocative reason here is the appearance in the quantum theory of the characteristic length L, (which we may also consider as the Schwarzschild length of a mass M):

$$L \sim (G\hbar c^{-3})^{\frac{1}{2}} \sim 10^{-33} \text{cm} \; ; \; M \sim (\hbar c G^{-1})^{\frac{1}{2}} \sim 10^{-5} \text{gm}. \quad (32)$$

It is of course, a very ridiculous number (below which the classical Einstein equations no longer make sense). However, it is comforting that $L = 10^{-33}$ cm does not intrude on us if it acts as a cutoff. On the other hand, the early speculations of Pauli, Landau, and Klein[20] expressed the hope that the cutoff comes in some logarithmic way, and, therefore, perhaps actually makes some observable difference. Their approach was the historical antecedent to those described by Professor Salam. In any case, it seems clear that if the Einstein theory is not just an averaged macroscopic phenomenological description (and there is no reason to think it is), then it must eventually be quantized as a local field. More important than the rigor questions attached to any non-free-field theory will be the physical consequences. These will either occur through dramatic emergence of 10^{-33} cm (as in a logarithmic cutoff or a la Salam), or, in quite different way, through the non-quantized parts of the field which, like $\underset{\sim}{E}^L$ in electrodynamics, are functions of the other dynamical variables. These, through non-linearity, may become effective sooner; the virtual "coulomb" effects may be more important than virtual transverse "photons".

Quite aside from its quantization, the existence of the gravitational field means that the energy-momentum tensor of any matter system does have a direct observable significance. It is not merely a quantity whose integrals are used for Poincaré generators, but a current which, being coupled to a measurable field, has to be taken seriously. For example, one has been able to obtain some information from the use of equal-time commutation relations involving the energy-momentum tensor with itself or with currents. Conversely, the Einstein equations make sense only so long as the notion of a stress-tensor makes sense, at least as a local operator if not a c-number. At the moment we cannot even be sure about this below 10^{-13} cm, which is some twenty orders of magnitudes above our other length!

A somewhat different application of general relativity is to an old idea -- consistency of local higher spin (>1) fields. It seems to be the case that such fields cannot couple consistently[21] to gravitation, even by adding non-minimal interactions. Now, whatever else such a field may or may not couple to (e.g. photons), it must couple to the gravitational field. This inconsistency would be a nice way of accounting for the apparent absence of this type of particles. (I am talking, of course, about elementary local fields).

The last subject I will mention is another topical and amusing illustration of interaction between gravitation and elementary particle physics. This rapprochement lies in the area of scale and conformal invariance. In the absence of dimensional constants and masses in a theory, the stress tensor should have the general property of vanishing trace. This is indeed the case for massless spin ½ and the electromagnetic field. However, the energy-momentum tensor of the free scalar field does not have vanishing trace (except in two-dimensional spacetime). The question arises, can one find a "new, improved" tensor which has vanishing trace? This was the motivation for the work of Callan, Coleman and Jackiw.[22] They wanted to improve the properties of the matrix elements of the renormalized energy-momentum tensor.

Now formally, there is always freedom to add, to the conventional symmetric energy-momentum tensor, any symmetric tensor which is identically conserved and which is linear in the second derivatives of some function of the field. The most general addition which will not spoil the Poincaré algebra looks like $\Delta^{\mu\nu} \equiv (H^{\mu\alpha\,\nu\beta})_{,\alpha\beta}$ where H has the indicated symmetries. The new improved energy-momentum tensor is:

$$T_{\mu\nu}(\phi) = (\phi_{,\mu}\phi_{,\nu} - \tfrac{1}{2}\eta_{\mu\nu}\phi_{,\alpha}^2) - \tfrac{1}{6}(\partial^2_{\mu\nu} - \eta_{\mu\nu}\Box)\phi^2. \qquad (33)$$

The trace now vanishes because ϕ still satisfies the free field equation

$$\Box\phi = 0. \qquad (34)$$

From the point of view of a relativist, one cannot add random terms to $T_{\mu\nu}$, the physical source of the gravitational field, with impunity. The piece we have added in Eq. (33) is in fact, nothing but a nonminimal coupling of the scalar field to gravitation. The action of such a theory is given by a famous equation of a conformally invariant scalar field:

$$I \sim \int [g(\partial\phi)(\partial\phi) - \tfrac{1}{6}R\phi^2]d^4x_o \qquad (35)$$

This means that in the absence of gravitation the curvature R won't appear in the field equations, but will change the stress. Eq. (33) is one which, under various guises, has been studied by many people on both sides of the gravitation-particles curtain. Gürsey[23] used (35) in a very beautiful paper to study whether Mach's principle can be accomodated within general relativity. Incidentally, if one replaced this coefficient 1/6 in Eq. (35) by an arbitrary constant one would obtain effectively the Brans-Dicke theory.[24] The particular value of this constant that Brans and Dicke favor is, however, quite different from the number we have here.

A number of people[25] have very recently been looking at gravitation starting from a form like (35). In my approach, there is no "bare" gravitational term, and we have a purely conformal theory which permits coupling only to systems with scale invariance:

$$T_\alpha{}^\alpha(\text{matter}) = 0. \tag{36}$$

Scale invariance is thus a consistency requirement in the same way conservation of current is a consistency requirement in electrodynamics. Next, in the spirit of field theory, one immediately goes ahead to break the symmetry. One simple way to break it is to add a term which involves a range parameter μ^{-1}:

$$I' \sim \tfrac{1}{2} \int \mu^2 \phi^2 d^4 x. \tag{37}$$

It is natural to make the range of the order to the radius of the universe, both to avoid experimental conflict and to tie scale invariance breaking to the finiteness of the "ultimate" comparison system. One then gets a theory at the cosmological level with roughly the right gravitational constant and the right cosmological term (if one were sure from experiment that there is a right cosmological term), these being determined in a Machian way from the average mass density of the universe. The model resembles that obtained within a purely general relativistic framework in Gürsey's paper.[23]

I cite this program not for its promise (it could be called "π in the sky" for several obvious reasons), but as our example of a motivation from elementary particles which leads to a theory on the cosmological level. I suppose this is particularly appropriate to the conformal group, which inverts large and small distances.

Predictions of future developments are notoriously useless, but there are certainly plenty of open problems in both classical gravitation and its relation to particles. For example, realistic quantum fields and classical general relativity have in common that in neither has a non-trivial, non-singular bounded solution been ever set forth. In general relativity one doesn't yet have an example of a system which starts out at $t = -\infty$ with some weak distribution of gravitational excitations and comes out with an analogous distribution at $t = +\infty$ (though there is little doubt that such solutions exist). The strongest results one has are of the more destructive kind -- collapse theorems. Their relevance to the real world depends, in my mind, on the extent to which quantum effects leave them applicable, as both $T_{\mu\nu}$ and $G_{\mu\nu}$ must begin to be treated as operators (at 10^{-13} and 10^{-33} cm?). Also these theorems become operative at incredible matter densities where we don't have any real idea about the validity of our usual "low-energy" $T_{\mu\nu}$'s. Some preliminary model gravitational quantization effects have just recently been looked at in this connection[26]. Of course, if gravitation has relevance as a universal regulator field, as Salam suggests, the two areas will become very directly linked. In any case, it is a safe bet that the interplay of gravitation with astrophysics is going to become very important as one is forced to deal in detail with very dense systems (whether neutron stars or the universe at birth), and as, for a change, experimental data really starts coming in from all those accelerators way out there.

REFERENCES

1. Deser, S., Revs. Mod. Phys., 29 417 (1957).
2. Weber, J., Phys. Rev. Letters, 22 1320 (1969).
3. Nauenberg, M. (U. C. Santa Cruz): Is this an announcement of the spin and parity of the new particle? Deser: Only Professor Weber can make such an announcement.

3a. Very recently the implications of finite gravitational range have revealed some extremely unexpected features [H. Van Dam and M. Veltman, Nucl. Phys. B22, 397 (1970), D. G. Boulware and S. Deser, to be published.

4. Einstein, A., Sitzungsben. Kon. Preuss. Akad. Wiss. Berlin 1917, 778.
5. There is a consistent non-local possibility [S. Deser and B. L. Laurent, Ann. Phys. (N.Y.) 50, 76 (1968)] but it only reinforces one's belief in Einstein's wisdom.
6. Kraichnan, R. H., M.I.T. Thesis (1947) and Phys. Rev. 98 1118 (1955).

   Papapetrou, A., Proc. Roy. Irish Acad. 52A 11 (1948).
   Gupta, E. N., Proc. Phys. Soc. London 65A, 608 (1952).
   Feynman, R. P., Proc. Chapel Hill Conf. (1956).
   Thirring, W., Fortschr. der Physik 7 79 (1963).

Halpern, L., Bull de L'Acad. R. de Belgique, Cl. des Sci. 49 226 (1963).

7. Deser, S., Gen. Rel. and Grav. 1, 9 (1970).
8. Nordström, G., Ann. d. Phys. 42 856 (1913); 43 533 (1914).
9. Freund, P. G. O., and Nambu, Y., Phys. Rev. 174 1746 (1968).
10. Deser, S., and Halpern, L, Gen. Rel. and Grav. 1, 131 (1970).
11. Deser, S., Phys. Rev. 187, 1931 (1969).
12. Dirac, P. A. M., Proc. Roy Soc. London A246 333 (1958).
13. A survey of and references to our work is provided by: Arnowitt, R., Deser, S., and Misner, C. in Gravitation: An Introduction to Current Research, ed. by Witten, L. (Wiley, New York 1962).
14. DeWitt, B. S., Phys. Rev. 162 1195, 1239 (1967).
15. Fadeev, L. D. and Popov, V. N., Phys. Letters 25B 29 (1967) and unpublished.
16. Feynman, R. P., Acta Phys. Polon. 24 697 (1963).
17. Mandelstam, S., Phys. Rev., 175 1604 (1968).
18. Weinberg, S., Phys. Rev. 138B 988 (1965).
19. Schwinger, J., Phys. Rev. 127, 324 (1962); 132, 1317 (1963).
20. See for example, Landau, L. D. in Niels Bohr and the Development of Physics (McGraw Hill, New York (1955)); Klein, O., and Pauli, W. in Helv. Phys. Aca Suppl. 4 (1956), and ref (1).

21. Aragone, C., Nuovo Cim. **64A**, 841 (1969) and Aragone, C., and Deser, S., Nuovo Cim. (in press).
22. Callan, C. G., Coleman, S., and Jackiw, R., Ann. Phys. (N.Y.) **59**, 42 (1970).
23. Gürsey, F., Ann. Phys. (N.Y.) **24**, 211 (1963).
24. Brans, C., and Dicke, R. H., Phys. Rev. **124**, 925 (1961).
25. e.g., Schwinger, J. -- <u>Particles, Sources and Fields</u> (Addison-Wesley 1970), Deser, S., Ann. Phys. (N.Y.) **59**, 248, (1970).
26. Misner, C. W., Phys. Rev. **186**, 1319 (1969); The most recent review article on quantization is: Brill, D. R. and Gowdy, R. H. -- Reports on Adv. in Phys. **33**, 413 (1970).
27. Brill, D. and Deser, S., Ann. Phys. (N.Y.) **50**, 548 (1968).
28. Brill, D., Deser, S., and Fadeev, L. D., Phys. Lett. **26A**, 538 (1968).
29. Deser, S., Nuov. Cim. **55B**, 593 (1969).

DISCUSSION OF PROFESSOR DESER'S TALK

F. C. Michel, (Rice): I wonder if I can get a comment from you. It seems curious that the spin of the gravitational field is not determined by the general principles which one ascribes to the field, in other words, equivalence and conservation of energy or what have you, but rather by experiment (and I might say, parenthetically, not by Weber's experiment, but by some rather more ancient ones). I believe this is in exact analogy to particle theory, to some extent. What speculations could you offer with regard to gravitation?

Deser: My comment would be that the spin determination does come both ways. Weber sees quadrupole and not monopole oscillations. Theory, it seems to me, also _does_ force a spin on you: There is the half-empirical, half-theoretical fact I stated that you cannot have a spin zero field or spin 1 field. Of course, there could logically be universes different from ours, (e.g., a Nördstrom one), but _we're_ necessarily in the **tens**or business. There need not be an electromagnetic field as far as we know. But there does have to be a geometry; that's perhaps a little bit more fundamental. If you say it that way then you've got the spin 2 in a more "forced-play" fashion.

There is incidentally another nice derivation of the Einstein equations which starts out in terms of small spin

2 excitations in a background geometry. The weak excitations cannot consistently constitute a dynamical system in a background field unless they are the variations of that background metric. Here you get to the Einstein equations as the result of the existence of a geometry and of a massless spin 2 excitation in its presence.

Salam, (Imperial College): We know of the existence of f^o particles of spin 2. One might expect, if there is quantum gravity, the f^o and gravitons will intermix like the ρ^o and the photon. This may presumably even have the effect of "fast" energetic gravitons interacting with matter, not through the volume, but through the surface (converting to f^o's and then the f^o's interacting on the surface). You may get a weaker gravity if one is dealing with "fast" gravitons at very high energies.

Deser: I think the idea of the f^o-graviton mixing is a very nice one. I must say that although I understand it at the level (which is sufficient, practically) of linear gravitons, I've never quite understood it at the point where the gravitons represent the full theory. But that's just because I haven't really spent enough time on it. [This has since been carried through further -- see Proc. 1971 Coral Gables Conference.

Segal, (MIT): What is the current situation on the positivity of the energy of the field?

Deser: Thank you for asking that question, which I should have taken up in the Hamiltonian discussion. It is one of the essential physical requirements on a dynamical system (and particularly if it is to be quantized) that it have positive energy and a timelike four-momentum ($m^2 \equiv -p^\mu p_\mu > 0$). For general relativity, a direct proof is probably impossible since the energy is such an implicit, non-linear, functional of the dynamical variables, but Brill and I[27] and Brill, Fadeev and I[28] gave two different variational derivations of $E > 0$, based on the behavior of the energy under small changes of the dynamical variables (which is complicated enough). Brill and I showed that the gravitational field (like any "good" dynamical system) has only one extremum; that extremum is a minimum; and that minimum occurs at vacuum, that is flat space (which has zero energy). We also showed that every extremum has zero energy. Furthermore, there are real theorems which insure that functionals with the above properties are indeed positive, aside from possibly, questions of whether the "manifold of geometries" is one to which the theorems apply as is. I mention also that, whenever one can calculate energy explicitly (this includes some rather general systems), it is positive. Likewise, a similar variational argument may be given[29] that $m^2 > 0$. So the four-momentum seems to have the good properties of being forward timelike, at least classically. Were this not true, quantization would be in real trouble as existence of a single negative energy state in relativity implies that there is no lower bound to energy at all (and there's no way to adjust the zero of energy in this game!).

<u>Nauenberg (Santa Cruz)</u>: This is a question to Professor Salam. As I understand it, there are questions in renormalizing--that is in making the quantized General Theory of Relatively finite. I don't recall that you discussed these problems in your talk. I mean, how do you solve the difficulties that Feynman and other pointed out about the quantized theory and how to make it finite?

<u>Salam</u>: I took the Feynman prescription by writing down matrix elements, that is, using his action principle. I think the difficulties in quantizing gravity arise because we are so fond of the canonical formalism. If you give that up and assume that there are other ways of quantizing theories, like the Feynman path integral method of doing it, then Feynman does give you a complete description of how to quantize the theory and write down matrix elements. The result of this (Fadeev and Popov have done this more carefully) is that you get a set of rules; those rules give

you the propagator for the graviton. Then in the graviton-graviton loops you have to add what are called "fictitious particles". This already happens for the Yang-Mills theory; this is nothing foreign--this is to preserve unitarity. And I wrote down the effective Lagrangian of those fictitious particles. Those fictitious particles are nothing very strange--they are just the extra degrees of freedom which have to be integrated out from the gravitational field if you only need two degrees of freedom. No matter how the other ones are floating about, they conspire to make it look as if you are introducing fictitious particles. Feynman, Dirac, Mandelstam, Fadeev, etc. all agree. And they have a very well defined set of rules for quantized gravity. There are no infinities at this stage. The infinities arise if you insist (as everybody else, like Feynman and Mandelstam and so on, have done) in not distinguishing between, the covariant and the contravariant tensors. If you distinguish between them then you get this sort of interaction: $\sqrt{-g}(g^{\mu\nu}...)$ in which $\sqrt{-g}$ is the crucial helpful factor. You express the covariant tensor, the other one, in terms of this one. This object then looks like

$$\frac{1}{1 + Kh} \, .$$

if I write

$$g^{\mu\nu} \sim 1 + Kh.$$

Therefore the effective Lagrangian (for example, for an electron-photon interaction) with the interposition of this "g" becomes

$$\frac{\bar{\psi}\gamma_\mu \psi A_\mu}{1 + Kh}$$

where h is the gravitational field, the field which I was talking about. The matrix element which I will compute, for example for electron-electron photon and millions of gravitons (h is going across):

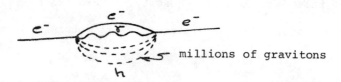

is very trivially written out in the form which I had. This is:

$$\int \frac{\frac{1}{x^5}}{1-K/x^2} d^4x,$$

so that the infinity at x = 0 has just disappeared. What I was trying to stress was that whereas in the past there were conjectures that this cutoff would help, we can now actually, by a very simple calculation show that these conjectures, to the second order, are absolutely right. One has now to check among the higher orders; and this is the program that one will hopefully undertake. This is the effective cutoff that comes in--it is not a cutoff just put in by hand, but just smears the place x^2 (where $x^2 = 0$ gives infinity). If this term wasn't there you'd have $\frac{1}{x^5}$ which would be the normal linear divergence. There's e^2 in front; I've done nothing to the e^2; all I've done is to the infinity which is smeared out visibly, not just by saying that the curvature is taken.

Freund: On the dominance of gravitational form factors and the non-linear aspect of the theory: The original way I formulated this point was to say that matrix elements of the energy momentum tensor are dominated by f-mesons. At that level you would say that the non-linear aspects do not appear very clearly in any way. You can formulate it in a field theoretic way, for instance taking the Coleman, et al tensor and constructing a theory where you get a field-current identity. The Coleman, et al tensor should be identical to the f-meson field. Then the source of the gravitational field will be just the f-meson field, the hadronic part, and then the non-linear aspects of course come clearly into play.

Deser: And the energy of the f-meson field itself; that is not a separate thing now? You have now the f-field and the ϕ-field, right?

Freund: That's right.

Deser: And so you have a stress tensor of the f-field besides, presumably?

Freund: That's right.

Deser: So you have both of them entering in. Incidentally this new improved stress tensor is already beginning to crack a little bit; one has to renormalize the 1/6 for example, so that...

Freund: Well, uh...

Deser: ...the diseases of field theory at small distances are not all cured yet.

Freund: There are besides, other ambiguities I understand, in this tensor which still have to be taken care of.

Rosenfeld: Well I think everybody's hungry so I'm closing this session and thanking all the speakers.

WEAK INTERACTIONS*

Jan Nilsson

Institute of Theoretical Physics

Goteborg, Sweden

Before reviewing the developments during the past decade in weak interaction theory it may be appropriate to outline briefly some of the major contributions which were made prior to 1960. In particular the exciting discoveries in the late 1950's merit special attention. The experimental verification of parity non-conservation came about in 1956. Its impact on weak interaction physics was immediate and profound. It opened up entirely new vistas which were thoroughly explored in the next couple of years. Already by 1958, the essential elements of our present knowledge with regard to weak interaction processes had been found and much of the past decade was devoted to further elaboration of the ori-

*Invited talk at the Symposium "The Past Decade in Particle Theory" on April 14-17, 1970, Austin, Texas.

ginal ideas put forth in the period 1956-1958.

1. A theory of weak interaction processes emerges.

Beta-decaying nuclei have been studied since the turn of the century. However, it was not until the 1930's that the first substantial contributions to a theoretical understanding of these processes were made. To explain why the energy spectrum of the emitted electrons from β-decaying nuclei is continuous Pauli[1] suggested that the energy released in the process is shared between the nucleus, the electron and a new hypothetical particle, the neutrino. It took almost thirty years until one verified Pauli's hypothesis experimentally[2]. Nevertheless, Fermi[3] made use of Pauli's suggestion and proposed a remarkably successful theory for beta decay in 1934. Drawing heavily on the analogy with electromagnetism he assumed that the beta interaction is local and that it may be expressed in terms of an interaction energy density proportional to the wave functions of the neutron, proton, electron and the neutrino. To obtain as close an analogy with electromagnetism as possible he further chose a vector type interaction. In a first approximation he considered the nucleons to be at rest so that only the analogue of the electrostatic interaction survives.

Also considering that the wave lengths of the emitted electron and neutrino are large compared to the size of the nucleus it follows that only the gross features of the nucleus are important. More precisely, only the analogue of the total electric charge enters the calculation while the details of the charge distribution are unimportant. Decays for which one may neglect the motion of the nuclear particles and the variation of the lepton wave functions over the nuclear volume are called "allowed Fermi transitions." If the "allowed" approximations are valid it is straightforward to calculate the shape of the electron spectrum taking into account also the Coulomb effects on the electron wave function. The remarkable agreement between the predicted "allowed" shape and the experimentally observed energy spectrum for a large class of beta decays provided the first quantitative verification of Fermi's theory.

To calculate the total decay rates one must also know the nuclear matrix elements. Assuming these to be constant (and of order unity) Fermi showed that for allowed decays the total decay rate is proportional to the square of the coupling strength, the Fermi coupling constant, multiplied by a known function of E_e^{max}, where E_e^{max} is the upper endpoint of the electron spectrum. It then follows that the product of this function and the life-time, the so-called ft-value, should be a constant for all allowed decays, and deviations from this rule must be due to variations in the nuclear matrix elements. Fermi's predictions have been eminently confirmed for a large class of decays.

In Fermi's original theory the nuclear spins are unaffected by the weak interactions. Of course, Fermi was aware that this is a consequence of his specific choice of a vector interaction and that other possible alternatives exist. The immediate success of his theory seemed to warrant the underlying assumptions, however, and there was no immediate need to proceed further. A couple of years later Gamow and Teller[4] found themselves faced with apparent difficulties which could most easily be resolved by assuming that there also exist decays where the nuclear spins are affected by the weak forces in a fashion similar to the way in which the magnetic part of the electromagnetic interactions affects nuclear spins, except that the interaction does not vanish for zero momentum transfer. It is now well established that there exist such Gamow-Teller transitions and that the same "allowed shape" of the electron spectrum is obtained both for Fermi and for Gamow-Teller transitions of the allowed type. One has later extended the analysis beyond the family of allowed transitions and established agreement between theory and experiment also for so called unique first-forbidden, unique second-forbidden etc. transitions for which the variation of the lepton wave functions over the nuclear volume is essential.

A more complete analysis of the beta decay process

$$n \to p + e^- + \bar{\nu},$$

where the nucleons may or may nor be bound in a nucleus, starts from an interaction energy density

$$H'(x) = \sum_i \left[C_i \bar{\psi}_p(x) \Gamma_i \psi_n(x) \bar{\psi}_e(x) \Gamma_i \psi_\nu(x) + h.c. \right]. \quad (1)$$

The summation extends over the five possible types of coupling consistent with a Lorentz invariant theory (including space reflection invariance). Depending on the transformation properties of the bilinerar Dirac covariants $\bar{\psi}(x) \Gamma_i \psi(x)$, where $\Gamma_i = \{1, \gamma_\mu, \sigma_{\mu\nu}, \gamma_\mu \gamma_5, \gamma_5\}$, one refers to the couplings as scalar (S), vector (V), tensor (T), axial vector (A) and pseudoscalar (P), respectively. In a non-relativistic limit for the nucleon wave functions one finds that S and V interactions correspond to Fermi transitions while T and A correspond to Gamow-Teller transitions. The pseudoscalar interaction is negligible in this limit. The obvious question at this stage of the game was which of these couplings do actually enter in beta decay. From the previous discussion it is clear that both S and/or V and T and/or A must be present since one had observed pure Fermi and pure Gamow-Teller transitions, with a common effective coupling constant. Now, the allowed shape of the electron spectrum provides little information with regard to the type of couplings that enter. Omitting many of the details of the analysis we note that the angular correlation of the electron and the neutrino (determined indirectly by observing the recoil effect on the nucleus) has a particularly simple form

$$\frac{d^2\Gamma}{dE_e d(\cos\theta_{e\nu})} \propto 1 + \underline{a}\frac{v_e}{c}\cos\theta_{e\nu} \qquad (2)$$

where the angular correlation coefficient $\underline{a}$ takes on the values $\underline{a} = -1, +1, +1/3, -1/3$, for S, V, T, and A interactions respectively. Hence, by measuring the electron-neutrino angular correlation one stands a good chance to determine the nature of the weak interactions in beta decay. The measurement of $\underline{a}$ in the mixed transitions of Ne^{19} and the neutron gave a value near zero suggesting a mixture of V and A or of S and T. For the almost pure Fermi transition A^{35}

one had found a large positive value indicating a V interaction. On the other hand the first measurement of an electron-neutrino angular correlation for a pure Gamow-Teller transition was made on He^6 and one obtained a positive value for $\underline{a}$ favoring a T interaction and, hence, the MS combination in eq. (1). This is the situation as it presented itself immediately prior to the discovery that parity is not conserved for weak processes. Conflicting evidence did not seem to allow for a universal interaction!

Attempts to understand the decays of K mesons into two or respectively three π mesons, the so-called $\theta\tau$ puzzle, had prompted Lee and Yang[5] to examine the question of parity conservation in 1956. For more than fifty years one had a accumulated information on the beta decay process. So deeply rooted was apparently the belief in space reflection invariance that previous to Lee and Yang no one had seriously examined its experimental status with regard to weak processes. On the basis of the careful analysis by Lee and Yang, Wu et al.[6] proceeded to test parity conservation in the beta decay of polarized Co^{60}. More precisely, they measured the up-down asymmetry of the electron emission with regard to the polarization direction of the Co^{60} nuclei. This correlation is of the type $<\vec{\sigma}>\cdot\vec{p}_e$, and it is of pseudoscalar nature. Its presence was confirmed, establishing beyond doubt that parity conservation does not hold in this decay. Shortly afterwards it was demonstrated that also in the weak decay of the μ particle space reflection invariance is violated[7]. These discoveries initiated a new phase in the development of a theory of weak interaction processes. Not only did parity non-conservation imply the presence of new correlations which could provide additional information about the nature of the interaction, but it did also make the theoretical analysis more complicated since the interaction energy density (1) now must be replaced by the more general expression (still assuming a local coupling)

$$H'(x) = \sum_i \left[\bar{\psi}_p(x) \Gamma_i \psi_n(x) \bar{\psi}_e(x) \Gamma_i (C_i + C'_i \gamma_5) \psi_\nu(x) + h.c. \right], \quad (3)$$

allowing for parity conserving as well as parity violating contributions. At that time it seemed natural to ask whether also time reversal invariance might be violated, in which case the ten coupling constants C_i and C'_i may be complex-valued.

Already before the experimental verification of parity violations in beta-decay and μ-decay it had been noted by Salam[8] that this effect could be related to the vanishing of the neutrino mass. For the photon, which mediates the electromagnetic interactions, gauge invariance provides a natural explanation why it should be massless. Peierls had pointed out that one lacked a corresponding principle for the neutrinos. Salam now made the observation that if one required the Hamiltonian density to be invariant under the transformation $\psi_\nu(x) \to \gamma_5 \psi_\nu(x)$ then one must conclude that (i) the neutrino is massless, and (ii) space reflection invariance in general must be violated. If further lepton conservation is assumed this restricts the interaction for μ-decay to be of the V and A type. For β-decay one obtains less restrictive conditions since only one neutrino participates in the process. Arguing along similar lines Landau[9] and Lee and Yang[10] somewhat later reached similar conclusions proposing the two-component neutrino hypothesis as a model for the weak interactions. This hypothesis amounts to putting $C_i = \pm C'_i$ in the equation (3). Although intimately connected with the theory of β-decay the two-component neutrino hypothesis did not resolve the outstanding question of which couplings actually occur in the beta interaction.

During that same period Sudarshan and Marshak[11] carried out an extensive analysis to investigate the possibility of a universal interaction responsible for all weak processes. They were faced with the following crucial pieces of experimental information:

(i) The electron-neutrino angular correlation data favored VA if one accepted the A^{35} measurements, in which case the results of the He^6 experiment had to be refuted. If, on the other hand, the He^6 result was accepted at the expense of the A^{35} findings one would conclude that the interaction is of the ST type.

(ii) The measured electron asymmetry in the decay of polarized nuclei as well as the measurement of the longitudinal polarization of beta electrons from Co^{60} (a pure Gamow-Teller transition) yielded a large negative value for the correlation parameter. This result is consistent with either A or T depending on the helicity of the emitted antineutrino.

(iii) The corresponding measurements for the Fermi transition of Ga^{66} was consistent with V or S, again depending on the sign of the antineutrino helicity.

(iv) For the mixed transitions of Sc^{46} and Au^{198} one has obtained results consistent with (ii) and (iii). Assuming that the neutrinos occuring in Fermi transitions are the same as those appearing in Gamow-Teller transitions one found the results consistent with VA or ST depending on the antineutrino helicity.

(v̈) Measurements of the beta-polarized gamma correlation in the mixed transition of Sc^{16} had given a nearly maximal effect of parity violation indicating approximately equal strength for the V and the A couplings or, alternatively, for the S and T couplings. Also, the coupling constants must have approximately the same relative phase. Again one must assume that the same kind of antineutrino participates both in the Fermi and in the Gamow-Teller transitions.

(vi) Finally, if one considers the $\pi \to e + \nu$ process on the same footing as nuclear beta decay, additional information was to be gained by studying this process.

Since the π meson is a pseudoscalar particle this process can only occur via an A or P interaction. An A type interaction would lead to a definite prediction for the branching ratio $\Gamma(\pi \to e + \nu)/\Gamma(\pi \to \mu + \nu)$ which exceeded the upper limit determined by experiments by one order of magnitude. In the case of a pure S interaction the disagreement between theory and experiment would be several orders of magnitude worse.

On the basis of their analysis Sudarshan and Marshak concluded that only a V-A interaction could provide a unified model for the weak processes known at that time. This conclusion made them refute the result of the He^6 experiment as well as the existing upper limit for the branching ratio $\Gamma(\pi \to e\nu)/\Gamma(\pi \to \mu\nu)$. Later experiments have confirmed their model. To account for the V-A interaction they proposed an elegant symmetry principle requiring chirality invariance. Other arguments were later presented by Feynman and Gell-Mann and by Sakurai[12] for a weak interaction theory of the V-A type (corresponding to $C_V = C'_V \equiv G_V$; $C_A = C'_A \equiv G_A$ and all other C_i's and C'_i's vanishing in equation (3)). Feynman and Gell-Mann[12] also noted that the V-A theory offered an opportunity to understand the near equality for the coupling constant G_μ in μ-decay and the Fermi coupling constant G_V in a beta-decay. This near equality had been pointed out much earlier[13], but it then seemed puzzling considering the fact that μ-decay is a purely leptonic process while the nuclear beta decay involves hadrons, and one would expect the strong interactions to "renormalize" the coupling strength in this latter case. The conserved vector current hypothesis (CVC) proposed by Feynman and Gell-Mann in their paper provides a very attractive explanation, as noted already earlier by Gerstein and Zeldovich[14].

The V-A interaction for leptonic and semi-leptonic processes as it was conceived in these pioneering papers may be written as a current interaction in the following way

$$H'(x) = \frac{G}{\sqrt{2}}\left\{ h_\lambda^+(x)\ell^\lambda(x) + h^\lambda(x)\ell_\lambda^+(x) + \ell_\lambda^+(x)\ell^\lambda(x) \right\} \quad (4)$$

where

$$\ell_\lambda(x) = \bar{\psi}_\mu(x)\gamma_\lambda(1+\gamma_5)\psi_\nu(x) + \bar{\psi}_e(x)\gamma_\lambda(1+\gamma_5)\psi_\nu(x) , \quad (5)$$

$$h_\lambda(x) = \bar{\psi}_n(x)\gamma_\lambda(1+\gamma_5)\psi_p(x) + \ldots \quad (6)$$

The CVC hypothesis imposes on the vector part $h_\lambda^V(x)$ of $h_\lambda(x)$ the condition

$$\partial^\lambda h_\lambda^V(x) = 0 , \quad (7)$$

$$h_\lambda^V(x) = \bar{\psi}_n(x)\gamma_\lambda\psi_p(x) + \ldots \quad (8)$$

We note that H'(x) in (4) describes $\beta^\pm$-decay, $\mu^\pm$-decay and $\mu^\pm$-capture on nuclei. In addition the "diagonal" terms correspond to scattering processes like $e + \nu_e \to e + \nu_e$. We shall return to some of these processes later.

In summary, the universal V-A theory of the current x current type together with the CVC hypothesis provided a most satisfactory frame-work for the understanding of a remarkably wide class of weak processes. Its basic features have remained unshaken by the later confrontations between theory and experiments. Of course, the extension to new phenomena has made it necessary to analyze the structure of the currents, in particular the hadronic current, more in detail. The situation is reminiscent of the developments in electrodynamics when one wanted to incorporate the strongly interacting particles into the theory. The very structure of the hadrons must be accounted for in terms of form factors etc. Recognizing that the fundamental tools in weak interaction theory were developed by 1958, we shall not pursue this historical expose' further but rather outline the theory in its present version so that we may examine its current status and the challenging problems which still remain unsolved.

2. <u>The current x current theory of weak interactions</u>.

It is consistent with our present knowledge of weak interaction phenomena to write the effective weak S-operator in the following way

$$S = 1 - \frac{iG}{\sqrt{2}} \int d^4x \; d^4x' J_\lambda^+(x) \rho^{\lambda\lambda'}(x - x') J^\lambda(x') , \qquad (9)$$

corresponding to a four-fermion interaction structure as indicated in figure 1.

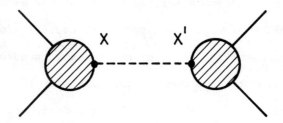

Figure 1: The structure of a weak interaction vertex.

It is an empirical fact that for small momentum transfers it is a good approximation to take

$$\rho_{\lambda\lambda'}(x - x') = g_{\lambda\lambda'}\delta^4(x - x') + \ldots ,$$

and in most of my talk I shall assume that we may couple the currents in this way. The self-coupled weak current $J_\lambda(x)$ is of the form

$$J_\lambda(x) = h_\lambda(x) + \ell_\lambda(x) , \qquad (10)$$

where, as before,

$$\ell_\lambda(x) = \bar{\psi}_\mu(x)\gamma_\lambda(1 + \gamma_5)\psi_{\nu_\mu}(x)$$
$$+ \bar{\psi}_e(x)\gamma_\lambda(1 + \gamma_5)\psi_{\nu_e}(x) \equiv (\bar{\mu}\nu_\mu) + (\bar{e}\nu_e), \quad (11)$$

and $h_\lambda(x)$ is the hadron current whose properties we shall discuss below. In (11) we have taken notice of the difference between the neutrino coupled to the electron and the one coupled to the muon. The basis for this distinction between ν_μ and ν_e is the observation of two separate conservation laws for the lepton numbers L_e and L_μ. The lepton current (11) also implies μe-universality, that is, apart from differences related to phase space weak interaction matrix elements are invariant under the simultaneous replacements $\mu \rightleftarrows e$ and $\nu_\mu \rightleftarrows \nu_e$. Finally, we note that $\ell_\lambda(x) \to h_\lambda(x)$ will be discussed below.

The hadron current has a much more complicated structure reflecting the effects of the strong interactions. The fact that hadrons have a finite extension makes the current operator $h_\lambda(x)$ non-local and we may more appropriately write it

$$h_\lambda(x) = \int d^4x_1 d^4x_2 \bar{\psi}_1(x_1) \Gamma_\lambda(x_1 - x, x_2 - x) \psi_2(x_2) + \ldots , \quad (12)$$

where Γ_λ is a vector-axial vector operator formed out of γ-matrices and differential operators. Since $h_\lambda(x)$ has

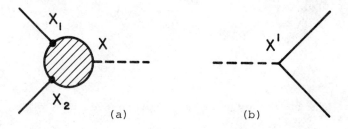

Figure 2: The hadronic part (a) and the leptonic part (b) of a weak vertex.

nonvanishing matrix elements between hadronic states, which are labeled by the quantum numbers of the strong interaction symmetries, it is natural to examine if there exist symmetry patterns with regard to hypercharge Y and isospin I. Indeed one has found that this is the case. With regard to hypercharge there is strong evidence for a selection rule limiting the change in hypercharge to $\Delta Y = 0, \pm 1$, where

$$\Delta Y = Y_f - Y_i , \quad (13)$$

Y_f and Y_i denoting the total hypercharge of the hadrons the final and the initial state respectively. On the is of this selection rule one separates the hadronic current into a hypercharge conserving current $j_\lambda(x)$ with $= 0$ and a hypercharge changing current $s_\lambda(x)$ with $|\Delta Y| = 1$ that

$$h_\lambda(x) = j_\lambda(x) + s_\lambda(x) . \qquad (14)$$

ce we cannot have terms with $\Delta Y = 1$ as well as $\Delta Y = -1$ present ultaneously without obtaining $|\Delta Y| = 2$ transitions in the leptonic term $h_\lambda^+ h^\lambda$ of the S-operator we must make a choice. is consistent with our definition of the S-operator and of the ton current to choose $\Delta Y = -1$.

Before we discuss the isospin selection rules it is th noting that the processes described by the S-operator naturally fall into three categories depending on which rents are involved

(i)   leptonic processes   — $\ell_\lambda^+(x)\ell^\lambda(x)$,

(ii) semileptonic processes — $h_\lambda^+(x)\ell^\lambda(x) + h.c.$,

(iii) nonleptonic processes — $h_\lambda^+(x)h^\lambda(x) .$

h regard to <u>semileptonic processes</u> there is evidence for following isospin selection rules

(i) $|\vec{\Delta I}| = 1$ for hypercharge conserving processes ,

(ii) $|\vec{\Delta I}| = \tfrac{1}{2}$ for hypercharge changing processes .

We can account for them by assuming the current $j_\lambda(x)$ to be an isovector current (I = 1) and the current $s_\lambda(x)$ to be an isospinor current (I = ½).

For the <u>nonleptonic processes</u> we are primarily interested in the hypercharge changing transitions since processes with $\Delta Y = 0$ and $\Delta Q = 0$ can take place by means of strong interactions and the weak contributions are then in general hard to observe. There are some exceptions, however, as discussed in the section on nonleptonic processes.[15] For the hypercharge changing processes one has found that the selection rule $|\vec{\Delta I}| = ½$ is accurate to within a few percent. Since these decays are represented by the term $(j_\lambda^+ s^\lambda + h.c.)$ our previous assumptions regarding the isospin properties of the currents in general would yield $|\vec{\Delta I}| = ½$ and $|\vec{\Delta I}| = \tfrac{3}{2}$ and there is no a priori reason why one or the other should be suppressed. Of course, one can arrange it so that in the coupling between $j_\lambda(x)$ and $s_\lambda(x)$ there is no $|\Delta I| = \tfrac{3}{2}$ contribution, but that would require the introduction of neutral currents (note that $\ell_\lambda(x)$ is a charged current). On the basis of a universal coupling scheme we would then expect also neutral lepton currents, but experimental evidence does not seem to permit neutral lepton currents. The question of the nonleptonic $|\vec{\Delta I}| = ½$ rule is discussed more in detail in the report by

Pakvasa and Rosen[15].

For nonleptonic processes one clearly has $\Delta Q = 0$, where Q is the electric charge quantum number. Since we have just discarded the possibility of neutral lepton currents on experimental grounds we must have $|\Delta Q| = 1$ for the semileptonic processes. For hypercharge changing semileptonic processes one may then have $\Delta Q = \Delta Y$ as well as $\Delta Q = -\Delta Y$. It is an empirical fact that processes with $\Delta Q = -\Delta Y$ are strongly suppressed. Hence one speaks about the $\Delta Q = \Delta Y$ rule. The status of this rule will be examined. Of particular interest in this context are the processes $K_L^0 \to \pi^{\pm} + \ell^{\mp} + \nu_\ell$ where ℓ stands for a muon or an electron. We note in passing that by virtue of the Gell-Mann-Nakano-Nishijima relation $Q = I_3 + \frac{1}{2}Y$ the semileptonic $|\vec{\Delta I}| = \frac{1}{2}$ rule implies the $\Delta Q = \Delta Y$ rule while the converse statement is not true.

During the past decade all these empirical selection rules have been incorporated into more elaborate symmetry schemes. We shall return to this question somewhat later in this review.

3. The experimental status of the empirical selection rules.

We have referred to a number of selection rules which have been deduced from the experimental findings. We shall briefly quote some relevant experimental results in support of these rules. No attempt is made to make this review complete. It is merely meant to give an idea of the precision by which these rules have been tested.

Lepton number conservation

The conservation laws for the electron lepton number L_e and the muon lepton number L_μ are based on the following assignments of these quantum numbers.

	e^-	ν_e	μ	ν_μ	all other particles
L_e	1	1	0	0	0
L_μ	0	0	1	1	0

For the antiparticles the lepton numbers have the opposite sign. It is seen that the lepton current, as we have previously given it, does not carry any lepton number and it is now assumed that these quantum numbers are strictly conserved. Evidence in support of these rules comes from

(i)   absence of neutrinoless double beta decay ,

(ii)   antineutrino capture processes ,

(iii)   high energy neutrino experiments ,

(iv)   absence of certain decay modes .

Double beta decay might occur as

$$(Z,A) \rightarrow (Z+2, A) + e^- + e^- , \qquad (15)$$

in which case it is referred to as neutrinoless, or as

$$(Z,A) \to (Z+2, A) + e^- + \bar{\nu}_e + e^- + \bar{\nu}_e \quad . \qquad (16)$$

The neutrinoless process violates L_e-conservation while in (16) one obviously satisfies this selection rule. If the conservation law for L_e holds, we then expect (16) to be seen while (15) should be absent. If L_e is not conserved the first process is enhanced by a phase space factor of order 10^5 over the second process. Experimentally the two processes can be distinguished by examining the effective mass of the electron pair. In Table 1 there are some experimental results given.

Decay Process	Experimental	Half life (years) Theoretical		Reference
		two neutrinos	neutrinoless	
$Te^{130} \to Xe^{130}$	$10^{21.34\pm0.12}$	$10^{22.5\pm2.5}$	$10^{16.3\pm2}$	16
$Se^{82} \to Kr^{82}$	$6\times10^{19\pm0.3}$	$10^{22.5\pm2.5}$	$10^{16\pm2}$	17
$Ca^{48} \to Ti^{48}$	$\geq (3.5\pm9)\times10^{19}$	$10^{21\pm2}$	-	18
	$\geq 10^{21}$	-	$3\times10^{15\pm2}$	19
$Ge^{76} \to Se^{76}$	$\geq 1,5\times10^{21}$	$10^{22\pm2.5}$	$10^{16\pm2}$	20

Table 1: Experimental and theoretical estimates for half-lifes of some double beta decays.

For comparison we have also given some theoretical estimates based on the assumption (i) that L_e is conserved and the process is of the type (16), and (ii) that L_e is maximally violated and the decay is neutrinoless. It is seen that all data are consistent with conservation of L_e although one can clearly not exclude a small contribution from L_e-violating interactions.

Nuclear reactors provide an intense flux of $\bar{\nu}_e$-particles. As a test of L_e-conservation one has studied the capture processes

$$\bar{\nu}_e + (Z,A) \rightarrow (Z+1, A) + e^- , \qquad (17)$$

$$\bar{\nu}_e + (Z,A) \rightarrow (Z-1, A) + e^+ . \qquad (18)$$

The first process violates L_e-conservation while the second one is L_e-conserving. Experimentally[22] one has determined the cross section for (18) in the process $\bar{\nu}_e + p \rightarrow n + e^+$ to be $(11 \pm 2.5) \times 10^{-44} cm^2$. A theoretical estimate with reasonable assumptions regarding the neutrino flux and the neutrino energy spectrum yields $(11 \pm 1.6) \times 10^{-44} cm^2$. For process (17) one determined[23] the cross section to be $(0.1 \pm 0.6) \times 10^{-45} cm^2$/atom consistent with a vanishing cross section. The quoted experiment dealt with the process $\bar{\nu}_e + Cl^{37} \rightarrow A^{37} + e^-$.

More recent experiments[24] with neutrino beams from multi-

BeV accelerators have resulted in more conclusive evidence for the two-neutrino hypothesis. In these experiments one has a neutrino beam containing predominantly ν_μ-particles. The energy spectrum is peaked around 1 BeV. If $\nu_\mu = \nu_e$ one would expect approximately equal number of electrons and muons in the reaction

$$\nu_\mu + n \rightarrow \ell^- + p \,..\qquad (19)$$

If $\nu_\mu \neq \nu_e$ one would expect less than one percent of electrons considering the purity of the beam. In the experiments one found a substantial number of muon events while the number of electron events was down by a factor of 10^2 at least. Similar experiments with antineutrino beams[24] have further confirmed the separate conservation of L_e and L_μ. Of course, these experiments are not unambiguous as tests of lepton number conservation. A gross violation of µe-universality (cf. below) could also explain the observed differences between the electron and the muon mode of (19). In view of the present status of µe-universality this does not seem to be a reasonable way out, however.

Separate conservation of L_e and L_μ also explains the apparent non-occurrance of several decay modes for the muon and the K meson. In Table 2 some relevant branching ratios are listed.

Process	Branching Ratio
$\mu^+ \to e^+ + \gamma$	$< 6 \times 10^{-19}$
$\to e^+ + 2\gamma$	$< 1.6 \times 10^{-5}$
$\to e^+ + e^- + e^+$	$< 1.25 \times 10^{-7}$
$\mu^- + Cu \to e^- + Cu$	$< 2.4 \times 10^{-7}$
$K_L^0 \to e^\pm + \mu^\mp$	$< 8 \times 10^{-6}$
$K^+ \to \pi^+ + e^- + \mu^+$	$< 3 \times 10^{-5}$

Table 2: Experimental tests of muon and electron lepton number conservation[26].

Although these processes violate the separate conservation of L_e and L_μ they do satisfy the conservation law for the total lepton number $L = L_e + L_\mu$. In fact the old conservation law for L was replaced by separate conservation laws for L_e and L_μ to explain why these decays did not seem to occur.

In conclusion we note that there is a good deal of support for the separate conservation of the lepton numbers L_e and L_μ but more accurate tests of possible violations would certainly be desirable[25].

Muon - electron universality

We have already stated the content of this principle; the muon component and the electron component of the lepton current $\ell_\lambda(x)$ couple with equal strength. The most accurate and unambiguous tests concern the branching ratios $\Gamma(M \to e + \nu_e)/\Gamma(M \to \mu + \nu_\mu)$ where M stands for K or π. These branching ratios are independent of the strong interaction effects. In the case of three-body decays of mesons or baryons one must introduce additional assumptions about the form factors to obtain unique predictions. In Table 3 some of the best tests of the μe-universality are listed. The theoretical values for the branching ratios of the two-body decays include small but significant corrections due to electromagnetic effects.

Of course, the most natural place to test μe-universality would be in purely leptonic processes like elastic $\ell \nu_\ell$-scattering. Experimentally this is not feasible yet, however.

Absence of neutral lepton currents

The absence of neutral lepton currents of the type $(\bar{\mu}\mu)$, $(\bar{e}e)$, or $(\bar{\nu}\nu)$ has previously been invoked as an argument against neutral components in the hadron current (to explain the nonleptonic $|\Delta \vec{I}| = \frac{1}{2}$ rule). The experimental situation

Branching Ratio	Experimental[26]	Theory
$\dfrac{\Gamma(\pi^+ \to e^+ + \nu_e)}{\Gamma(\pi^+ \to \mu^+ + \nu_\mu)}$	$(1.24 \pm 0.03) \times 10^{-4}$	1.23×10^{-4}
$\dfrac{\Gamma(K^+ \to e^+ + \nu_e)}{\Gamma(K^+ \to \mu^+ + \nu_\mu)}$	$(1.86 \pm 0.46) \times 10^{-5}$	2.34×10^{-5}
$\dfrac{\Gamma(K^+ \to \pi^0 + \mu^+ + \nu_\mu)}{\Gamma(K^+ \to \pi^0 + e^+ + \nu_e)}$	0.67 ± 0.02	$0.65^{(x)}$
$\dfrac{\Gamma(\Sigma^- \to n + \mu^- + \bar{\nu}_\mu)}{\Gamma(\Sigma^- \to n + e^- + \bar{\nu}_e)}$	0.42 ± 0.06	$0.45^{(x)}$

Table 3: Tests of the μe-universality. Theoretical estimates marked by star have been obtained assuming constant form factors and SU(3) invariance for the strong interactions.

is summarized in Table 4.

Process	Rates[26]
$K^0_S \to \mu^+ + \mu^-$	$< 3 \times 10^{-7}$
$K^0_L \to \mu^+ + \mu^-$	$< 2 \times 10^{-6}$
$\to e^+ + e^-$	$< 2 \times 10^{-5}$
$K^+ \to \pi^+ + e^+ + e^-$	$< 4 \times 10^{-5}$
$\to \pi^+ + \mu^+ + \mu^-$	$< 2.4 \times 10^{-6}$
$\to \pi^+ + \nu + \bar{\nu}$	$< 7.7 \times 10^{-7}$ *
$K^0_S \to e^+ + e^-$	$< 2 \times 10^{-7}$

Table 4: Tests of the absence of neutral lepton currents. Rates are given in terms of the total decay rates.

* R. Steining et al., Berkely (unpublished)

Local action of the lepton current

The fact that the lepton current is assumed to be a local operator leads to a number of explicit predictions in say, high energy neutrino reactions[27]. Clearly the most informative tests must be performed at high energies providing large momentum transfers. So far no experimental information is available. Of course, one may take the apparent point structure of the μ-decay process as indirect evidence for the local structure of the lepton current (see below). This is not a very crucial test, however, due to the small energy release in this decay.

The selection rule $|\Delta Y| < 2$

The most accurate test of this selection rule is the very small measured difference for K_L^0 and K_S^0 [28]. If $|\Delta Y|=2$ contributions were present the mass difference $\Delta m = m_{K_L^0} - m_{K_S^0}$ can be estimated to be of order

$$\Delta m \sim 10^5 \tau_S^{-1},$$

where τ_S is the life time of K_S^0. If the selection rule $|\Delta Y| < 2$ holds then the mass difference must be a second-order effect in which case one finds

$$\Delta m \sim \tau_S^{-1}.$$

Experimentally one has determined Δm to be [26]

$$\Delta m = (0.473 \pm .014) \cdot \tau_S^{-1},$$

indicating that the $|\Delta Y| = 2$ matrix element is down by at least a factor 10^5 as compared to the $|\Delta Y| = 1$ matrix element. Of course, it could be an accidental effect that the $|\Delta Y| = 2$ contribution is strongly suppressed in the mass difference Δm [29]. The first estimate above assumes this not to be the case.

Less accurate tests of the $|\Delta Y| < 2$ selection rule refer to the absence of decay modes for which $|\Delta Y| = 2$. Some relevant branching ratios are given in Table 5.

Decay Process	Rate
$\Xi^0 \to p + \mu^- + \bar{\nu}_\mu$	$< 1.3 \times 10^{-3}$
$\to p + e^- + \bar{\nu}_e$	$< 1.3 \times 10^{-3}$
$\to p + \pi^-$	$< 0.9 \times 10^{-3}$
$\Xi^- \to n + \mu^- + \bar{\nu}_\mu$	—
$\to n + e^- + \bar{\nu}_e$	$< 1 \times 10^{-2}$
$\to n + \pi^-$	$< 1.1 \times 10^{-3}$

Table 5: Branching ratios for some processes with $|\Delta Y| = 2$[26]. All rates are expressed in terms of total decay rates.

Isospin selection rules

The experimental evidence for the $|\vec{\Delta I}| = 1$ rule governing the hypercharge conserving semi-leptonic weak processes is essentially indirect. The decay processes are of the type

$$h \to h' + \ell + \nu_e , \qquad (20)$$

and whether or not the hadrons h and h' belong to the same iso-multiplet the $|\vec{\Delta I}| = 1$ rule does not possess much predictive power. This is so because of the peculiar distribution of quantum numbers among the weakly decaying hadrons. However, there are two cases for which the rule at least in principle may be tested. Assuming $h_\lambda(x)$ to be an isovector density one concludes from the Wigner-Eckart theorem that the matrix elements for the two processes

$$\Sigma^- \rightarrow \Sigma^0 + e^- + \bar{\nu}_e$$

and

$$\Sigma^0 \rightarrow \Sigma^+ + e^- + \bar{\nu}_e$$

are the same. Unfortunately, the latter process is a very rare decay mode and one does not know the experimental value for the corresponding decay rate. For the beta decay of $\pi^\pm$ one similarly finds that in the limit of exact isospin invariance for the strong interactions the isovector property of $h_\lambda(x)$ together with the Bose statistics for the π-mesons imply that the form factor conventionally denoted $f_-(q^2)$ must vanish. However, one arrives at the same result from the assumption that the vector current is conserved which weakens the conclusions one may draw from the fact that $f_-(q^2)|_{exp} \simeq 0$.

We have already pointed out that the available evidence for the $|\vec{\Delta I}| = 1$ rule is indirect. However, since the lepton current carries away one unit of electric charge it follows from the Gell-Mann-Nakano-Nishijima relation that $|\Delta I_3| = 1$ so that $|\vec{\Delta I}| = n$, where $n \geq 1$. The simplest choice is $n = 1$. The success of the CVC theory, in which the hypercharge conserving weak vector current is a component of the isospin current and hence trivially satisfies $|\vec{\Delta I}| = 1$, lends support to this assumption. Also recent success with the SU(2) x SU(2) current algebra and the octet nature of the weak currents (the Cabibbo theory) should be recalled in this context. Of course, many other elements enter into the picture and all one can safely say is that the $|\vec{\Delta I}| = 1$ rule for the hypercharge conserving semileptonic processes is part of an overall picture which seems to be consistent with observations.

More crucial tests of the $|\vec{\Delta I}| = 1$ rule must await the availability of high energy neutrino beams. With access to suitable neutrino beams one may study processes like

$$\nu_\ell + p \rightarrow \ell^- + p + \pi^+,$$

$$\nu_\ell + n \rightarrow \ell^- + n + \pi^+,$$

$$\nu_\ell + n \rightarrow \ell^- + n + \pi^0.$$

On the basis of the $|\vec{\Delta I}| = \frac{1}{2}$ rule for hypercharge changing semileptonic processes rests on more solid evidence. The existence of the decays $K^+ \to \ell^+ + \nu_\ell$ and $\Lambda \to p + e^- + \bar{\nu}_e$ implies that $s_\lambda(x)$ at least has an $I = \frac{1}{2}$ component even if it does not exclude that other component exist with higher isospin content. Studies of K_{ℓ_3} and K_{ℓ_4} decays have a more direct bearing on the $|\Delta \vec{I}| = \frac{1}{2}$ rule. On the basis of this rule one easily establishes the following relation between the decay rates

$$\Gamma_L(\ell^-): \quad K_L^0 \to \pi^+ + \ell^- + \bar{\nu}_\ell,$$

$$\Gamma_L(\ell^+): \quad K_L^0 \to \pi^- + \ell^+ + \nu_\ell,$$

$$\Gamma(K_{\ell 3}^-): \quad K^- \to \pi^0 + \ell^- + \bar{\nu}_\ell,$$

$$\Gamma_L(\ell^-) + \Gamma_L(\ell^+) = 2\Gamma(K_{\ell 3}^-) = 2\Gamma(K_{\ell 3}^+). \quad (21)$$

The experimental results are given in Table 6, and they are seen to agree with the prediction above in the case of $K_{\mu 3}$ while the K_{e3} data may indicate a small violation of the $|\Delta \vec{I}| > \frac{1}{2} \to |\Delta \vec{I}| = \frac{1}{2}$ rule.

l	$\Gamma_L(\ell^-) + \Gamma_L(\ell^+)$	$2\Gamma(K_{\ell 3}^-)$
e	$(7.22 \pm 0.29) \times 10^6$	$(7.87 \pm 0.11) \times 10^6$
μ	$(4.98 \pm 0.22) \times 10^6$	$(5.17 \pm 0.18) \times 10^6$

Table 6: Experimental tests of the $|\Delta \vec{I}| = \frac{1}{2}$ rule for hypercharge changing semileptonic processes[26].

Phase space corrections and radiative corrections are of order one per cent and do not change this conclusion. For the $K_{\ell 4}$ decays similar relations may be deduced, but the experimental situation does not yet permit any meaningful tests of the predictions.

The presence of $\Delta Y = -\Delta Q$ transitions would imply that $|\Delta I| > \frac{1}{2}$. Tests of the $\Delta Y = \Delta Q$ rule (see below) thus furnish indirect evidence in support of the $|\Delta \vec{I}| = \frac{1}{2}$ rule for the hypercharge changing semileptonic processes. Note, however, that the $\Delta Y = \Delta Q$ rule is consistent with the $|\Delta \vec{I}| = \frac{1}{2}$ rule but does not necessarily imply its validity.

For a discussion of the nonleptonic $|\Delta \vec{I}| = \frac{1}{2}$ rule we refer to the review paper by Pakvasa and Rosen[15].

The $\Delta Y = \Delta Q$ rule

The apparent absence of decay modes for which $\Delta Y = -\Delta Q$ were noted quite early and provided the first indication of a selection rule at work. More recently one has found alternative methods to test the rule in the decays of neutral K-mesons. The charge asymmetry in the $K_L^0{}_{,\ell 3}$ decay may be written

$$\frac{\Gamma_L(e^+)}{\Gamma_L(e^-)} \simeq 1 + 4\mathrm{Re}\varepsilon \times \frac{1-|x|^2}{|1+x|^2} , \qquad (22)$$

where ε is the conventional parameter measuring the CP-violation[30] in these decays and

$$x = \frac{h_+(0)}{f_+(0)} \qquad (23)$$

is the ratio between form factors defined by

$$<\pi^0(p')|s_\lambda(0)|K^-(p)> = \frac{1}{(2\pi)^3} \cdot \frac{1}{\sqrt{4p_0 p_0'}} \left[f_+(q^2)(p+p')_\lambda + f_-(q^2)(p-p')_\lambda \right]$$

$$<\pi^+(p')|s_\lambda(0)|K^0(p)> = \frac{1}{(2\pi)^3} \cdot \frac{1}{\sqrt{4p_0 p_0'}} \left[h_+(q^2)(p+p')_\lambda + h_-(q^2)(p-p')_\lambda \right] ,$$
$$(24)$$

where $q = p' - p$. In the case of $K^0_{L,e3}$ decays one can neglect the terms proportional to f_- and h_- since they are multiplied by the electron mass, and in that case x is the ratio between the amplitude for which $\Delta Y = -\Delta Q$ and the amplitude with $\Delta Y = \Delta Q$. The charge asymmetry above does not yield x directly but rather a functional relation between ε and x. To determine x separately one must consider the time dependence of the charge asymmetry in a regeneration experiment[31]. The result is[26]

$$\mathrm{Re}\, x = 0.021 \pm 0.036$$
$$\mathrm{Im}\, x = -0.099 \pm 0.047$$

or

$$|x| = 0.10 \pm 0.06$$

4. The Cabibbo model for the hadron current

The symmetry properties of the strong interactions attracted considerable attention during the late 1950's and the early part of the 1960's. From these discussions the SU(3) symmetry emerged[32] incorporating the invariance under isospin rotations and hypercharge gauge transformations within the same theoretical framework. The possible connection between these invariance properties of the strong interactions and the underlying currents provided a link between the strong and the weak interactions. This line of thought is already present in the CVC hypothesis which we shall discuss more in detail later.

It was noted by Cabibbo[33] and others that the empirical selection rules for the weak interactions can be accomodated within an SU(3) scheme in a very natural way by assuming that the hadronic currents $j_\mu(x)$ and $s_\mu(x)$ belong to the same octet of SU(3) current densities. One had further noted that the rates for hypercharge changing semileptonic baryon decays were suppressed by at least an order of magnitude as compared to the hypercharge conserving processes (the beta decay of the neutron). Both features are accounted for in the Cabibbo model in which we may write the currents

$$j_\mu(x) = \cos\theta \left[h_\mu^1(x) + i\, h_\mu^2(x) \right]$$
$$s_\mu(x) = \sin\theta \left[h_\mu^4(x) + i\, h_\mu^5(x) \right] \tag{25}$$

or alternatively

$$h_\mu(x) = \cos\theta \left[h_\mu^1(x) + i\, h_\mu^2(x) \right] + \sin\theta \left[h_\mu^4(x) + i\, h_\mu^5(x) \right] \tag{26}$$

where $\{h^i(x)\}_{i=1}^8$ denotes the unitary octet of current densities and where we have adopted Gell-Mann's conventions for the SU(3) indices[32]. The parameter θ is known as the Cabibbo angle. Decomposing the octet currents into vector and axial vector components, $v_\mu^i(x)$ and $a_\mu^i(x)$ respectively, so that

$$h_\mu^i(x) = v_\mu^i(x) + a_\mu^i(x) , \tag{27}$$

we can now establish the relation between the weak currents and the generators F_i for the SU(3) symmetry group of strong interactions. By hypothesis we have

$$F_i = \int d^3x\, v_0^i(x) , \tag{28}$$

and the statement that the h_μ^i's form a unitary octet is expressed by the following commutation relation

$$\left[F_i, h_\mu^j(x) \right] = i\, f_{ijk}\, h_\mu^k(x) , \tag{29}$$

where the f_{ijk}'s denote the structure constants of SU(3). The identification (28) implies that in the limit of exact SU(3) symmetry all the vector currents are conserved and the corresponding generators F_i are then time independent. This is the SU(3) generalization of the CVC hypothesis. Since SU(3) is only an approximate symmetry of the strong interactions one must in general expect rather large corrections to predictions derived from this principle. However, the fact that the strong interactions do conserve isospin and hypercharge implies that F_1, F_2, F_3, and F_8 are time independent and that the corresponding vector currents are conserved (neglecting electromagnetic and weak corrections). Furthermore, in view of the Ademollo-Gatto theorem[34] many predictions which follow from the generalized CVC hypothesis are good to second order in the SU(3) violation. The theorem states that to first order in the SU(3) breaking the matrix elements of the vector currents $v_\mu^i(x)$ between single particle states belonging to the same SU(3) multiplet are not affected by the SU(3) violations in the limit of vanishing momentum transfer.

Before proceeding to the explicit predictions of the Cabibbo model we note that the current (26) by no means is the most general one. Retaining the idea of octet current densities one could, for example, have introduced different angles for the vector and the axial vector components, and one could also have chosen a more general normalization replacing $\cos\theta$ and $\sin\theta$ by arbitrary parameters. However, Cabibbo's particular choice of current enjoys a kind of universality property which makes it attractive. More precisely, consider the following lepton charges constructed out of the weak lepton current

$$L_+ = \int d^3x \left[\bar{\psi}_{\nu_e}(x)\gamma_0\psi_e(x) + \bar{\psi}_{\nu_\mu}(x)\gamma_0\psi_\mu(x) \right] ,$$

$$L_- = \int d^3x \left[\bar{\psi}_e(x)\gamma_0\psi_{\nu_e}(x) + \bar{\psi}_\mu(x)\gamma_0\psi_{\nu_\mu}(x) \right] .$$
(30)

From the anticommutation relations of the lepton fields one finds

$$\left[L_+, L_- \right] = 2L_3 ,$$
(31)

with

$$L_3 = \int d^3x \left[\bar{\psi}_{\nu_e}(x)\gamma_0\psi_{\nu_e}(x) - \bar{\psi}_e(x)\gamma_0\psi_e(x) \right.$$
$$\left. + \bar{\psi}_{\nu_\mu}(x)\gamma_0\psi_{\nu_\mu}(x) - \bar{\psi}_\mu(x)\gamma_0\psi_\mu(x) \right]$$
(32)

and

$$\left[L_3, L_+ \right] = L_+$$
$$\left[L_3, L_- \right] = -L_- ,$$
(33)

which are the commutation relations for an SU(2) algebra. If we now in an analogous fashion form

$$H_{\pm} = a(F_1 \pm i F_2) + b(F_4 \pm i F_5), \quad (34)$$

where the F_i's are related to the vector part of the weak hadron currents as given in (28), then one finds by straightforward computation that

$$[H_+, H_-] = 2H_3 \quad (35)$$

with

$$H_3 = (a^2 + \frac{b^2}{2})F_3 - abF_6 + \frac{b^2\sqrt{3}}{2} F_8. \quad (36)$$

In order that also the hadronic charges $H_\pm$ and H_3 should close under commutation to form an SU(2) algebra the parameters a and b must satisfy the following condition

$$a^2 + b^2 = 1, \quad (37)$$

and we may write $a = \cos\theta$ and $b = \sin\theta$. A similar argument involving both the vector and the axial vector charges leads to the same conclusion ensuring a universal strength for the leptonic current and the hadronic current as given in the Cabibbo model.

Before considering the experimental status of the Cabibbo model we note that starting from the commutation relations (29) it is easily established that the hyperchange changing semileptonic processes must satisfy the $|\vec{\Delta I}| = \frac{1}{2}$ rule and, hence, also the $\Delta Y = \Delta Q$ rule. Conversely, any contribution from $|\vec{\Delta I}| = \frac{3}{2}$

requires a departure from the idea that the weak hadronic current is contained within a unitary octet. The validity of the $|\Delta \vec{I}| = \frac{1}{2}$ rule is, therefore, a crucial test of the model.

In applications of the Cabibbo model we are interested in matrix elements of the type $<B_k|v_\mu(0)|B_n>$ and/or $<B_k|a_\mu(0)|B_n>$ where B_k and B_n are members of the baryon octet. The fact that the vector currents are directly related to the SU(3) generators implies the following simple structure for the matrix elements of the vector current (in the limit of vanishing momentum transfer)

$$<B_k(p')|v_\mu(0)|B_n(p)> = \frac{1}{(2\pi)^3}\sqrt{\frac{M_k M_n}{p_0 p_0'}}\,\bar{u}(p')\gamma_\mu \left[\cos\theta\, f_{1+i2,kn} + \sin\theta\, f_{4+i5,kn}\right]u(p) \,. \tag{38}$$

For the matrix element of the axial vector current we must recognize that in the decomposition of a tensor product of two SU(3) octets the octet representation occurs twice corresponding to so called F and D type couplings. Hence, in the limit of zero momentum transfer one finds

$$<B_k(p')|a_\mu(0)|B_n(p)> \frac{1}{(2\pi)^3}\sqrt{\frac{M_k M_n}{p_0 p_0'}} \times$$

$$\times \bar{u}(p')\gamma_\mu\gamma_5 \left[(\cos\theta\, f_{1+i2,kn} + \sin\theta\, f_{4+i5,kn})\,F + (\cos\theta\, d_{1+i2,kn} + \sin\theta\, d_{4+i5,kn})\,D\right]u(p) \,, \tag{39}$$

where F and D denote unknown reduced matrix elements. The
structure constants f_{ijk} and the related symmetric constants
d_{ijk} are tabulated for example in Gell-Mann's original paper
on the eightfold way[32]. The two matrix elements (38) and (39)
contain three unknown parameters θ, F, and D. The parameters
F and D must be real if time-reversal invariance holds. The
most recent experimental fits give the following values[35]

$$\theta = 0.242 \pm 0.004 \, ,$$
$$F = 0.460 \pm 0.015 \, , \quad (40)$$
$$D = 0.771 \pm 0.016 \, .$$

On the basis of (38) and (39) we may now express all the
semileptonic decays of baryons ($B_n \to B_k + \ell + \bar{\nu}_\ell$) in terms
of the three parameters θ, F, and D. Some results are listed
in Table 7 where we have also given (i) the numerical values
obtained from (40) and (ii) the experimental values reported
in the literature. Since the A/V ratio for the neutron
usually is used to determine the Cabibbo parameter it is seen
that much experimental work remains until one can convincingly
argue that the Cabibbo model has been confirmed for the baryon
decays. In making the comparison between theory and experiments one should bear in mind that (i) the theoretical predictions in Table 7 have been derived under the explicit assumptions that effects of SU(3) breaking are negligible in the

	vector current matrix element (V)	axial vector current matrix element (A)	A/V "theoretical"	A/V experimental
$n \to p + e^- + \bar{\nu}_e$	$\cos\theta$	$(F+D)\cos\theta$	1.23	1.23 ± 0.01
$\Sigma^- \to \Lambda + e^- + \bar{\nu}_e$	0	$2/\sqrt{6}\, D\cos\theta$	V/A=0	$V/A = 0.35 \pm 0.18$
$\Xi^- \to \Xi^0 + e^- + \bar{\nu}_e$	$-\cos\theta$	$(D-F)\cos\theta$	-0.31	—
$\Lambda \to p + e^- + \bar{\nu}_e$	$-3/\sqrt{6}\,\sin\theta$	$-1/\sqrt{6}(D+3F)\sin\theta$	0.71	0.63 ± 0.08^{35}
$\Sigma^- \to n + e^- + \bar{\nu}_e$	$-\sin\theta$	$(D-F)\sin\theta$	-0.31	$|A/V| = 0.21 \pm 0.12^{35}$
$\Sigma^0 \to p + e^- + \bar{\nu}_e$	$-1/\sqrt{2}\,\sin\theta$	$1/\sqrt{2}(D-F)\sin\theta$	-0.31	—
$\Xi^- \to \Lambda + e^- + \bar{\nu}_e$	$3/\sqrt{6}\,\sin\theta$	$1/\sqrt{6}(3F-D)\sin\theta$	0.21	—
$\Xi^- \to \Sigma^0 + e^- + \bar{\nu}_e$	$1/\sqrt{2}\,\sin\theta$	$1/\sqrt{2}(F+D)\sin\theta$	1.23	—
$\Xi^0 \to \Sigma^+ + e^- + \bar{\nu}_e$	$\sin\theta$	$(F+D)\sin\theta$	1.23	—

Table 7: The hadronic part of the matrix elements for semileptonic baryon decays within the Cabibbo model. The "theoretical" A/V ratio is obtained by inserting the latest fits for the Cabibbo parameters F and D.

matrix elements and (ii) terms proportional to the momentum transfer can be neglected. Both assumptions merit special attention if any serious disagreement should arise.

Accepting the theoretical predictions of Table 7 regarding the hadronic part of the matrix elements for the semileptonic baryon decays one may proceed to compute the decay rates. The predictions so obtained agree quite well with the experimental results, but they are not particularly sensitive tests of the Cabibbo model and we refrain from quoting any numbers.

Additional information regarding the Cabibbo model is to be gained from the pseudoscalar meson decays; to wit $K_{\ell 2}$, $K_{\ell 3}$ and the corresponding decays for the π mesons. For $K_{\ell 2}$ and $\pi_{\ell 2}$ only the axial vector current contributes while in the $K_{\ell 3}$ and $\pi_{\ell 3}$ decays only the vector current yields a nonvanishing hadronic matrix element.

Considering first the $K_{\ell 2}$ and $\pi_{\ell 2}$ decays we deduce from the transformation properties of the hadronic matrix elements the following general expressions

$$<0|a_\lambda(0)|K^-> = \frac{1}{(2\pi)^{\frac{3}{2}}} \frac{1}{\sqrt{2E_K}} f_K p_\lambda^K ,$$

$$<0|a_\lambda(0)|\pi^-> = \frac{1}{(2\pi)^{\frac{3}{2}}} \frac{1}{\sqrt{2E_\pi}} f_\pi p_\lambda^\pi ,$$

(41)

where f_k and f_π are constants. In the limit of exact SU(3) invariance for the strong interactions and with the assumed SU(3) transformation properties for the axial vector currents one has $f_\pi = f_k$. If we for the moment accept this result we obtain the following expression for the branching ratio

$$\frac{\Gamma(K^- \to \mu^- + \bar{\nu}_\mu)}{\Gamma(\pi^- \to \mu^- + \bar{\nu}_\mu)} = \tan^2\theta \, \frac{m_K}{m_\pi} \, \frac{(1 - m_\mu^2/m_K^2)^2}{(1 - m_\mu^2/m_\pi^2)^2} \,. \tag{42}$$

Inserting the experimental value for the branching ratio[26] one finds

$$\theta \equiv \theta_A^M = 0.269 \pm 0.001, \tag{43}$$

which is to be compared with the result (40) obtained from the baryon decays. To account for the discrepancy it is most natural to question the assumption $f_\pi = f_k$. For the axial vector current matrix elements there is no Ademollo-Gatto theorem and the effects of SU(3) violations may be appreciable. Adopting this attitude we shall later use the result (43) to estimate the effects of the SU(3) violations.

Turning next to the $K_{\ell 3}$ decays it follows from the transformation properties of the matrix element under SU(3) and the Lorentz group that the hadronic part may be written

$$<\pi^0(p')|v_\lambda(0)|K^-(p)> = \frac{1}{(2\pi)^3} \frac{1}{\sqrt{4p_0 p_0'}} f_{4+i5,jk} \, [\tilde{f}_+(q^2)(p+p')_\lambda +$$
$$+ \tilde{f}_-(q^2)(p-p')_\lambda] \,, \tag{44}$$

where $f_{4+i5,jk}$ is the appropriate SU(3) structure constant and $q = p' - p$. In the limit of exact SU(3) where the masses are degenerate and the current is conserved one finds

$$\tilde{f}_+(0) = 1 \,,$$
$$\tilde{f}_-(0) = 0 \,. \tag{45}$$

More often one expresses the matrix element (44) in terms of the form factors $f_\pm(q^2)$ (cf. the eq.(24)), where

$$f_\pm(q^2) = f_{4+i5,jk} \tilde{f}_\pm(q^2) \,. \tag{46}$$

Exact SU(3) symmetry then implies

$$f_+(0) = 1/\sqrt{2} \,, \tag{47}$$

and it follows from the Ademollo-Gatto theorem that this result is good to second order in the SU(3) breaking. The form factors $f_\pm(q^2)$ are not expected to vary strongly over the physical range of q^2- values in these decays. Furthermore, one may use the energy-momentum conservation and the Dirac equation for the leptons to conclude that in the full matrix element for $K_{\ell 3}$ the term proportional to $f_-(q^2)$ is multiplied by the lepton mass. For the electronic mode one may then neglect this term in comparison with the term proportional to $f_+(q^2)$. If we neglect the q^2 dependence of $f_+(q^2)$ and consider the K_{e3} decay rate one then deduces the following experimental result for the Cabibbo angle

$$\sqrt{2}\, f_+(0)\sin\theta = 0.255 \pm 0.006, \qquad (48)$$

and inserting (47)

$$\theta \equiv \theta_V^M = 0.277 \pm 0.006, \qquad (49)$$

which agrees quite well with the result from the baryon decays.

If we accept the result (48) as the true value for the Cabibbo angle we may estimate the SU(3) breaking as measured by the violation of the relation $f_\pi = f_K$. Retaining these as separate constants we deduce

$$\frac{\Gamma(K^- \to \mu^- + \bar{\nu}_\mu)}{\Gamma(\pi^- \to \mu^- + \bar{\nu}_\mu)} = \frac{f_K^2}{f_\pi^2} \tan^2\theta \, \frac{m_K}{m_\pi} \frac{(1 - m_\mu^2/m_K^2)^2}{(1 - m_\mu^2/m_\pi^2)^2}, \qquad (50)$$

which replaces (42). From (48) and (50) we obtain

$$\frac{f_K}{f_\pi} \frac{1}{\sqrt{2}\,f_+(0)} = 1.21 \pm .13 \qquad (51)$$

If we take $f_+(0) = 1/\sqrt{2}$ we conclude that

$$\frac{f_K}{f_\pi} = 1.21 \pm .13, \qquad (52)$$

which does not seem unreasonable since there is no Ademollo-Gatto theorem operating in this case as we have previously noted.

This concludes our discussion of the Cabibbo model. At present one may safely state that there is no serious disagreement between the predictions of the model and experiments, but there is neither any compelling argument speaking in its favor. We shall have to wait for more decisive tests before a firm decision can be made.

5. The current algebra of SU(3) x SU(3)

The vector charges F_i have previously been given in terms of the weak vector currents. In the same fashion we may define eight axial vector charges F_i^5 by

$$F_i^5 = \int d^3x \, a_0^i(x) \tag{53}$$

One may then replace the commutation rule (29) by the following set of cummutation relations

$$[F_i, F_j] = i \, f_{ijk} \, F_k ,$$
$$[F_i, F_j^5] = i \, f_{ijk} \, F_k^5 . \tag{54}$$

It has been suggested[36] that the algebra of vector and axial vector charges be closed by specifying also the commutation relations for the F_j^5's. For example we may choose

$$[F_i^5, F_j^5] = i \, f_{ijk} \, F_k . \tag{55}$$

Since the operators F_i and F_i^5 in general are time dependent all commutation relations are to be interpreted as equal time commutation relations.

If we now define

$$F_i^L = (F_i + F_i^5)/2,$$
$$F_i^R = (F_i - F_i^5)/2, \qquad (56)$$

it is easily verified that

$$[F_i^{L,R}, F_j^{L,R}] = i f_{ijk} F_k^{L,R},$$
$$[F_i^L, F_j^R] = 0, \qquad (57)$$

from which we learn that $\{F_i^L\}$ and $\{F_i^R\}$ each separately span an SU(3) algebra. Since the two algebras are mutually commuting we here deal with the direct sum, that is, the vector and the axial vector charges from the algebra SU(3) x SU(3). Provided we use the same Cabibbo angle for the vector currents, as prescribed by the original Cabibbo model, it is clear that the charges appearing in the theory of weak interactions all are of the type F_i^L. We may also express this fact by stating that the weak charges belong to the representation (8,1) of the algebra SU(3) x SU(3). An unequal mixture of vector and axial vector contributions to the weak current would invalidate this assignment.

Assuming the commutations relations (54) and (55) to be correct one may ask for the corresponding commutation relations for the underlying current densities $v^i_\mu(x)$ and $a^i_\mu(x)$, etc. One must then recognize that the commutation properties of the current densities are not uniquely specified by the commutation rules for the charges. One has tried various approaches to circumvent this inherent arbitrariness. For example, one has worked out the commutation rules for the current densities in different field theory models starting from the canonical quantization rules for the fields. The currents are then identified with the source terms in the field equations. However, great care must be exercised in this approach to avoid inconsistencies[37]. In particular, the appearance of gradient terms, the so-called Schwinger terms, in the commutation relations between space and time components of current densities requires special attention. Since the explicit form of the Schwinger terms is model dependent it is important to notice that in many applications of current algebra the results are insensitive to the appearance or non-appearance of Schwinger terms.

A more direct approach to remedy the arbitrariness in the commutation relations is to identify the current densities directly with vector fields[38]. The current densities then appear as the dynamical variables of the theory subject to the canonical quantization rules. In this approach the Schwinger terms are given once and forever, and one deals with an algebra of fields rather than with a current algebra. It has been shown that all the successful results of current algebra in fact can be obtained also in this way. This should be no real surprise since we have observed that many of the current algebra results are independent of the Schwinger terms, and in a sense the algebra of fields approach amounts to a specific choice for these terms.

We end this brief outline of the current algebra approach by noting that the cummutation rules are non-linear relations and, hence, they fix the relative scale of the quantities which appear in them. To the extent that one can compute matrix elements of these relations between physical states important information is to be gained. However, in practice one usually has to involve additional assumptions or approximations and the relevance of the results then also reflects the validity of these additional assumptions. In the next few sections we shall examine some of these aspects of the theory.

6. The conserved vector current hypothesis and the partial conservation of the axial vector current

We have already several times referred to the conserved vector current (CVC) hypothesis. In its most general form it amounts to the following set of conservation laws

$$\frac{\partial}{\partial x_\mu} v^i_\mu(x) = 0 , \qquad (58)$$

where $\{v^i_\mu(x)\}^8_{i=1}$ is the vector current density octet out of which the infinitesimal generators $\{F_i\}$ of the strong interaction SU(3) symmetry group are constructed. It is then part of the CVC hypothesis that the weak currents and the isovector part of the electromagnetic current belong to the same unitary octet. This in turn implies that there are relations between weak and electromagnetic processes, which can be used to test the soundness of the hypothesis. Originally the CVC idea was restricted to the $\Delta Y = 0$ weak processes and it is clear, that it is most reliable within this domain. This is so since we are then exploiting only the isospin invariance which is a better symmetry for the strong interactions than SU(3) as, for example, mass differences indicate. From the experimental point of view the CVC predictions for $\Delta Y = 0$ processes are very well satisfied. Tests refer to the following:

(i) The non-renormalization of the weak charge, that is, the form factor $F_1(q^2)$, defined by

$$<p(p')|v_\mu(0)|n(p)> = \frac{1}{(2\pi)^3}\sqrt{\frac{m_N^2}{p_0'p_0}}\;\bar{u}(p')[\gamma_\mu F_1(q^2)$$

$$+ i\sigma_{\mu\nu}q^\nu F_2(q^2) + q_\mu F_3(q^2)]\,u(p), \qquad (59)$$

satisfies

$$F_1(0) = 1. \qquad (60)$$

It follows from this relation, eq. (9) and the Cabibbo model for the hadronic current that the experimentally observed effective coupling constant in nuclear beta decay is

$$G_\beta \equiv G\,F_1(0)\,\cos\theta = G\cos\theta, \qquad (61)$$

which should be compared with the coupling constant for, say, μ-decay

$$G_\mu \equiv G. \qquad (62)$$

Experimentally one has (after correcting for electromagnetic effects)

$$G_\beta\big|_{0^{14}} = (1.3995 \pm 0.0094) \times 10^{-49}\;\text{erg cm}^3,$$

$$G_\mu\big|_{\text{exp}} = (1.4351 \pm 0.0011) \times 10^{-49}\;\text{erg cm}^3.$$

Accepting $\cos\theta = 0.972$, we obtain from (61)

$$G\big|_{O^{14}} = (1.4399 \pm .0094) \times 10^{-49} \text{ erg cm}^3 \ .$$

in excellent agreement with the result from μ decay.

(ii) The relation between the weak and electromagnetic form factors in, say, nuclear beta decay. More precisely, the CVC hypothesis implies that

$$F_1(q^2) = F_1^V(q^2) \ ,$$

$$F_2(q^2) = \frac{\mu_n - \mu_p}{2M_n} F_2^V(q^2) \ , \quad (63)$$

$$F_3(q^2) = 0 \ ,$$

where F_1^V and F_2^V are the isovector nucleon electromagnetic form factors, and where μ_p and μ_n denote the proton and the neutron anomalous magnetic moments. The coefficient $(\mu_p - \mu_n)/2M_n$ in front of $F_2^V(q^2)$ has been checked experimentally by measuring the slope of the electron spectrum in the weak decays of N^{12} and B^{12} which are members of an isotriplet with C^{12*} as the third member. The CVC prediction is well borne out by the experimental findings[39].

(iii) The beta decay of the π meson, that is,

$$\pi^- \to \pi^0 + e^- + \bar{\nu}_e \ ,$$

for which the rate is uniquely predicted by CVC. The analysis is very similar to the analysis of the K_{e3}^- decay previously discussed. For the hadronic part of the matrix element one finds

$$<\pi^0(p')|v_\mu(0)|\pi^-(p)> = \frac{1}{(2\pi)^3}\sqrt{\frac{1}{4p_0 p_0'}} \; f_{1+i2,jk}$$

$$\times [\tilde{f}_+(q^2)(p+p')_\mu + \tilde{f}_-(q^2)(p-p')_\mu]. \tag{64}$$

Defining

$$f_\pm(q^2) = f_{1+i2,jk} \, \tilde{f}_\pm(q^2) ,$$

we deduce from CVC

$$f_+(q^2) = \sqrt{2}\, F_\pi(q^2) ,$$
$$f_-(q^2) = 0 , \tag{65}$$

where $F_\pi(q^2)$ is the electromagnetic form factor of the π meson normalized so that $F_\pi(0) = 1$. Since the energy release is very small in this process it is reasonable to neglect the q^2-dependence of $f_+(q^2)$ and one obtains

$$\Gamma(\pi_{e3}^-) = \frac{G^2}{30\pi^3} \Delta^5 \; \phi(\frac{m_e}{\Delta}) , \tag{66}$$

with

$$\phi(x) = \sqrt{1-x^2}\,(1 - \frac{9}{2}x^2 + 4x^4) + \frac{15}{2}x^4 \ln(1 + \frac{\sqrt{1+x^2}}{x})$$

and $\Delta \equiv (E_e)_{max} = (4.52 \pm 0.10)$ MeV. From this then follows (including isoactive corrections)

$$\left.\frac{\Gamma(\pi_{e3})}{\Gamma(\pi_{\mu 2})}\right|_{th.} = 1.05 \times 10^{-8},$$

which should be compared with[40]

$$\left.\frac{\Gamma(\pi_{e3})}{\Gamma(\pi_{\mu 2})}\right|_{exp} = (1.023 \pm .069) \times 10^{-8}.$$

(iv) The supression of the vector contribution in the decay $\Sigma^{\pm} \rightarrow \Lambda^0 + e^{\pm} + \nu_e$. We have previously found that in the Cabibbo model the vector contribution vanishes in the limit of zero momentum transfer. More generally we obtain

$$<\Lambda(p')|v_\mu(0)|\Sigma^-(p)> = \frac{1}{(2\pi^3)}\sqrt{\frac{m_\Lambda m_\Sigma}{p_0' p_0}} \bar{u}(p')[\gamma_\mu \hat{F}_1(q^2) + i\sigma_{\mu\nu} q^\nu \hat{F}_2(q^2) + q_\mu \hat{F}_3(q^2)] u(p) \quad . \quad (67)$$

The term $\hat{F}_3$ is negligible in this decay since after contraction with the lepton part it is proportional to m_e. From CVC it then follows that $(M_\Sigma - M_\Lambda)\hat{F}_1(q^2) = 0$ or $\hat{F}_1(q^2) = 0$. Hence only the weak magnetism term survives here. It is expected to be small since it is proportional to q_μ. Exploiting CVC further it may be related to the Σ-Λ transition magnetic moment, which governs the $\Sigma^0 \to \Lambda + \gamma$ decay. Rough estimates yield a vector contribution several orders of magnitude smaller than the axial vector contribution to the decay rate. Of course, a better place to look for a possible vector contribution is in the electron-nuetrino angular correlations or in parity-violating effects, which are due to the interference between vector and axial vector contributions. Due to poor experimental accuracy the findings so far are inconclusive, but consistent with the CVC predictions.

The success of the conserved vector current hypothesis makes it tantalizing to try the same prescription also for the axial vector current densities. However, a conserved axial vector current density hypothesis immediately leads to predictions which cannot be reconciled with the experimental findings. For example, it predicts zero rate for the decay $\pi \to \mu + \nu_\mu$ which happens to be the dominant decay mode for the charged π mesons. Also, there is no exact symmetry observed in nature that calls for a conserved axial vector current. Noting that the divergence of the axial vector current has the same quantum numbers as the π-meson for the hypercharge conserving current $a_\mu^{(0)}(x) \equiv a_\mu'(x) + ia_\mu^2(x)$ and as the K-meson for the hypercharge changing current $a_\mu^{(1)}(x) \equiv a_\mu^4(x) + ia_\mu^5(x)$ it has been suggested[41] that the following relations hold

$$\partial^\mu a_\mu^{(0)}(x) = f_\pi m_\pi^2 \phi_\pi(x) , \qquad (68)$$

$$\partial^\mu a_\mu^{(1)}(x) = f_K m_K^2 \phi_K(x) . \qquad (69)$$

These two equations constitute the hypothesis of a partially conserved axial vector current (PCAC). In the limit $m_\pi(m_K) \to 0$ the currents are conserved reflecting a symmetry which, however, is broken in the real world. It is generally assumed that the mechanism by which this symmetry is broken also is responsible for the mass of the $\pi(K)$ meson being non-vanishing in which case the mass is a measure of how badly the symmetry is broken. One may rephrase the PCAC hypothesis in dispersion-theoretic language by insisting that all matrix elementa of $\partial^\mu a_\mu(x)$ obey unsubtracted dispersion relations in the momentum transfer variable, and that these relations are dominated by the $\pi(K)$ meson pole contribution.

There is no easy way to verify the PCAC hypothesis. Mostly one must invoke additional assumptions and/or approximations to reach physical predictions which can be tested. Some applications will be considered in the next section.

Adler[42] has suggested a consistency condition for the iso-spin symmetric πN scattering amplitude derived solely from PCAC as a clean test. However, the resulting condition involves the πN amplitude at a point in the non-physical region, and one must rely on approximations before one may compare with experimental results. A fair estimate yields agreement between theory and experiment to within 10%.

7. The Goldberger-Treiman relation and the Adler-Weisberger sum rule

The Goldberger-Treiman relation reads

$$f_\pi = \frac{\sqrt{2}\, m_N g_A}{g_r}, \qquad (70)$$

and it expresses the π meson decay constant

$$f_\pi \equiv \frac{1}{\sqrt{2q_0}} <\pi^+(q) | \partial^\mu a_\mu^{(0)}(x) | 0> \bigg|_{x=0} \times \frac{1}{m_\pi^2}, \qquad (71)$$

in terms of the πN coupling constant g_r ($g_r^2/4\pi \simeq 14.6$) and the ratio

$$g_A \equiv \frac{G_1(0)}{F_1(0)} \qquad (72)$$

between the effective coupling strength of the axial current and the vector current in beta decay. We have previously defined $F_1(q^2)$ in eq. (59) and $G_1(q^2)$ is defined analogously by the following equation

$$<p(p')|a_\mu(0)|n(p)> = \frac{1}{(2\pi)^3}\sqrt{\frac{m_N^2}{p_0'p_0}}\ \bar{u}(p')[\gamma_\mu G_1(q^2) +$$

$$+ q_\mu G_3(q^2) + i\sigma_{\mu\nu}q^\nu G_2(q^2)]\gamma_5 u(p) \ . \tag{73}$$

The first derivation [43] of the GT relation was based on dispersion-theoretic arguments and required rather drastic approximations. Nevertheless, it turned out to be accurate to within 10%, and PCAC hypothesis was originally introduced[41] to put the derivation on firmer ground. We recall the normalization of the π meson field

$$<\pi^+(q)|\phi_\pi^+(x)|0>\bigg|_{x=0} = \frac{1}{\sqrt{2q_0}} \ , \tag{74}$$

and the definition of the πN coupling constant

$$<p(p')|(\Box^2 + m_\pi^2)\phi_\pi^+(0)|n(p)> = (-q^2 + m_\pi^2)<p(p')|\phi_\pi^+(0)|n(p)> \equiv$$

$$\equiv \frac{1}{(2\pi)^3}\sqrt{\frac{m_N^2}{p_0'p_0}}\ i\sqrt{2}\ g_r(q^2)\bar{u}(p')\gamma_5 u(p) \ , \tag{75}$$

with $g_r \equiv g_r(m_\pi^2)$ and $q = p' - p$. Taking the divergence of (73), recalling the PCAC relation (68) and equating this with (75) one finds in the limit $q^2 \to 0$

$$f_\pi = \frac{\sqrt{2}\, m_N\, G_1(0)}{g_r(0)} \quad . \tag{76}$$

Assuming $g_r(q^2)$ to be a slowly varying function so that $g_r(0) \simeq g_r(m_\pi^2) \equiv g_r$ and further recalling that $F_1(0) = 1$ if CVC holds we finally arrive at the GT relation (70). From this brief outline of the derivation it is clear that the accuracy by which this relation holds is a rather direct check of the PCAC hypothesis for the hypercharge conserving current. As one would expect the corresponding relation for the hypercharge changing current is less accurate.

In many applications of current algebra the PCAC hypothesis plays a vital role. As a matter of fact this hypothesis is absolutely essential in order to evaluate the matrix elements of the commutator relations which current algebra prescribes[44]. The first successful application of current algebra in combination with the PCAC hypothesis along these lines is the Adler-Weisberger relation[45]. The derivation of this relation starts from the commutator (cf. eq. 55)

$$[F_1^5 + iF_2^5,\ F_1^5 - F_2^5] = 2F_3 \equiv 2I_3 \tag{77}$$

of the SU(3) x SU(3) integrated current algebra. Considering the eq. (77) taken between proton states of equal momenta one introduces a complete set of intermediate states on the left hand side. The only single-particle contribution to the sum over intermediate states comes from the neutron and the corresponding term is easily evaluated. It is clearly proportional to $G_1(0)$ or g_A. To obtain the contribution from multiparticle intermediate states the PCAC hypothesis is invoked and after some formal computations one can relate this sum to the total cross section for πN scattering. More precisely one finds

$$1 - \frac{1}{g_A^2} = \frac{4m_N^2}{g_r^2(0)} \cdot \frac{1}{\pi} \int_{m_N+m_\pi}^{\infty} \frac{\omega d\omega}{\omega^2 - m_N^2} \times [\sigma_0^+(\omega) - \sigma_0^-(\omega)], \quad (78)$$

where $\sigma_0^\pm(\omega)$ is the total cross section for scattering of a zero mass $\pi^\pm$ on a proton at center of mass energy ω. If the Pomeranchuk theorem holds, as present experimental results seem to indicate, the integrand vanishes for large ω. In any case we know from experimental results concerning real π_p scattering that the difference between the two cross sections is largest for energies in the N* region, which is easily accessible for experimental investigations. Of course, the experimental data on $\pi_p \to \pi p$ scattering must be extrapolated to zero mass for the π-mesons. This introduces uncertainties, but a careful analysis yields

$$|g_A| = 1.24 \pm 0.03,$$

which should be compared with the experimental value[40]

$$g_A|_{exp} = 1.23 \pm 0.01.$$

Analogous results for the hypercharge changing current have been obtained[46] starting from

$$[F_4^5 + iF_5^5, F_4^5 - iF_5^5] = F_3 + \sqrt{3}\, F_8 , \qquad (79)$$

relating the axial vector coupling constants in the beta decay of hyperons to the corresponding scattering processes. The lack of accurate experimental data makes the check less stringent. Also, one expects the PCAC hypothesis for $a_\mu^{(1)}(x)$ to be less accurate than in the case of $a_\mu^{(0)}(x)$.

8. Low energy theorems and current algebra

There is a vast literature on applications of the PCAC hypothesis in evaluating matrix elements of current or current density commutation relations which follow from SU(3) x SU(3) or any of its subgroups. Time does not permit me to give more than a very brief account of the accomplishments in this field of research. Results fall broadly into two categories; (i) low energy theorems for soft π-meson emission and low energy theorems for the emission of a photon or a virtual lepton pair

($\bar{\mu}\nu_\mu$ or $\bar{e}\nu_e$) of zero four-momentum and (ii) sum rules. The low energy theorems express the matrix element for the emission of, say, a π-meson of zero four-momentum in terms of the matrix element for the same process without a π-meson emitted. As an example let me outline the derivation of the Callan-Treiman relation[47]. Consider the quantity M_μ defined by

$$M_{\mu\nu}^{(ij)} = i\int d^4x \, \exp[-iqx]\,\theta(x_0)<0|\left[a_\mu^{(i)}(x), h_\nu^{(j)}(0)\right]|K^+>. \quad (80)$$

If we integrate by parts and neglect the surface terms we find

$$iq^\mu M^{(ij)} = i\int d^4x \, \exp[-iqx]\,\theta(x_0)<0|\,[\partial^\mu a_\mu^{(i)}(x), h_\nu^{(j)}(0)]|K^+> +$$

$$+ i\int d^4x \, \exp[-iqx]\,\delta(x_0)<0|\,[a_0^{(i)}(x), h_\nu^{(j)}(0)]|K^+>. \quad (81)$$

PCAC and the Goldberger-Treiman relation yield

$$\partial^\mu a_\mu^{(i)}(x) = \frac{m_N m_\pi^2 g_A}{g_r} \phi_\pi^{(i)}(x). \quad (82)$$

Next, let $q \to 0$ in eq. (81). One can then show that the left hand side vanishes and one remains with

$$i\int d^4x\,\delta(x_0)<0|\,[a_0^{(i)}(x), h_\nu^{(j)}(0)]|K^+> =$$

$$= -\frac{ig_A m_N}{g_r} \int d^4x\,(m_\pi^2 - \square^2)\,\theta(x_0)<0|\,[\phi_\pi^{(i)}(x), h_\nu^{(j)}(0)]|K^+>. \quad (83)$$

Now choose for $a_\mu^{(i)}(x)$ the neutral, hypercharge conserving axial current ($i = 3$) and for $h_\nu^{(j)}(0)$ the full hypercharge changing current ($j = 4 + i5$). On the left hand side we further make use of the prescribed current density commutation relation

$$[a_0^{(3)}(x), h_\nu^{(4+i5)}(0)]_{x_0=0} = \frac{1}{2} \delta^3(\vec{x}) h_\nu^{(4+i5)}(0).$$

(84)

It is now seen that inserting (84) in (83) the left hand side is proportional to the hadronic part of the matrix element for $K^+ \to \ell^+ + \nu_\ell$ and the right hand side is proportional to the corresponding matrix element for $K^+ \to \pi^0 + \ell^+ + \nu_\ell$ evaluated at zero four-momentum for the π^0-meson. By choosing other current densities in (80) and/or other states when we take the matrix element, analogous relations are obtained pertaining to other processes. Clearly one must exercise care in the formal manipulations. We have also mentioned previously the appearance of so-called Schwinger terms when one considers commutation relations between current densities. We shall not discuss these technical points further but rather summarize some of the results with regard to semi-leptonic processes. For a review of the non-leptonic processes we refer to the paper by Pakvasa and Rosen[15].

(i) Callan-Treiman type relations

We have already outlined the derivation of the Callan-Treiman relation in which the amplitude for $K_{\ell 3}$ decay is related to the corresponding amplitude for $K_{\ell 2}$ decay. The final result reads

$$f_K = \frac{2g_A m_N}{g_r} \cdot [f_+(0) + f_-(0)], \qquad (85)$$

where f_K is the decay constant for $K_{\ell 2}$ decay defined in eq. (41) and f_+ and f_-, which pertain to $\bar{K}_{\ell 3}$ decay, have been defined in eqs. (44) and (46). Experimentally one has

$$f_K \Big|_{\exp} = (0.070 \pm .001) m_K$$

To evaluate the right hand side we take notice of the experimental value for $f_+(0)$ determined in K_{e3} decay[40]

$$f_+(0) \Big|_{\exp} = 0.16 \pm 0.01,$$

and we deduce the following value for

$$\xi(0) \equiv \frac{f_-(0)}{f_+(0)}, \qquad (86)$$

$$\xi(0) \Big|_{th} = 0.36 \pm 0.09.$$

This result should be compared with the experimental value[40]

$$\xi(0)\Big|_{exp} = -0.72 \pm 0.21 \quad .$$

A word of caution is in place here. The experimental situation with regard to the $K_{\ell 3}$ decay parameters is still unclear with different methods for the determination of the parameters yielding conflicting results[48]. However, recent experiments seem to indicate a rather large and negative value for $\xi(0)$ while the Callan-Treiman relation favors a positive and possibly small value of $\xi(0)$. This discrepancy has lead to many speculations[49] taking into account finer details of the form factors dependence on q^2 etc. Some improvements in the relation between theory and experiment has been accomplished but the question whether the two sides can be reconciled is still an open one.

One can, of course, obtain the analogous relation for the $\pi_{\ell 2}$ and $\pi_{\ell 3}$ decays. It reads[47]

$$f_\pi = \frac{g_A m_N}{g_r} f_+(\pi) , \qquad (87)$$

and $f_{\pi\,exp} = 0.94\, m_\pi$ while the experimental results for

the quantities on the right hand side give the value
0.85 m_π.

Applying the same line of reasoning one can also derive relations for the form factors in $K_{\ell 4}$ decays in terms of the $K_{\ell 3}$ decay parameters. As shown by Weinberg[50] the predictions so obtained are very well satisfied by the experimental results.

(ii) Sum rules

The derivation of sum rules involves many technical details which we shall not go into[51]. Among the sum rules there is one that has attracted particular attention, namely Adler's sum rule[52] concerning the total cross sections for high energy neutrino reactions. The sum rule follows rather directly from time-time component commutation relations and reads

$$\lim_{E_\nu \to \infty} \left[\frac{d\sigma(\nu p)}{dq^2} - \frac{d\sigma(\bar\nu p)}{dq^2} \right] = - \frac{G^2}{\pi} (\cos^2\theta + 2 \sin^2\theta) , \quad (88)$$

where $G \simeq 10^{-5}/m_N^2$ and θ is the Cabibbo angle. With the advent of new and powerful particle accelerators high energy neutrino beams may soon be available to test this sum rule. Of course, the question at what energy E_ν one may expect it to be satisfied cannot be answered with certainty even if there are indications that it may in fact be rather small (at least for small q^2-values).

8. Some open questions

Most of the information we have about the weak interactions has been derived from the limited phenomenology of low energy decay processes. The greater freedom afforded by collision processes is yet to be gained when high energy neutrino beams become more readily available. The new large machines under construction therefore will greatly enhance our possibilities to study the weak interactions. In that respect the future looks bright. We shall not forget, however, that the study of low energy phenomena continues to produce new insight and that we still live with many questions, which are not answered by our present understanding of the weak interactions. I shall here only mention some of these open questions.

CP and T violations in weak processes

It is now soon ten years since Fitch and Cronin reported the first indication of CP violation in a weak process, namely the decay $K_L^0 \to 2\pi$. A great deal of ingenuity both from experimentalists and theorists has gone into the question of where to put the blame on the failure of this invariance principle, and yet we still do not know the answer. Noting that CP and T violating effects so far only have been observed in processes involving the neutral K-meson system I shall not discuss the subject further since it will be reviewed by Professor Wolfenstein[53].

Elastic electron-neutrino scattering

The current x current model for the weak interactions leads to a very definite prediction for elastic electron-neutrino scattering as well as for elastic muon-neutrino scattering. It follows from the eqs. (9) and (11) that the processes

$$e^- + \nu_e \rightarrow e^- + \nu_e$$

$$e^- + \bar{\nu}_e \rightarrow e^- + \bar{\nu}_e$$

$$\mu^- + \nu_\mu \rightarrow \mu^- + \nu_\mu$$

$$\mu^- + \bar{\nu}_\mu \rightarrow \mu^- + \bar{\nu}_\mu$$

should occur with the strength of coupling equal to that of μ-decay. However, processes like

$$e^- + \nu_\mu \rightarrow e^- + \nu_\mu$$

$$\mu^- + \nu_e \rightarrow \mu^- + \nu_e$$

should not occur, at least not in lowest order. At the present time little is yet known about these processes. The one which has been examined most closely is the $\bar{\nu}_e + e^- \rightarrow \bar{\nu}_e + e^-$ where very intense but low energy antineutrino beams from a nuclear reactor have been used. So far one has only established an upper limit for the cross section[54]

$$\sigma(\bar{\nu}_e e \rightarrow \bar{\nu}_e e)\Big|_{exp} < 4\sigma_{th}$$

These processes are of particular interest with regard to the model of Gell-Mann et al [55] in which higher order effects for "diagonal" processes may change the effective coupling strength appreciably.

Second class currents

Long ago Weinberg[56] introduced the concept of first and second class hypercharge conserving currents on the basis of their G-transformation properties. A first class vector current is a current which is even under the G-transformation ($G = C \exp[i\pi I_2]$). A first class axial vector current is odd under the G-transformation. For the second class currents the opposite transformation properties hold. The existence of first class hypercharge conserving currents is established by the occurrance of the decays $\pi^+ \rightarrow \pi^0 + e^+ + \nu_e$ and

$\pi^+ \to \mu^+ + \nu_\mu$ since second class currents cannot contribute to these processes. But for semileptonic $\Delta Y = 0$ processes involving baryon states of indefinite G-parity both classes of currents contribute. A closer examination shows for example that the terms with $F_1(q^2)$ and $F_2(q^2)$ in (59) as well as those with $G_1(q^2)$ and $G_2(q^2)$ in (73) are contributions from first class currents while the terms with $F_3(q^2)$ respectively $G_3(q^2)$ come from second class currents. Hence, by investigating the distinctive spectrum and spin correlation effects due to F_3 and G_3 one may reach a conclusion of whether second class currents exist or not. A possible source of information may be high energy neutrino reactions like $\nu_\mu + n \to p + \mu^-$. Recently Wilkinson and Alburger[57] have reported results from beta decays of mirror nuclei indicating discrepancies as large as 20% in their ft-values. If we assume CPT and T invariance one can show that these discrepancies must come from second class currents. Since known T-violating effects are negligible with regard to effects of this size one may be inclined to interprete the data of Wilkinson and Alburger as verification of the presence of second class currents. However, one deals with nuclear states which contain isospin impurities due to electromagnetic effects and it is not inconceivable that the observed discrepancies can be explained as results of such isospin impurities of the nuclear states. Certainly this proposition should be much more carefully investigated and the fate of second class currents should also be examined in terms of high energy neutrino reactions, where the effects can be expected to be much larger.

K meson decays

We have already mentioned the persisting uncertainties and possible inconsistencies in our understanding of the $K_{\mu 3}$ decay. This is an area where both experimental and theoretical inadequacies are apparent. Experiments yield conflicting results for the decay parameters depending on the method used for their determination (muon polarization, the Dalitz plot or the $K_{e3}/K_{\mu 3}$ branching ratio). On the theoretical side Dashen-Weinstein type computations must be improbed so that the approximations are better controlled.

Recently[58] the K-mesons have given rise to another apparent discrepancy. One has determined the branching ratio for the decay $K_L^0 \rightarrow \mu^+ + \mu^-$ to be smaller than 1.8×10^{-9}. However, on theoretical grounds[59] one expects this branching ratio to be larger than 6×10^{-9} in violation of what experimental findings tell us. So far there is no explanation for this discrepancy.

The structure of weak interactions

When we introduced the current x current picture in eq. (9) we allowed for an intrinsic structure in terms of a structure function $\rho_{\lambda\lambda'}(x - x')$ in the effective S-operator. We noted that low energy phenomenology was consistent with

$$\rho_{\lambda\lambda'}(x - x') \simeq g_{\lambda\lambda'}\delta^4(x - x') ,$$

However, from simple unitarity considerations we also know that this approximation must break down if not before at least at energies of the order of 300 Bev. The past decade has seen many attempts to tackle the problem of this intrinsic structure of weak interactions. Inadequate experimental facilities have provided few opportunities to reach any conclusions so far and it still remains essentially virgin land to be explored in the years to come.

Most of the considerations regarding structure effects have centered around the idea that the weak interactions are mediated by intermediate vector bosons[60]. These particles are so far purely hypothetical and hence they can be asked to do many things including for example to provide for CP violations. Accordingly there are many versions of the intermediate vector boson model. For low energies where $q^2 \ll m_w^2$ the net result of such a model is the replacement of the weak interaction coupling constant G by a dimensionless coupling constant g characterizing the interaction of the boson field with the weak currents. More precisely one obtains

$$\frac{G}{\sqrt{2}} = \frac{g^2}{m_w^2} \quad . \qquad (89)$$

The effects on low energy phenomenology of the intermediate boson strongly depends on its mass. The larger its mass is the harder it will be to see any effects. Of course, the most direct proof of its existence would be to produce it and see it decay. Many attempts to do so have failed possibly indicating that its mass in fact is quite large. Present lower limit for m_w is set at somewhere around 2BeV. Speculative estimates based on theoretical considerations vary from 5BeV to about 40BeV[61]. If the latter value turns out to be right it will be rough going before anyone can establish its existence. The hunt for the intermediate vector boson is the weak counterpart to the strong search for quarks! In both cases, of course, the hunted objects may prove themselves very elusive if not altogether non-existing.

9. Conclusion

In summary we may state that for leptonic and semi-leptonic weak processes the V-A current x current theory provides a most satisfactory framework for understanding at least low energy phenomena. Any other intrinsic structure is not expected to alter this judgment. The past decade has seen much progress in our understanding of the properties of the hadronic current revealing interesting interrelations between the different types of interactions in nature, but not basically changed our view of the weak interactions themselves. This is certainly in part due to the lack of high energy and high intensity neutrino beams and it is not surprising that much attention has been paid to possible neutrino experiments in the discussions related to the new large accelerators that are under construction. With these new facilities coming into operation I am sure weak interaction physics will once more prove itself a most exciting field of research.

Reference List

1. W. Pauli, unpublished (1933).
2. F. Reines and C. L. Cowan, Phys. Rev. 113, 273 (1959).
3. E. Fermi, Nuovo Cimento 11, 1 (1934) and Zeitschrift für Physik 88, 161 (1934).
4. G. Gamow and E. Teller, Phys. Rev. 49, 895 (1936).
5. T. D. Lee and C. N. Yang, Phys. Rev. 104, 254 (1956.
6. C. S. Wu, et. al. Phys. Rev. 105, 1413(1957).
7. R. Garwin, L. Lederman and M. Weinreich, Phys. Rev. 105, 1415 (1957).
   J. I. Friedman and V. L. Telgdi, Phys. Rev. 105, 1681 (1957).
8. A. Salam, Nuovo Cimento 5, 299 (1957).
9. L. Landau, Nuclear Physics 3, 127 (1957).
10. T. D. Lee and C. N. Yang, Phys. Rev. 105, 1671 (1957).
11. E. C. G. Sudarshan and R. E. Marshak, Proc. Padua-Venice Conference on Mesons and Newly Discovered Particles 1957 and Phys. Rev. 109, 1869 (1958).
12. R. P. Feynman and M. Gell-Mann, Phys. Rev. 109, 193 (1958)
    J. J. Sakurai, Nuovo Cimento 7, 649 (1958).
13. O. Klein, Nature 161, 897 (1948)
    J. Tiomno and J. A. Wheeler, Rev. Mod. Phys. 21, 144 (1949)

T. D. Lee, M. Rosenbluth and C. N. Yang, Phys. Rev. 75, 905 (1949).

14. S. S. Gerstein and Y. B. Zeldovich, Soviet Phys. JETP 2, 576 (1956).

15. S. Pakvasa and S. P. Rosen, Fields and Quanta, 2, 93(1971).

16. T. Kirsten, O. A. Schaeffer, E. Norton and R. W. Stoenner, Phys. Rev. Letters 20, 1300 (1968).

17. T. Kirsten, W. Gentner and O. A. Schaeffer, Zeitschrift fur Physik 202, 273 (1967).

18. cf. review article by H. Primakoff and S. P. Rosen, Rept. Prog. Phys. 22, 121 (1959); Phys. Soc. (London) 78, 464 (1961).

19. R. K. Bardin, P. J. Gollon, J. D. Ullman and C. S. Wu, Phys. Letters 26B, 112 (1967).

20. E. Fiorini, reported to the XVth International Conference on High Energy Physics, Kiev 1970.

21. E. Fiorini et al., reported by L. Wolfenstein, Proc. Heidelberg Int. Conf. on Elementary Particles, North Holland Pub. Co. (1968).

22. F. Reines and C. L. Cowan, Phys. Rev. 113, 273 (1959) F. Reines, Ann. Rev. Nucl. Sci. 10, 1 (1960)

R. E. Carter et al., Phys. Rev. 113, 280 (1959).

23. R. Davis, Proc. Int. Conf. on Radio-Isotopes Sci. Res., Paris 1957, Pergamon Press (1958).

24. G. Danby et al., Phys. Rev. Letters 9, 36 (1962)

    J. L. Bienlein et al., Phys. Letters 13, 80 (1964).

25. B. Pontecorvo, Proc. Heidelberg Int. Conf. on Elementary Particles, North-Holland Pub. Co. (1968).

26. Particle Properties, Particle Data Group, April 1971.

27. T. D. Lee and C. N. Yang, Phys. Rev. 126, 2239 (1962)

    A. Pais, Phys. Rev. Letters 9, 117 (1962)

    A. Pais, Theoretical Physics, I. A. E. A., Vienna (1963).

28. L. B. Okun and J. B. Pontecorvo, Soviet Phys. JETP 5, 1297 (1957).

    See also R. E. Marshak et al., Theory of Weak Interactions in Particle Physics, Interscience, New York (1969), p. 536.

29. S. L. Glashow, Phys. Rev. Letters 6, 196 (1961).

30. See for example, R. E. Marshak et al., Theory of Weak Interactions in Particle Physics, Interscience, New York (1969), p. 472.

31. See for example, S. Bennett et al., Phys. Letters 27B, 244 (1968).

32. M. Gell-Mann, The Eightfold Way: A Theory of Strong Interaction Symmetry, Caltech preprint CTSL-20 (1961)
    Y. Ne'eman, Nucl. Physics 26, 222 (1961)
    A. Salam and J. C. Ward, Nuovo Cimento 20, 419 (1961).
33. N. Cabibbo, Phys. Rev. Letters 10, 531 (1963).
34. M. Ademollo and R. Gatto, Phys. Rev. Letters 13, 254 (1964).
35. Reported to the XVth International Conference on High Energy Physics, Kiev (1970).
36. M. Gell-Mann, Phys. Rev. 125, 1067 (1962)
    M. Gell-Mann and Y. Ne'eman, Annals of Physics 30, 360 (1965).
37. See for example, G. Kallen, Gradient Terms in Commutators of Currents and Fields in Particles, Currents and Symmetries, (ed. P. Urban), Springer Verlag, Wien (1968).
38. T. D. Lee, S. Weinberg and B. Zumino, Phys. Rev. Letters 18, 1029 (1967).
    T. D. Lee and B. Zumino, Phys. Rev. 163, 1667 (1967).
39. See for example, C. S. Wu, Rev. Mod. Physics 36, 618 (1963).
40. Rev. of Particle Properties, Rev. Mod. Physics 43, 51 (1971).
41. M. Gell-Mann and M. Lévy, Nuovo Cimento 16, 705 (1960)
    J. Bernstein et al., Nuovo Cimento 17, 757 (1960).
42. S. L. Adler, Phys. Rev. 137, 1022B (1965).
43. M. L. Goldberger and S. B. Treiman, Phys. Rev. 110, 1178 (1958).

44. A. P. Balachandran and H. Pietschmann, Nucl. Phys. $\underline{43}$, 321(196

45. S. L. Adler, Phys. Rev. Letters $\underline{14}$, 1051 (1965) and
    Phys. Rev. $\underline{140B}$, 736 (1965)
    W. I. Weisberger, Phys. Rev. Letters $\underline{14}$, 1047 (1965) and
    Phys. Rev. $\underline{143}$, 1302 (1966).

46. W. I. Weisberger, Phys. Rev. $\underline{143}$, 1302 (1966).
    D. Amati, C. Bouchiat and J. Nuyts, Phys. Letters $\underline{19}$, 59 (1965).

47. C. G. Callan and S. B. Treiman, Phys. Rev. Letters $\underline{16}$, 153 (1966).
    V. S. Mathur, S. Okubo and L. K. Pandit, Phys. Rev. Letters $\underline{16}$, 371 (1966).

48. For a rather complete report we refer to M. K. Gaillard and L. M. Chounet, Cern Yellow Report 70-14 (1970).

49. R. Dashen and M. Weinstein, Phys. Rev., $\underline{183}$, 1261(1969).
    R. Acharya and L. Brink, to be published.

50. S. Weinberg, Phys. Rev. Letters $\underline{17}$, 336 (1966).

51. For an excellent review of these questions see S. L. Adler, and R. F. Dashen, Current Algebras, Benjamin, New York (1968).

52. S. L. Adler, Phys. Rev. $\underline{143}$, 1144 (1966).

53. L. Wolfenstein, Fields and Quanta, $\underline{2}$, 293(1971).

54. F. Reines and H. S. Gurr, Phys. Rev. Letters 24, 1448 (1970).
55. M. Gell-Mann, M. L. Goldberger, N. Kroll and F. Low, Phys. Rev. 179, 1518 (1969).
56. S. Weinberg, Phys. Rev. 112, 1375 (1958).
57. D. H. Wilkinson, Phys. Letters 31B, 447 (1970).
    D. H. Wilkinson and D. E. Alburger, Phys. Rev. Letters 24, 1170 (1970).
    D. H. Wilkinson and D. E. Alburger, Phys. Letters 32B, 190 (1970).
58. A. R. Clark, et. al. Phys. Rev. Letters 26, 1667 (1971).
59. L. M. Seghal, Nuovo Cimento 45, 785 (1966) and Phys. Rev. 183, 1511 (1969)
    C. Quigg and J. D. Jackson, UCRL-18487.
60. T. D. Lee and C. N. Yang, Phys. Rev. 108, 1611 (1957).
    See also R. Marshak et al., Theory of Weak Interactions in Particle Physics, Interscience, New York (1969).
61. J. Schechter and Y. Ueda, Phys. Rev. D2, 736 (1970).

CP VIOLATION[†]

L. Wolfenstein

Department of Physics

Carnegie-Mellon University

Pittsburgh, Pennsylvania 15213

One of the most striking events of the last decade was the experimental discovery of CP violation in the decay of the K-meson. This discovery occured in 1964 in the famous experiment of Fitch, Cronin, and Christenson.[1] Let me just remind you what was involved.

The K-meson has two states which in the strong interactions are called K^0 and $\bar{K}^0$. These particles have strangeness +1 and -1, respectively. Since it violates strangeness the weak interaction mixes these together; in particular the standard weak interaction connects these states in second order. Since the standard weak interaction theory is CP conserving, the eigenstates of interest for the weak inter-

---
[†] Invited talk presented at the Symposium "The Past Decade in Particle Theory" at the University of Texas in Austin on April 14-17, 1970.

actions should be the CP eigenstates. Defining $|\bar{K}^0\rangle = CP|K^0\rangle$, the CP eigenstates are the states $|K_1\rangle = (|K^0\rangle + |\bar{K}^0\rangle)/\sqrt{2}$ and $|K_2\rangle = (|K^0\rangle - |\bar{K}^0\rangle)/\sqrt{2}$, and these should be the actual eigenstates or decaying states of the system. In fact two different decay modes were observed: a short lived K_S meson which decayed into two pions and a long lived K_L meson which decayed into, among other things, three pions. The K_S was identified as being the CP + 1 state; that is, the K_1. The K_L was identified with the CP - 1 state since $3\pi^0$ have CP = -1. Therefore, things seemed to fit very well until the discovery in 1964 that the long lived K_L in fact also upon rare occasions, namely one time in a thousand, decayed into two π's, which is a state of CP = +1.

It is, of course, a somewhat ironic thing that it was the fact that a K meson decayed into 2π's and also 3π's which was the origin of the discovery of parity violation in the famous τ-θ puzzle. That was for the charged K meson, and now some years later the discovery that the neutral K meson has these two decay modes should get us into another hassle, this time concerning CP violation. However, there is a striking difference in the subsequent history. Once Lee and Yang suggested in the paper in 1956, that the τ-θ puzzle meant that parity was violated, within a year and a half countless examples of parity violation were discovered. The whole theory of weak interaction was illuminated by that discovery.

The situation with respect to CP violation is very different. Since the original discovery of this decay mode (K_L into 2π) in 1964 essentially nothing has happened. And after some 6 years, we know almost nothing more than we knew in 1964. So let me start by summarizing all the experimental information that exists concerning CP violation.[2]

There are essentially three experiments in which CP violation has been observed. Mainly it has been observed in the decay of the long lived K-meson into π^+, π^-. It has been observed that the long lived K-meson also decays into $2\pi^0$'s. This is hardly a surprise, since (π^+, π^-) and $2\pi^0$, are intimately connected by the strong interactions, and it would be very strange if something that decayed into $\pi^+ \pi^-$ were not to decay into $2\pi^0$'s. Finally, there is one other observation concerning the K^0 system, again the K_L. The decays $K_L \to \pi + e + \nu$ and $K_L \to \pi + \mu + \nu$ mentioned by Dr. Nilsson show a charge asymmetry. If you compare the rate for $\pi^+ e^- \bar{\nu}_e$ with the charge conjugate mode $\pi^- e^+ \nu_e$, this ratio is not one as it should be if CP were good and if the K_L were a CP eigenstate. In fact it differs from one by some three parts per thousand. All these effects are the order of a few parts per thousand.

Now what has been learned from the quantitative studies which have been made of these parameters in the last six years? Well, I am afraid the answer is not very much. One thing, of course, is that the observed CP-violating amplitudes are of the order of parts per thousand, which was known from the first experiment. Since that time the only very accurate result is the measurement of the phase ϕ_{+-} of the amplitude ratio,

$$\eta_{+-} = \frac{A(K_L \to \pi^+ \pi^-)}{A(K_S \to \pi^+ \pi^-)} = 2 \times 10^{-3} e^{i\phi_{+-}}$$

for the CP-violating K_L decay to the CP-conserving K_S decay. One can measure by interference experiments not only the magnitude of that amplitude ratio but also its phase and after many vicissitudes this phase is settling down around $45°$.

There are two things of interest that one can conclude from this numerical result. The first is that while this is directly evidence for CP violation, it is also indirectly evidence for time reversal violation. Now, of course, we tend to believe for well-known reasons in the CPT theorem, so that CP violation implies T-violation on theoretical grounds. But one might nevertheless say our theories don't always work, and therefore, let us consider the alternative hypothesis that time-reversal invariance is a good symmetry and that CPT is violated. If one then follows through on that, one can make some deductions about various phases. As you know time reversal arguments commonly give you conditions on phases. If one takes then the experimental results that this phase is $45°$ coupled with a rather limited information that exists on η_{00}

$$\eta_{00} = \frac{A(K_L \to \pi^0 \pi^0)}{A(K_S \to \pi^0 \pi^0)},$$

then one can show that the hypothesis of time reversal invariance in fact does not work, and therefore, this experiment indirectly by analysis of the numbers involved is inconsistent with time reversal invariance. Arguments of this sort have been given by Kabir and Casella.[3]

The second point that can be made is that all the data that does exist so far is consistent with a one-parameter theory of the CP violation. The one-parameter theory is the theory that says that all the effects of CP violation are associated with a mixing in the K^0 system. Consider the states K_1 and K_2 as basis states, K_1 is CP-even, and K_2 is CP-odd; the whole effect in the one-parameter theory is due to the fact that the decaying states, K_S and K_L have an admixture of the wrong type of CP. Defining the basis vectors

$$K_1 = \begin{bmatrix} 1 \\ 0 \end{bmatrix} \quad ; \quad K_2 = \begin{bmatrix} 0 \\ 1 \end{bmatrix}$$

then the most general admixutre, assuming CPT invariance, is

$$K_S = \begin{bmatrix} 1 \\ \varepsilon \end{bmatrix} \quad ; \quad K_L = \begin{bmatrix} \varepsilon \\ 1 \end{bmatrix}.$$

The whole effect is due to the admixture ε. The quantities η_{+-} and η_{00} can be calculated by assuming that when the K_L decays, its K_1 component decays into 2π's as expected and the admixture of K_1 in the K_L state is the source of all CP violation; then $\eta_{+-} = \eta_{00} = \varepsilon$. ε is a complex quantity, but we shall now try to express it in terms of a single real parameter. If we look at the self-energy matrix of the K^0 system again in the $K_1 - K_2$ representation, then the CP violation shows up as off-diagonal elements that mix K_1 and K_2. The self-energy matrix is a combination of a dispersive part M and an absorptive part Γ:

$$M - \frac{i\Gamma}{2} = \begin{bmatrix} m_1 & im' \\ -im' & m_2 \end{bmatrix} - \frac{i}{2} \begin{bmatrix} \gamma_1 & i\gamma' \\ -i\gamma' & \gamma_2 \end{bmatrix}$$

The matrices M and Γ are Hermitian matrices and by CPT invariance they are anti-symmetric. The one-parameter theory is based upon the assumption that γ' is 0; that is, that there are no absorptive contributions. Absorptive contributions correspond to real intermediate states and therefore to CP violation occurring in the transition amplitudes from K_1 and K_2 to intermediate states. A one-parameter model is based upon the idea that all the CP violations occur only in the dispersive part of the mass matrix so there is only one parameter which is m'. This parameter determines the phase ϕ_ε of ε in terms of the masses and widths

$$\phi_{+-} = \phi_\varepsilon = \tan^{-1} \frac{m_L - m_S}{\gamma_S/2} \simeq 45°$$

All the data that does exist is consistent with this one-parameter model.

A simple realization that I will talk about this one-parameter model is the so-called super-weak theory. The super-weak theory also says that CP violation will not be found in any other process except K^0 unless one goes to orders of magnitude of 10^{-9}. Now at present there is no other clear-cut evidence for CP violation or for T violation, but there are very few experiments which are more accurate than 1%. There are only two experiments that may be cited that have an accuracy of the order of a few parts per thousand. There are the experiments on studies of time reversal in nuclear physics which study nuclear reactions and inverse reactions. They have been studied to a few parts per thousand but involve complex nuclei and involve a search for T violation in strong interactions. The only other experiment that perhaps is of this order is the comparison of the electric dipole moment of the neutron with the magnetic dipole moment. The ratio is found to be less than 10^{-9}. Now in order to get an electric dipole moment for the neutron, one has to violate parity. We know that parity is violated in weak interactions. If this were the only thing that inhibited an electric dipole moment then this ratio might be 10^{-6}. Thus you might say that somehow this experiment tests an extra factor of 10^{-3} due to time reversal violation. This is a rather difficult argument to make quantitative. There are few tests, then, that test for CP violation or T violation at a level of better than 1%. So the fact that CP or T violation has not been found in any other process than K^0 decay is not a very strong argument to say that no such violation exists.

Now I want to turn to the question of what are the theoretical understandings of this CP violation? Since there is little data, there is much room for theoreticians to play. If you have only one good experimental number then the number of one-parameter theories that will explain that one experimental number is some class of infinity. I want to limit my discussion very severely and apologize to numerous members of the audience who have their own theories of the CP violation and limit myself to theories which have at least the following properties. Firstly, I'm going to assume that CP violation is not some kind of a mean trick being played on us by some perverse god, but in fact it is some basic feature of the elementary particle interactions, an intrinsically big effect which just happens to show up small in the place we have seen it. I assume that CP violation has something to tell us about the nature of elementary particle interactions, that the answer to the CP violation question is also the answer to some other question.

I want to consider three other questions that may be related to CP violation: (1) What is the dynamical structure of the weak interactions? (2) Does there exist an intermediate vector boson which mediates the weak interactions? and (3) Does there exist an interaction much weaker than the weak interactions, intermediate in strength between the weak and gravitational interactions?

Now by the dynamical structure of the weak interactions I mean: there is a good deal of feeling that the effective interaction which Professor Nilsson reported (previous paper) is not the final answer about the weak interaction, that in some sense it is a phenomenological interaction which describes successfully or reasonably successfully what goes on at low momentum transfer at which nearly all the processes discussed by Professor Nilsson occur. Perhaps the real weak interaction is quite different. As an example, and not a very happy example, of a model of this sort, there is the model of Kummer, Segre, and Christ[4] which explains the weak interactions as being due to a box diagram where there are in fact four fictitious intermediate particles, two scalar bosons and two intermediate fermions, all heavy undiscovered particles.

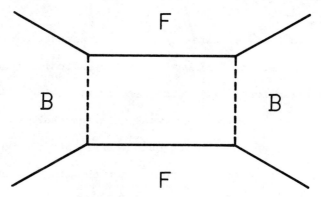

Figure 1: Box Diagram for Weak Interactions.

The advantage of this theory is that it involves scalar bosons, and therefore is renormalizeable. My only point in bringing it up lies in the fact that in this model by suitable choice of the vertex operators, one can reproduce the effective Hamiltonian of the standard theory in the limit where the momentum transfer is low compared to the intermediate masses. The true interaction in this theory has a more complex structure. If there are theories of this sort, then there will be corrections of the order of $\frac{Q^2}{M^2}$ where M^2 is the intermediate mass in the dynamical structure and Q^2 is some momentum transfer defined in terms of the external momenta. If one has the intermediate masses of the order of 5 GeV or so, then effects of this type would be of the order of parts per thousand and this is suggestive that perhaps CP violation is also an effect of this sort. I know of no real model that has this feature, it is just sort of a schematic idea for somebody's model.

Let me turn to the question of the intermediate vector boson. Let me remind you that with the intermediate vector boson W the basic interaction is a semi-weak interaction of the form $g J_u W_u$. The interaction in which the ordinary weak interaction occurs is order g^2, involving a diagram with a virtual intermediate vector boson which transmits the weak interaction from one semi-weak vertex to the other. Now what might the intermediate vector boson have to do with CP violation?

There are at least two kinds of possibilities here. Once we have interactions of order g^2, one may think that to get something which is in effect one part per thousand below that, one might consider terms of order $g^2 \alpha$ where α is the electromagnetic coupling. These terms would be electromagnetic corrections to the weak interaction. Effects of $g^2 \alpha$ might come into play if there is a CP violating electromagnetic interaction.

The first theory of this sort that I know of was a proposal by Salzmann and Salzmann[5] that the intermediate vector boson might have an electric dipole moment. Now that particular theory and many theories of this sort face a difficulty in the fact that, if there is a CP violating vertex involving a photon and the intermediate vector boson, then there exists the possibility of a diagram such as in Figure 2 contributing to the electric dipole moment of the neutron.

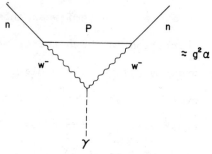

Figure 2: Weak Boson Contribution to the Neutron Electron Dipole Moment.

In this case the photon instead of being virtual is the real zero frequency photon or electromagnetic field which is involved in making the measurement of the dipole moment. This has a difficulty then that the electric dipole moment is measured to be so small that the dipole moment of the neutron due to this diagram would tend to rule out this model. However, like any good theory, it will die slowly and it is possible by various methods to suppress the electric dipole moment of the neutron.

Let me turn to the other suggestion which is that perhaps the CP violating effect might occur in order g^3. Once you have the semi-weak coupling and the ordinary weak interaction is of order g^2, it becomes interesting to ask might you make observations of things of order g^3. Now in order to do that it is necessary to have a tri-linear coupling f between three intermediate bosons or, equivalently, to have a non-vanishing vacuum expectation value of the product of 3 intermediate boson fields. Schematically, a g^3 contribution to a weak process is described by Figure 3, where I and I' are intermediate states, one of which may be the vacuum. This theory has been worked out in considerable detail by Professor Okubo[6] who describes this tri-linear interaction between the 3 intermediate boson fields as being of a Yang-Mills variety. Of course, in order that this be of order g^3 it is necessary that f, the tri-linear interaction coupling be of order unity. This requires that the intermediate vector boson have a strong interaction of the tri-linear variety.

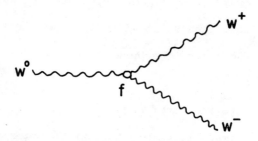

Figure 3: Tri-Linear Weak Boson Interactions.

The idea that intermediate bosons might have strong interactions has been an idea which has been propagated and advertised by Professor Marshak for many years. Now in order to have a nonvanishing expectation value and to have an interaction involving a charged vector boson one must have in fact, one positive, one negative, and one neutral boson coming into the vertex. This thus requires that this model involves neutral intermediate vector bosons as well as charged intermediate vector bosons. Furthermore, in order that there be no effects of order g, it is necessary that the positive and negative particle not be anti-particles to one another. Thus it requires a triplet of intermediate vector bosons.

A simple realization of this model involves a semi-weak interaction of a form suggested independently by Segre.[7]

$$H_{W\ meson} = g\{W_2^- \left[J_\pi + \cos\theta\ L\right]$$
$$+ \overline{W_3^+} \left[J_K + +\sin\theta\ L\right]$$
$$+ W_1 \left[J_{K^0} + J_\eta\right]\}$$
$$+ f\varepsilon_{abc} W_a W_b W_c$$

In this realization the Cabibbo angle is thrown on the lepton current L. Here one type of charged intermediate boson interacts with a strangeness changing current J_K plus $\sin\theta$ times the lepton current, while the other interacts with the strangeness conserving J_π plus $\cos\theta$ L. A neutral boson, which is the source of all strangeness-changing non-leptonic processes, interacts with a strangeness changing current J_{K^0} plus a strangeness conserving current J_η. To that, one adds the tri-linear coupling.

Where does CP violation come in? Well CP violation can come in very amusingly by simply putting an i in front of the semi-weak coupling constant g, in other words, by letting the semi-weak coupling be imaginary, so that in this case there is a very large CP violation. The whole weak interaction is CP violating. Of course, everyone who plays these games knows enough to be very suspicious. This is because you know that the i can easily be absorbed into redefinitions of fields. In this case it can't because if you reabsorb this i in redefining the W fields, the i pops up in the trilinear strong interaction. This is what T. D. Lee calls a symmetry clash, a clash between the strong interaction of a W and its weak interaction. This model has been investigated by Marshak and others[8] for the possibility of finding large effects in certain processes. Particularly they have looked at the process where a K-meson decays into a π meson and two leptons such as $K^+ \rightarrow \pi^+ e^+ e^-$, (Figure 4) a decay which has not yet been observed.

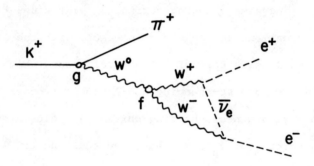

Figure 4: Decay of $K^+ \rightarrow \pi^+ e^+ e^-$

Well, let me turn now to the third of the possibilities I have outlined; namely, there might exist an interaction much weaker than the weak interaction, intermediate in strength between the weak and gravitational interactions. It is, of course, very striking that there exist a rather large range of orders of magnitude of interactions. And since we at the moment have no understanding of why interactions occur with different strengths we have no reason certainly at all to believe that there are no interactions in this huge range between weak and gravitational.

We want to ask about the possibility of interactions much weaker than the weak, say, interactions which might lie somewhere between 10^{-12} to 10^{-20} (See Table 1).

Electromagnetic	$\frac{e^2}{4\pi}$	$\sim 10^{-2}$
Weak	$\frac{Gm_p^2}{4\pi}$	$\sim 10^{-6}$
"Missing" = Super-weak		$\sim 10^{-12} - 10^{-20}$
Gravitation	$\frac{G_{Grav} m_p^2}{4\pi}$	$\sim 10^{-38}$

Table 1: Possible Interaction Strengths

If we ask about such interactions, and I want to limit myself to short-range interactions like the weak interactions, you suddenly come across the question, how would you ever observe such very weak interactions? Well, you can go back and ask, how do you ever manage to observe the weak interaction itself? The answer is it's observed because of distinctive qualitative features. One feature is that neutrinos, which have no other interactions, have weak interactions. Particles like the neutron and π meson must decay via the weak interactions since they must decay into neutrinos in order to conserve energy. The second feature of the weak interactions is that they violate strangeness, so that they are responsible for the decays of strange particles. And the third feature is that they violate parity. In principle, one might have discovered the weak interactions by finding small parity violations in nuclear forces and being wise enough to say that was due to the fact there is a new weak interaction; that, of course, was not the way history went. So how can one look and find a much weaker interaction? Well, it must in some way have special properties. One can look for what symmetries it might violate. Unfortunately there are not many left by the time you get through the weak interactions. What symmetries are left to be violated? Well, first there is charge and baryon number. The violation of these, of course, would show up in the fact that the proton and electron would not be stable. Their stability has been measured to very high accuracy so we have to look for other violations. Well, if you look at the standard weak interaction theory which Professor Nilsson discussed, about all that is really left is CP and T and the two lepton numbers, the electron number and the muon number. That is, we are not likely to see a very weak interaction unless it has the properties of violating CP and T or violating the conservation of one of the two lepton numbers. Another possibility which might help us to observe it would be if it violated some other conservation laws to a greater degree than ordinary weak interactions do. For example, suppose it changes strangeness by more than one unit. Of course, in second order the usual weak interactions change strangeness by 2. This is how you get a mass difference between the two K-mesons. But if this new interaction changes strangeness by 2 then it would at least be competitive with second order weak interactions.

Now you try to find what possible phenomena might show up, what different experimental consequences there are. One possibility is to look at places where one does see some very weak effect; for example, places where you do observe a second order in the weak interactions. One of these is the mass matrix of the K^0, which as we noted does involve second order in the weak interactions. So one possibility to see this very weak interaction would be if it made a contribution to the mass matrix in first order. To do that, it has to change strangeness by 2. It would then show up in the mass matrix but how would you distinguish it? Well, if it changed strangeness by 2 and also violated CP then you would have a chance, because then it would show up as a CP-violating term in the mass matrix, namely as an off diagonal element in the $K_1 K_2$ representation. So the quantity called m' earlier when we looked at the mass matrix could come from the first order of a super-weak interaction provided it had those two properties. If we, in fact, explained the K^0 CP violation in this way, then the effective interactions that we need, is equivalent to one part in a thousand of a second order weak interaction, which means that it is something like 10^{-9} times the weak interaction or 10^{-15}. Where else could you then hope to see such an interaction? You might look at other $\Delta S = 2$ phenomena. For example, Ξ decay into nucleon and pion. That is really very difficult because the branching ratio is a second order weak effect. This branching ratio is going to be one part in 10^{12}. Then you have to look for a 1 part in 1,000 effect in a one part in 10^{12}. I would say that is the kind of experiment which the experimentalists don't pay you for advising them about.

If you ask where else you might hope to find it, another possibility is in the violation of lepton conservation. That is if this interaction also by chance violates the lepton conservation, then again you might hope to look for some effect. Where is the most hopeful place to look for a very small lepton non-conservation? Again, it is the place where you might compete with second order weak interactions. Now one of the few places where second-order weak interactions have been observed is in the process of double beta-decay, in which a nucleus, stable with respect to the usual first-order weak interactions, may decay by emitting two electrons. According to the standard lepton-conserving weak-interaction theory this decay involves the emission of two neutrinos

$$Z \to (Z+2) + e^- + e^- + \bar{\nu}_e + \bar{\nu}_e$$

and presently-observed rates of double beta-decay are not in disagreement with this interpretation. However, it was pointed out by Pontecorvo[9] that if a superweak interaction in the range we are considering were to violate lepton number by 2 and so allow in first order the process

$$Z \to (Z+2) + e^- + e^-$$

then the superweak interaction might indeed show up. This is particularly true for transitions involving small energy releases for which the phase space factors strongly favor the three-body final state over the five-body final state. Beyond these two places (K^0 decay and double beta-decay) I know of no other practical suggestions for observing a super-weak interaction.

Professor Telegdi told me that this is going to be my last talk on CP violation because he says that next year everyone will accept the super-weak interaction and, therefore, the whole subject of CP violation will be dead. In fact, since I first presented the idea of the super-weak interaction[10], I have spent a good deal of my time warning people not to believe it. The point is the following: the super-weak theory is very easy to disprove, one good experiment might disprove it, but it is extremely hard to prove. There are many other theories which give very similar predictions. At the moment it is true that there are no experiments which rule out the super-weak interaction theory and there is a good deal of prejudice, I think, in favor of it, particularly among theoreticians. This is because the super-weak interaction simply wouldn't bother anything else they've been doing all this time. That is perhaps an unfortunate aspect. What we have been doing all this time has not been very successful. We need to be bothered. So that I think the hope really should be the reverse. The situation then I would summarize as follows: although there are no experiments that disprove the super-weak interaction theory, the present experiments are so limited that as an argument in favor of the super-weak theory, it is a super-weak argument.

References

1. J. H. Christenson et al., Phys. Rev. Letters 13, 138(1964)
2. For a recent summary see J. Steinberger, CERN Topical Conference Weak Interactions (1969), p. 291.
3. P. K. Kabir, CERN Topical Conference on Weak Interactions (1969);R. C. Casella, Phys. Rev. Letters 21, 1128(1968).
4. N. Christ, Phys. Rev. 176, 2086(1968), and references therein.
5. F. Salzmann and G. Salzmann, Phys. Letters 15, 91(1965).
6. S. Okubo, Nuovo Cimento 54A, 491(1968); 57A, 794(1968); Annals of Physics 49, 219(1968).
7. G. Segre, Phys. Rev. 173, 1730(1968).
8. R. E. Marshak et al., Nuclear Physics B11, 253(1969).

DISCUSSION OF PROF. WOLFENSTEIN'S TALK

G. B. Yodh, University of Maryland: Can one observe effects of super-weak theory in the study of double beta decay of strangeness changing decays like

$$K^- \to \pi^+ + e^- + e^-?$$

L. Wolfenstein, Carnegie-Mellon University: Well, I am not sure what to say. I mean, if from the point of view I mentioned of it being super-weak, it clearly would have an extremely low probability, so that you wouldn't look for that. But there are other models in which it is suggested that the lepton violation might occur at the level of middle-weak, that it might occur not at the level of 10^{-9} type of interaction but there might be a lepton violation parameter of the order 10^{-3}. This would be interesting to look for.

R. Hofstadter, Stanford University: Can the super-weak interaction be observed in super-high energy experiments where phase space factors might help one see a small effect?

Wolfenstein: I don't think that it will help very much as far as the super-weak interactions are concerned. I think they help primarily from the point of view that the CP violation effect that may be not really super-weak but simply an effect as I mentioned in the first possibility. That is an effect which is suppressed by the low momentum transfers which allow the present phenomenological theory to be a good one and that when one does go to very high energies which are energies comparable to the masses that define the structure of the weak interactions, that there one may see very large effects. I think that is not really the super-weak philosophy, that is a different philosophy. That is certainly an interesting possibility.

<u>Y. Ne'eman, University of Texas and Tel-Aviv</u>: What is the status of the Gürsey-Pais cosmological super-weak model?

<u>Wolfenstein</u>: Yes, as far as we know Gürsey and Pais actually at the same time that I wrote this paper about the super-weak, wrote a preprint which was never published describing essentially exactly the same thing as the super-weak theory. They suggested that it was due to a cosmological effect of a scalar cosmological field. As far as I can see, that suggestion is totally equivalent to the super-weak with the exception of saying that if the experiment was done outside of the cosmos that the effect would disappear. They, in fact, even make the comment at the end of that note that even the ordinary weak interactions are cosmological in origin and that ordinary weak interactions would also disappear outside of the cosmos. So I find it becomes a kind of philosophical

question of whether the coupling constants in nature in some way are induced by the surrounding matter of the universe and that is a question relevant to problems of cosmological models, but it doesn't really affect particle theory.

W. Thirring, CERN: I would like to comment on the cosmological explanation of CP violation. If this field is scalar there is no reason why its mass should be zero and CP violation may be an effect of the local surroundings. I made various estimates of the range and first got a few Km. This was good because this explained why CP violation was a few % stronger at CERN than at Brookhaven where the ground is somewhat less dense. Another estimate gave a range of the order of the solar system. This was not bad either since it explained why data taken at different times of the year differed where the distance to the sun was different. However, the main diffculty of these attempts is to produce a static field under time reversal. Since the interaction should conserve T the source has to distinguish the arrow of time.

Relativity and Quantum Mechanics[†]

P. A. M. Dirac

Florida State, Physics Department
Tallahassee, Florida 32306

I was asked by the organizers of this conference to give an evening talk and I began to plan what I should say and then I began to think of a suitable title for it and I suggested the title Relativity and Quantum Mechanics. I shall be talking on the subjects of relativity and quantum mechanics, but I shall be dealing with these questions from my own personal point of view, expressing the opinion that I think one should have with regard to the difficulties.

Let us begin by starting from the position of physics in the last century. One had then a very definite and well-defined scheme built from Newton's equations and Maxwell's equations for the electromagnetic field and most physicists at that time were perfectly happy with this situation and believed that they had the essential basis for the understanding of the whole of nature. However, towards the end

---

[†]This talk was presented under the joint auspices of the Program on the Public Understanding of Science and the Symposium on the "Past Decade in Particle Theory", April 14-17, 1970.

of that century there were difficulties appearing. One of the first persons to realize the seriousness of these difficulties was Lord Kelvin, who discovered a scale of temperature, the absolute scale of temperature. It was reported that he was extremely unhappy, extremely depressed in the last years of his life because of the difficulties then facing science. He died in 1907. There was really no need to be depressed by the state of science. It wasn't his fault in any way, but still he did have the foresight to appreciate that very fundamental changes would be needed and he did not share in the general complacency of physicists at that time.

There were indeed very profound changes to be made which shook the whole basis of physics, and which appeared at the beginning of this century. Those changes are relativity and quantum theory. Relativity required us to adopt a different picture of the universe, a space-time picture, a four dimensional picture in which there was no absolute time. Various observers moving relative to one another would have different ideas of time. They would use different time axis and all of them would be equally right, and nature does not favor one of these times axis at the expense of the others. The idea was first properly understood by Einstein, although many other people were approaching that idea. It has been found to fit experiment in very many applications. Some of them are rather unexpected and difficult for the ordinary person to understand. For example, if one is moving very rapidly, one's own personal watch gets slowed up and the passage of time relative to this moving observer goes very slowly as would be observed by a stationary observer. This aspect of relativity theory has been affirmed in a very striking way by properties of a certain particle called the mu meson, or the muon for short, which is a particle with a very short lifetime, about a hundred millionth of a second. When you have a mu meson here, in a hundred millionth of a second it changes into something else. Now these mu mesons are produced at the top of the atmosphere come down and are detected at sea level. Now how can they possibly do that when they live for a hundred millionth of a second. They travel with very nearly the speed of light, but even at that speed you can only go a few centimeters in that short lifetime. The answer to that problem is of course that this lifetime of the meson refers to their own proper time and it is very much extended. A person traveling with these mesons would indeed seem to take only a hundred millionth of a second to get from the top of the atmosphere down to sea level. In that way this drastic departure from ordinary ideas of space and time is directly verified by experiment. There are many of these verifications of relativity and as a result it is universally accepted by physicists. Though there may be some that think very small modifications may be needed, but the general idea is universally accepted.

Now the other great change of the present century is quantum mechanics, a new kind of mechanics, a modification of the standard mechanics of Newton which is needed for the atomic world. This new mechanics has two forms, Heisenberg's and Schrödinger's. It was discovered independently by these two workers in about 1925 and they started out with two theories which looked very different but were later found to be equivalent. Now both these forms of quantum mechanics work from equations of motion. The idea with equations of motion is that you describe the state of the world at a certain time with mathematical variables and you set up equations which tell you how these variables change with the time; then integrating these equations you can see what the state of the world is at a future time starting from the initial state, and you can compare your results with experiment. Now you will see that here I have used the concept of time in an absolute sense, in a way which relativity says one should not. Here we have the beginning of the great difficulty. We have relativity and quantum theory--two theories which are very well established--each is very certain in its own domain, and yet they resist being reconciled to each other. If two theories are both right, then you would think they ought to be immediately fitted into a single scheme, but that is not the case with relativity and quantum mechanics. There is a definite conflict between them. This conflict has been the main problem of physics for the last forty years.

One can say that all the main efforts of physicists have revolved around this question of reconciling relativity and quantum theory. An enormous amount of work has been done on the subject but there is no solution in sight. Of course, with all the work that has been done, thre has been a lot of progress in subsidiary directions which don't really face up to the real difficulty. One can do very well if one considers problems just involving a single particle, even a particle moving under the influence of fields and sets up equations involving the three coordinates of the particle and also involving the time. These equations involve the four variables and one can bring in the necessary symmetry between these four variables which relativity requires and thus one can set up a satisfactory relativistic quantum theory for one particle, but that doesn't go very far.

Suppose we have two particles. Well then one would think that one would need to work with mathematical equations involving eight variables, because each of the particles will have its three coordinates and its own time giving you altogether eight, but then how do you bring in interaction between the particles. The nonrelativistic quantum mechanics provided quite a satisfactory solution for the problem of two interacting particles, two interacting electrons, assuming that there is the instantaneous Coulomb interaction between them. Then

Then one just needs seven variables, one time for both particles, and the three position variables of the two coordinates, and one has the Coulomb interaction expressible in terms of those variables. One got a very satisfactory theory for the quantum theory of two particles with quite a good degree of accuracy, but of course it is completely foreign to relativity. One has made a complete break between time variables and the space variables. If you want to have a relativistic theory, you must not have instantaneous action at a distance. You must replace that by action through the medium of a field, action which can be propagated only with the velocity of light. People have tried to set up a quantum theory in which particles interact through the medium of a field and the interaction proceeds with the velocity of light. They have made quite a lot of progress on this problem. They have set up equations which we call right. They seem to satisfy the general properties that one would require of the equation in order to be relativistic and to conform to the basic ideas of quantum mechanics, but still when one tries to solve these equations, one finds that they have no solutions. The trouble arises because one has gone to an infinite number of degrees of freedom. The fields which conveys the interaction from one particle to the other is looked upon as a dynamical system and it has an infinite number of degrees of freedom, and whenever you have an infinity occurring in a theory there is a question that arises whether the sums and integrals associated with this infinity will converge or whether they will just lead to divergences -- lead to infinities appearing in the equations which will spoil the applications of those equations and it turns out that that is the case with the equations which one gets when one sets up a relativistic quantum mechanics with particles interacting through the medium of a field. Of course, the problem is not solved at all if your equations do not have solutions and you are thrown back again at the beginning.

These difficulties were realized some forty years ago and there has been a big lapse of time since then. An enormous number of theoretical physicists have set to work on these problems and a great deal of development has been made. Various methods have been used. Some of these methods involve setting up rules for discarding the infinities from the equation. Your equations give you infinities and the infinities ought not to be there. Let us find a way for discarding those infinities and try to see if one can get on setting up a scheme, working from some modified equations.

Well, there has been a big development of the subject of quantum electrodynamics, that is to say, the quantum theory of interacting particles where the interaction is conveyed by the electromagnetic field following these rules. The surprising thing is that people have been able to find well-defined rules for discarding the infinities in such that when they use these rules they get results which are in excellent agreement with experiment. Now most theoretical physicists are very happy with the situation. They say that if you have got rules for carrying out calculations and the results of these calculations agree with experiments, what more can you want? Well, I feel that one should not be satisfied at all with this situation. I feel that it is inexcusable to neglect infinities which appear in your equations. It is quite common to neglect small quantitites. Engineers are doing it all the time and what characterizes a good engineer or a good physicist is to be able to appreciate which are the small quantities which one can afford to neglect and not get into trouble. But here one is doing just the opposite -- neglecting something which is infinitely great simply because one doesn't like to have it in the equation.

My own way of looking at this question is to simply say that one simply must not depart from the standard practice of mathematics. It would be better to satisfy oneself from the beginning with a theory which is only approximately relativistic. One can do that. One can get rid of the infinities by changing them into finite quantities, by making approximations. If you do that you get a quantum theory which is only approximately relativistic, only approximately in agreement with relativity, but you may arrange it so that the theory is in pretty good agreement with relativity so far as concerns low energy processes and the disagreement is serious only for the high energy processes. If you do that you will get a theory which I must confess is not at all inspiring, but I believe that it should be considered as the best one can do at the present time. To have a theory which is not inspiring, which is founded on logic is sounder than departing from logic and one has at any rate the basis for a program of work, trying to develop this theory and making it accessible to higher and higher energies. However, the situation remains that we are working with a theory which has an unsatisfactory basis in it and it seems pretty clear to anyone who has worked for some time in this field that some very drastic change is needed.

Perhaps we ought to feel rather like Lord Kelvin felt, that our present theory is not all as good as one is liable to think it is. One should not be at all complacent about it. One should realize that there are some very profound changes which will have to be made in it. One cannot say when these changes will come, but one must be not too closely attached to the present formalism because one knows that such changes will have to come before the present difficulties can be surmounted.

To understand the present general situation I think one could get a good picture by imagining how theoretical physics would have developed if it had not been for the genius of Heisenberg and Schrödinger to think of quantum mechanics. Before 1925, people were working with Bohr orbits. Quite a lot of developments had been made about Bohr orbits. They provided a pretty satisfactory nonrelativistic picture for atoms in which one electron essentially was the important one. Bohr had introduced the idea of equations of motion to atoms and he had found that in addition to the Newtonian equations one has to bring in some supplementary conditions called quantum conditions and it was possible in that way to fix the orbits, each electron having its own Bohr orbit. But there were great difficulties in understanding how two electrons would interact. These difficulties of course are not the same as the difficulties I was referring to earlier in this talk when I mentioned the difficulties of fitting them into relativity. These difficulties occur already in the nonrelativistic theory and they showed up most clearly when people tried to account for the spectrum of helium. The young people in those days were trying to account for the spectrum of helium by setting up a theory for interaction of Bohr orbits, and there is no doubt that they would have continued along those lines if it had not been for Heisenberg and Schrödinger. They would have probably continued for decades on those lines working out the interactions between Bohr orbits, and with a large number of people working on the problem and continually modifying their assumptions and comparing the results of their calculations with experimental progress which has been made. People would have found that some assumptions gave results in tolerable agreement with experiment and they would have become accepted more or less and people would have then gone on to build up more elaborate theories dealing in greater detail with this interaction. Many people would have gotten their Ph.D. degrees just by this kind of work.

Well with decades of work on those lines, you can imagine how it would have looked. I think that the result would have been very much the same as the present situation in quantum theory. Maybe they would have even been led to infinities which they would have been led to discard by certain rules. But you can see that such work would have been very unlikely to lead to any really important developments. You can imagine that there would have been some people who would have followed the axiomatic method, studying all the assumptions made in the theory of Bohr orbits and then criticizing those assumptions, maybe trying to modify them, trying to work out their consequences by development of illogical ideas. Such work would perhaps have some analogy to the present axiomatic development of quantum field theory.

However, the one fundamental idea which was introduced by Heisenberg and Schrödinger was that one must work with noncommutative algebra. Noncommutative algebra means working with a mathematical scheme in which $x \cdot y$ is not the same as $y \cdot x$. Now people studying Bohr orbits from the axiomatic method would have never thought of introducing that kind of mathematics into their work. It seems to me that we are now needing some further developments as drastic as the one that Heisenberg and Schrödinger introduced, some development so unexpected that it is hopeless to try to think of it from a direct enumeration of the axioms of the present theory, even a critical enumeration of these axioms. A person would lack the necessary intelligence to make such a big jump as we need to get out of the present difficulties.

Can one perhaps get any help in the present situation by thinking over how the last big development in atomic theory was made? Can one perhaps draw any lessons? Well, there is the lesson that one cannot expect a direct attack on the problem to succeed. The answer to the difficulties of people who were working with Bohr orbits consisted in introducing quite a new mathematics. From the earliest times, people studying nature had used ordinary algebra where $x \cdot y$ is the same as $y \cdot x$ and it is obvious that they should do so because they wanted equations which could fit observations. Observations were always expressed in terms of ordinary numbers and so far as ordinary numbers are concerned $x \cdot y$ is always the same as $y \cdot x$. One is going over a kind of abstract mathematics with a non-commutation, and such a jump into abstraction one cannot make by a direct attack on the problem. Let us try and figure out how this big abstraction of introducing noncommutative algebra actually came about in atomic physics.

Well, Heisenberg started with the idea that one should build up an atomic theory entirely in terms of observable quantities. That is a very sound idea philosophically. It is rather ironic that such a sound, practical down to earth idea let to such a revolution in thought, to such an abstract idea as noncommutative algebra. And it was only possible because Heisenberg didn't keep too strictly to his idea of working entirely in terms of observable quantitites. He kept to it only partially. Of course it is obvious that if you could make discoveries just by sticking to observable quantities then it will be pretty easy to make discoveries. There must be unobservable quantities coming into the theory and the hard thing is to find what these unobservable quantities are. The way Heisenberg proceeded was to criticize previous use of the Bohr orbits. What people were doing then was to make a Fourier analysis of the Bohr orbits and of course those coefficients of the Bohr orbits are not connected at all with experiment. Heisenberg had the idea of replacing these coefficients by quantities more closely connected with experiment, quantities which should be associated with radiative transitions, transitions in which the atom absorbs or emits radiation and he introduced therefore these new quantities which are each associated with two states because radiative transitions according to Bohr's idea are always connected with two states. He had these new quantities associated with two states replacing the Fourier coefficients of the Bohr orbits. Now when you consider these quantities each connected with two states and you want to fit them together, the natural way of doing it is in the form of a matrix array. And thus Heisenberg was led to the idea of matrices and bringing matrices into equations he found that he got an algebra in which x.y is not the same as y.x.

Heisenberg was in that way led very reluctantly to noncommutative algebra. He was very reluctant because it was so foreign to all ideas of physicists at that time and when it first turned up he thought there must be something wrong with his theory and tried to correct it but he was just forced to accept it. So you see that noncommutation came about in an indirect way. First of all, the matrix elements came in and then the matrices were found to satisfy noncommutative algebra. The matrix elements of course are not so closely connected with observation. They are not completely connected with observation, but one takes the squares of their moduli and uses them to give probabilities of transition processes in which photons are emitted or absorbed.

Now Schrödinger had an entirely different approach. He was working entirely independently. He had the idea that an atom should be treated as some kind of oscillating system and that it should have energy levels, which will come out as the eigenvalues of this oscillating system, by methods which were current in other problems in mathematical physics. By thinking on these lines, he was led to the wave equation to describe atomic states. This wave equation involved various operators, operators of differentiation, and when one studied the wave equation and tried to see how to connect it with physical reality one saw that these operators were related to dynamical variables. Now operators are usually noncommuting quantities and in that way Schrödinger was led to a theory involving noncommutative multiplication. In both cases the noncommutation appeared only as the end result of a chain of reasoning which started on a very different line. I think that the great new idea which will be needed to solve the difficulties of present day atomic physics will likewise come about, not by people who are seeking directly this idea but by people who are following some path which will lead them in a round about way to the great goal.

The great new idea of quantum mechanics is this noncommutative algebra which dominates the equations. It is really very surprising that one can take Newton's equations and make rather small modifications in them so as to bring in this noncommutation in a neat and elegant way. Of course, when you have got this theory involving the noncommutation, the question arises of how you are to use it to get experimental results. You cannot immediately say that these noncommuting quantities represent things that are observable because observable things are always expressed in terms of numbers and numbers certainly satisfy the ordinary algebra.

The general interpretation of quantum mechanics was worked out a few years after the original discovery of the equations and it was found that it has to be of a statistical character. One can calculate probabilities for atomic processes taking place and one can compare these probabilities with results of observation. And it is the fit that one then has which gives one so much confidence in the correctness of the nonrelativistic quantum mechanics. This is of course not altogether satisfying to have only a statistical interpretation after the complete connection between the mathematics and the physical observation which we had with the deterministic theory of the last century. We have to go over to a new scheme of things where we have to be content with probabilities. And many people are not really satisfied with this situation. I must confess that I belong to the school that is not completely satisfied with it, but one has to accept that it is certainly the best that can be done with the present framework of quantum mechanics. It has been very much studied. People have tried all possible alternatives for getting some closer connection and such alternatives just won't work. One has to reconcile oneself to this statistical connection.

The question arises whether the noncommutation is really

the main new idea of quantum mechanics. Previously I always thought it was but recently I have begun to doubt it and to think that maybe from the physical point of view, the non-commutation is not the only important idea and there is perhaps some deeper idea, some deeper change in our ordinary concepts which is brought about by quantum mechanics. If you refer to Heisenberg's theory, then you see that the basic quantities in his formulation were the matrix elements. The matrix elements are not directly observable, but one can take the square of the modulus of the matrix element, the matrix element is usually a complex number. We can take the square of its modulus and get a real number and this square of the modulus determines a probability of an emission or an absorption process. Now if you pass over to Schrödinger's picture, there you have the wave equation. It is an equation for a wave function. The wave function is again a complex number for values of its variables and it is not directly observable. But we can take the square of its modulus and get a probability. We get the probability for the electrons being in certain positions or having certain momentum values by forming the square of the modulus of the Schrödinger wave function. We see there the beginning of a new idea. The probabilities which we have in atomic theory appear as the square of the modulus of some number which is a more fundamental quantity. Now that holds with all the later developments of quantum mechanics. One is continually calculating probabilities but before one gets a probability one has some other quantity of which one takes the square of the modulus. This other quantity is called a probability amplitude. I believe that <u>this concept of the probability amplitude is perhaps the most fundamental concept of quantum theory</u>.

A probability amplitude is a complex number and we must take the square of its modulus to get the probability of any event in the atomic world. That gives one a very different idea of probability from the ordinary everyday notion of probability. You get this ordinary notion by examples such as throwing dice. The result of any particular throw you can work out mathematically, but your mathematics gives you the probability directly. It doesn't give you some quantity of which you have to take the square of the modulus. Here we have a new idea which results in the atomic world departing very drastically from our everyday world. The probabilities which we calculate in the atomic world are always given by this special mathematical process and there is nothing corresponding to it with the ordinary everyday probabilities. The immediate effect of the existence of these probability amplitudes is to give rise to interference phenomena. If some process can take place in various ways, by various channels, as people say, what we must do is to calculate the probability amplitude for each of these channels. Then add all the probability amplitudes, and only after we have done this addition do we form the square of the modulus and get the total result for the probability of this process taking place. You see that that result is quite different from what we should have if we had taken the square of the modulus of the individual terms referring to the various channels. It is this difference which gives rise to the phenomenon of interference, which is all pervading in the atomic world. You get the simplest example if you just suppose a stream of particles shot at a screen which contains two holes and then those which go through the holes are allowed to fall on another screen behind the first one. There is a probability amplitude for a particle going through each of the two holes. We add together these probability amplitudes to get the total probability amplitude for a particle striking the second screen in a certain place and as a result of our having to add the probability amplitudes, and not the probabilities, an interference pattern appears.

So if one asks what is the main feature of quantum mechanics, I feel inclined now to say that it is not noncommutative algebra. It is the existence of probability amplitudes which underly all atomic processes. Now a probability amplitude is related to experiment but only partially. The square of its modulus is something that we can observe. That is the

probability which the experimental people get. But besides
that there is a phase, a number of modulus unity which can
modify without affecting the square of the modulus. And
this phase is all important because it is the source of all
interference phenomena but its physical significance is obscure. So the real genius of Heisenberg and Schrödinger,
you might say, was to discover the existence of probability
amplitudes containing this phase quantity which is very well
hidden in nature and it is because it was so well hidden that
people hadn't thought of quantum mechanics much earlier.

If you go over to present day theory to see what people
are doing you find that they are retaining this idea of probability amplitude. The fundamental thing which nonrelativistic quantum theory has taught us is the existence of probability amplitude and no one is inclined to criticize it.
People think that they do form the building blocks needed for
atomic theory. Now there are a good many physicists who take
Heisenberg's original standpoint--that one ought to build a
theory entirely in terms of observable quantities, but they
don't keep rigorously to that. They say let us build it up
entirely in terms of probability amplitudes which are connected with observable probabilities. The difficulties in
present day theoretical physics are concerned with high
energy processes, processes in which extremely small distances
play an important role, and for these experiments with high

energies one takes a target and shoots some particles at it and sees what comes out from the target. The experimental results consist in observing the probability for various things to come out when you have various things going in. The experimental results are expressed in terms of probabilities connecting particular outgoing particles with particular ingoing particles. There is the school which thinks one should deal only with these observable probabilities, or rather with probability amplitudes connected with these observable probabilities, and they try to see what one can do with these probabilities. If they are considered altogether they fit into a mathematical scheme which is called the S-matrix. Now the S-matrix really contains all the information that one needs for describing these transition processes where you have some ingoing particle hitting targets and leading to outgoing particles. People have made a big development of this S-matrix theory, making various assumptions which appear reasonable. They assume that the theory must be relativistic. They assume that it must satisfy principles of causality, unitarity, and they assume the really over-riding principle that the S-matrix must be an analytic function of the variables that occur in it, something which one cannot explain very well in common terms but which is a very powerful mathematical tool. They have gone a long way in developing the S-matrix theory, but of course they are continually being helped by experimental results. When experiments are found to disagree with something, well just change the assumptions and find a new basis to carry on with. There is this S-matrix school. I feel however that they are not working on altogether the right line. It must be only an incomplete theory which they are striving for. The S-matrix should appear as the result of some calculations. One should have something like equations of motion at the basis of one's theory, and one should carry out some integrations of these equations and have the S-matrix appear right at the end of the calculation.

What will this calculation involve? One might make a shrewd guess that these quantities occurring in these calculations will all be of the nature of probability amplitudes. The justification for this guess is that in the nonrelativistic quantum mechanics, whether you look at it from Heisenberg's point of view or Schrödinger's point of view you are dealing all the time with mathematical quantities of the nature of probability amplitudes. Now it is true that the probabilities connected with these probability amplitudes are usually not things which can be observed directly. They are probabilities for the results of hypothetical experiments. You take the Schrödinger wave function and form the square of its modulus, you interpret that as the probability for the various electrons in the atom

being at certain positions at a certain time. Now there is
no experiment which will enable you to tell just what is the
probability for the particles being at certain positions at a
certain time. These probabilities which were imagined as the
building bricks of a complete theory refer to hypothetical
probabilities which are beyond the range of practical experiments, but still one does get some sort of picture of them.

I feel that what is needed now is to set up a relativistic quantum theory involving probability amplitudes. They
should be fitted together in some way which conforms to the
basic ideas of relativity. One should be able to picture them
as referring to the probabilities of hypothetical processes,
imagining some experimenters who are able to perform really
individual experiments--quite impractical--but still things
which one can imagine. We should fit them together into a
complete and unified scheme. Then the S-matrix will appear
at the end of this scheme. The hypothetical probability amplitudes are on the inside. On the outside we have those
which refer to probabilities which one can observe and which
enable one to compare the theory with experiment. If one
proceeds on those lines it is, I should think, a rather less
speculative method than trying to guess what is to replace the
general mathematical scheme which we are now using. After
one has set up these probability amplitudes and found the
correct relations between them there will presumably be some

general features of the theory appearing. Maybe some very simple symbolic equations which will form the groundwork of the new theory which everyone will start to learn when the next development has been made.

Of course one might try to short circuit this work and to make a guess at what the new equations will be like and various people have attempted this, of course without success. I have joined in it myself quite a bit. One of the first suggestions that was made was that people should use non-associative multiplication. A certain generalization of ordinary algebra was found to be very useful in describing nonrelativistic atomic theory. Why not make a further step in the same direction? It was first proposed by Jordan in the 1920's, but it didn't lead to anything. It might ultimately turn out to be correct, but there are many gaps which will have to be filled in before one can see that it is the correct idea. I tried myself to set up some new equations based on the existence of an ether. The idea of an ether was rejected when relativity was established, because if one has an all pervading ether, that ether will have a definite velocity. This velocity would provide a preferred time axis. One might think that an observer moving with this velocity of the ether would see the laws of nature in a simpler form from any other observer. That idea would contradict the basic principle of relativity, that all observers must be on the same footing. However, quantum mechanics introduces an uncertainty--that maybe there could be an ether whose velocity is uncertain. The velocity of the ether may have a definite probability for having a definite value. Under those conditions the idea of an ether of a definite velocity does not contradict relativity. The ether velocity would just be a new kind of field. Now physicists have been using all sorts of fields for describing the new particles of physics but so far they haven't found any use for the kind of the velocity field which one would be led to if one had postulated the idea of an ether. The idea of an ether is not dead, but it is just an idea which has not yet been found useful and so long as the fundamental problems remain unsolved, one must bear in mind this is a possibility.

Well, I think I would like to conclude at this point. I might just say that one should keep a completely open mind for the future. I feel pretty sure that the changes which will be needed to get over the present difficulties facing quantum theory and appearing as a resistance between the quantum theory and relativity will be very drastic just as drastic as the change from Bohr orbits to the quantum mechanics of Heisenberg and Schrödinger and therefore one should not become too much attached to the present quantum mechanics. One shouldn't build up ones whole philosophy as though this present quantum mechanics were the last word. If one does that, one is on very uncertain ground and one will in some future time have to change one's standpoint entirely.

PARTONS*

Richard P. Feynman

Department of Physics
California Institute of Technology
Pasadena, California 91109

Many of us are working on the same problem, which is to understand the behavior of hadrons, the strongly interacting particles. Among the various attempts to understand it in the past decade, there is a recent one that goes under the name of "partons". It sounds very mysterious, but I will try to show you that you knew it all the time, you just called it something else. Work on partons has been done by myself, Paschos, Drell, Bjorken, and others.

Almost everybody supposes that the strongly interacting particles obey a number of principles like the quantum-mechanical superposition of amplitudes, relativistic invariance, unitarity, and analyticity in some form or other. The problem is to

---

*An invited talk presented at the Symposium "The Past Decade in Particle Theory" at the University of Texas in Austin on April 14-17, 1970.

make a theory which is consistent with all of these principles at the same time. It would be very convenient if we could write our equations in a mathematical form such that we do not have to keep checking them for unitarity, relativistic invariance, quantum mechanics, or something else. We would then not get new equations if we add some of these features, for the special mathematical form would automatically have all the equations in it. The thing to do is to find a model which satisfies all of these conditions simultaneously. The only model that we have is quantum field theory. The idea is to take quantum field theory, despite its divergences and difficulties, and try to see what kinds of mathematical formulations it would have in it. We will start with something which satisfies all the conditions and later on perhaps we can modify some parts.

What is implied by quantum field theory? Of course you would have to make for field theory a specific model, yet in every quantum field theory there are fundamental fields and operators which create and annihilate some kind of particles that underlie the theory. We need a name for the objects which the fundamental operators a^* and a create and annihilate, objects that are not the final states of the system,

such as a complete proton, which may have parts inside. We
need a name for what one would call the "elementary particles"
of the theory. We do not know if there are any such particles
in the end, but we will start by supposing that there are because otherwise we would have no field theory at all. I will
call these things "partons". In the case of electrodynamics,
for example, the partons are the ideal, what one sometimes
calls "bare", electron and the ideal proton. The creation
and annihilation operators create and annihilate these. But,
a physical electron is an ideal electron with some photon
field around it, and is therefore a combination of various
partons.

What does a field theory look like? In quantum field
theory, instead of representing things by scattering amplitudes and Regge poles, we use a wave function, and it gives
the amplitude for finding a parton moving this way or that,
or two partons going in a certain direction, etc. So we
shall see if we can get anywhere by discussing what the character of the wave functions of the protons, the neutrons,
the pions might be, and shall try to represent the experimental results in terms of properties of the wave functions. Ultimately, maybe we will see some special properties of the

wave functions and perhaps get some idea of what the partons are, whether they have spin one-half, or they are quarks, if in fact they are anything. Also, maybe by using the wave function, one might be led, psychologically, to suggest a certain principle or find a general proposition that can be deduced outside the realm of abstract field theory, without talking about the objects inside. For example, one might be able to deduce it by looking at the commutation law of currents, or something like that, which does not say anything about the underlying machinery. In such a case one would establish something that is more satisfactory than if it were model dependent. But, it does not bother me too much if, at first, it is model dependent.

The characteristic of a wave function is that if you have an amplitude for finding a lot of particles or pieces around, the amplitude is given by

$$\text{Amp} = \frac{N}{E_o - \sum_i E_i},$$

where E_o is the state energy, say of a proton, and E_i are the constituent energies of the partons inside. The numerator is some kind of matrix element. Now an interesting thing about a wave function is that it does depend on the coordinate system in which one looks at it. It is an unrelativistic idea, really, because it cuts space-time at a given moment and asks, "What partons do you see?" If, say from a Lorentz transformation, it cuts space-time at a different angle, one would see other combinations. If one has never tried these things, the fact that the wave function is not a relativistic invariant may be bothersome for a moment. But you know that the sum of the momenta of the partons inside the wave function is always the same as the total momentum of the state, and that the momentum is conserved, but that this is not so for the energy. It is the lack of energy conservation which gives the strength of a certain amplitude, so it is not relativistically invariant.

It is not easy to transform a wave function, because one has to know the Hamiltonian. Therefore, there may be one system or another in which it is most convenient to look at the wave function and the usual suggestion is to look at the wave function in the rest system of the entire state. I prefer, instead, to look at the wave function in a system in which the particles are moving very fast, all extremely relativistic, in fact, in the limit. Thus, I would like to look at the wave function of a proton when it is moving extremely fast in the z-direction.

Now if such an object moves very fast, then in order to understand what happens it is convenient to rewrite the same expression as

$$\text{Amp} = \frac{N}{(E-P_{oz}) - \sum(E_i - P_{iz})} ,$$

since $\sum P_{iz} = P_{oz}$ where P_{oz} is the center-of-mass momentum in the z direction. When the thing is moving at extreme velocities in the z-direction, let me suppose that

$$P_{iz} = x_i W .$$

For example, W may be the energy in the center-of-mass of the original state, in which case x_o would be one. Or W could just be a scale, and we could leave x_o open. Then the energy of a particle (denoting the transverse component of a particular particle by Q_i) is

$$E_i = \sqrt{P_{iz}^2 + Q_i^2 + M_i^2} = \sqrt{x_i^2 W^2 + Q_i^2 + M_i^2} .$$

It has appeared experimentally in collision after collision of hadrons with each other, that the side-ways momenta are all limited -- they seem to be of the order of 340 MeV. As the energy of the collision rises to the higher and higher GeV's, there is no increase in the transverse momenta of the outgoing particles in the collision and they are finite. I am therefore going to guess that inside the wave function, it is also true of the partons. Therefore, as W goes to infinity, Q stays the same. Let us try it. Using $E - P_z \simeq \frac{M^2 + Q^2}{2Wx}$, we get

$$\text{Amp} \simeq \frac{2WN}{\frac{M_o^2 - Q_o^2}{x_o} - \sum \frac{M_i^2 - Q_i^2}{x_i}}$$

with $\sum x_i = x_o$. Now it turns out that as you take the limit for certain matrix elements, 2WN falls and the amplitude of those states is very low. But for various states 2WN approaches a constant, at most, and therefore, interestingly enough, the wave function has a definite limit as long as one stays away from x's near zero. That is, if you go to infinity with W, everything will be all right if the x's remain finite. There is a small technical point to straighten out as to what happens as x approaches zero.

However, we have already discovered something interesting. It is that if we represent transverse momenta in absolute units, and longitudinal momenta in the scale of the energy of the collision, the wave functions that come in are invariant as W goes to infinity. This leads to a suggestion about the energy behavior of highly inelastic collisions in general, namely, that expressed in terms of the variables Q's and x's, the laboratory experiments might approach invariance at extremely high energies.

A little more can be said about the wave function. I want to represent this fast moving photon by a number of particles moving at different speeds, sharing the total x_o, the total longitudinal momentum in scale, and let us say $x_o = 1$. All these particles share the total momentum of the system. How about some of them becoming negative? No, they cannot! If x were negative, when I take the square root of E, which has to be positive, it would be -xW, and so for particles going backwards, $E-P_z$ is very large, not very small. It is the $E-P_z$ in the denominator that makes the amplitude of that term negligible. So the wave function of a fast moving hadron consists of a lot of little particles moving forward, sharing the longitudinal momentum and having a finite, confusing, transverse momentum.

I have tried to use this concept of the wave function to understand a number of phenomena. I would ultimately like to be able to understand, as much as possible, all the characteristics of the things that we observe from such a point of view, such as where the Regge expressions come from, etc. So far, I have had but a limited success. However, even the crudest situation leads you to certain ideas. Consider the elastic form factor of a proton. That means that we hit the bunch of forward-moving partons transversely with a tremendous momentum by a photon, and ask with what amplitude do we get a proton back again. It would be represented by a number of partons moving in the direction of the final proton. Now the electromagnetic field, I am going to suppose, interacts with only one parton at a time, because most field theories have a propagator for the parton obtained by the substitution $p \rightarrow p-eA$. So what we are really asking is this: Suppose we have a number of these partons and we kick one of them sideways, what is the probability that the rest of them look like they were moving in the final direction? Now, for each of these initial partons there is a certain amplitude that it was not moving exactly in the forward direction, that it had a little transverse momentum, and there is some amplitude for the final proton state that its corresponding partons look the same way, so there is an overlap integral between the two. But the higher the momentum transfer is, the harder it is to get the overlap, and, therefore, the form factor should fall off with Q depending on how likely it is that the non-kicked partons can be seen to be going in the final direction. I would guess that there is a kind of universal function for such problems of high momentum transfer. There is, of course, a matrix element for the photon on one parton that happens not to be Q dependent, but the main Q-dependence, the rapid fall-off, must come from trying to get the fast moving particle going in one direction to look like a very quiet simple thing, a proton.

Since we know experimentally that this form factor falls off toward zero as Q goes toward infinity, perhaps as Q^{-3}, we learn immediately that there is no amplitude to find only one proton-like parton inside of a proton. Because if there were, there would be a certain finite amplitude, A let us say, to find a proton consisting of a proton-like parton all alone. Then when you kick it, it would be a deflected proton-like parton, and the amplitude of that for the proton is also A, so A^2 would be the ultimate limit of the form factor as Q goes to infinity. So we have to look for more complicated wave functions. This amplitude could be zero for several reasons, one of which would be that there is no parton in the fundamental field theory which has the same quantum numbers of charge and spin as the proton. Or it may be for reasons somewhat like the fact that the electron is never found with absolutely no photons around it. There is always a field around it. But I do not think that this case is realistic here because in QED it depends on the zero mass of the photon.

The next application which I want to describe is called "inelastic electron-proton scattering." In this experiment, we take a proton and hit it with an electron, and there is an exchange of a photon. There is also a small correction for the exchange of two photons, but that is removed by the theoretical people who analyze these experiments in order to have an interaction of first order in e^2 between the electron and proton. We can control the variable $q^2 = -Q^2$ and the variable $P_o \cdot q = M\nu$. The momentum transfer, q, is spacelike, and ν is the energy loss of the electron in the proton rest system. M is the mass of the proton. In principle, we can measure the probability amplitude to have different final

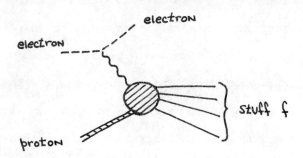

Figure 1

Inelastic Electron-Proton Scattering

states as a function of these two variables. In fact, however, at the present time, without looking at the final states of the proton we observe the total cross section by looking only at the final electron. So we have to discuss the total probability that a proton, being hit by a photon, absorbs a certain energy and a certain momentum.

First, we will describe the experiment. It turns out that the differential cross-section for inelastic electron-proton scattering can be expressed in terms of the angles in the laboratory and two functions, W_2 and W_1, which are functions of Q^2 and ν.

$$\frac{d^2\sigma}{dE'd\Omega} = \frac{e^4}{4E'^2 \sin^4\theta/2} \left[W_2 \cos^2\frac{\theta}{2} + 2W_1 \sin^2\frac{\theta}{2} \right]$$

Here E' is the final electron's energy. The amplitude for the scattering is

$$\text{Amp} = 4\pi\alpha \frac{(\bar{u}_2 \gamma_\mu u_1)}{q^2} (f|j_\mu(q)|P) .$$

With the known factors all factored out, we can express everything in terms of the quantity

$$K_{\mu\nu} = \sum_f (P|j_\nu(-q)|f)(f|j_\mu(q)|P), \quad \nu > 0 .$$

From relativistic invariance, we have

$$K_{\mu\nu}(q) = P_\mu P_\nu W_2(q^2, \nu) - \delta_{\mu\nu} W_1(q^2, \nu),$$

where both W_2 and W_1 are positive, and

$$\left(1 + \frac{\nu^2}{Q^2}\right) W_2 \geq W_1 .$$

Furthermore, since they are both positive, there is another relationship.

Because the angles used were small, the results are sensitive only to W_2, and we will plot the complete result as though we were plotting W_2, although this is a little bit erroneous. Let us look at the complete scattering, first at

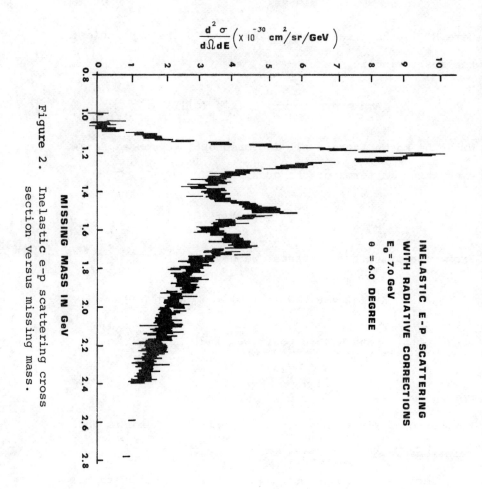

Figure 2. Inelastic e-p scattering cross section versus missing mass.

low values of momentum transfer and energy.

The missing mass is simply another way of expressing the variable ν,

$$M_f^2 = (p+q)^2$$
$$= M^2 + 2M\nu - Q^2.$$

But it measures the total four-momentum squared of all the products added together, such that if the object which is made in the collision is not a lot of pieces, but happens by accident to be one little ball of slightly excited "goop", there will be a resonance if that "goop" has a definite mass. In Figure 2 there are the famous resonances like the 1238 and there is a 1535 and something higher. The resonance at 1410 does not appear but it is very interesting that these resonances appear so beautifully.

Now if we hit the proton a little harder, it will be a little bit harder to get all those pieces to come back together again to form a resonance. We should still see the resonances but without so much glory.

In Figure 3, as expected, the resonances are much weaker and there is a rather large smear in the back which is sometimes called the "deep inelastic scattering." It is obvious

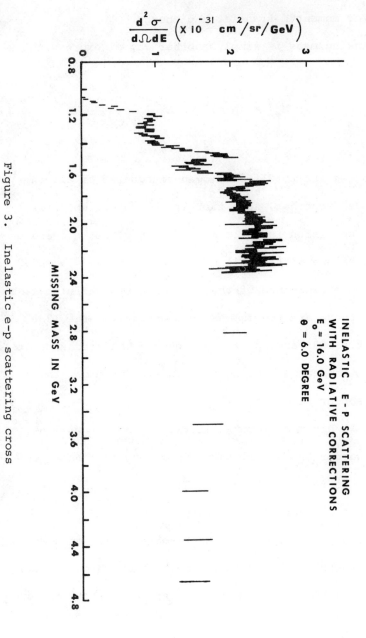

Figure 3. Inelastic e-p scattering cross section versus missing mass.

what that is: that is when you hit the proton and the pieces do not stay together, they just fly into a lot of pieces, but the chance that they are resonances becomes less and less.

Incidentally, it is very interesting that the way these resonances fall as Q^2 increases is almost exactly the same as the way the elastic scattering form factor goes out. There is a certain amplitude, that is not on these curves, which should be represented by a δ-function at exactly the mass squared of the proton, but that has been taken off so as to make this curve look good, otherwise this curve's elastic term is very big, and the size of it is the elastic scattering, which is the ordinary form factor. The funny thing is that all these fall off in proportion to the elastic form factor. In other words, the probability that the thing holds together to form a proton and the probability that it holds together, as you give it a higher momentum transfer, to form some excited state is a ratio roughly independent of Q^2. As Q^2 gets harder and harder, it is difficult to line things up, but if they are more or less lined up, they might as well be an N* as a proton.

Now let us consider deep inelastic scattering. Plotted together on Figure 4 are the data for several different

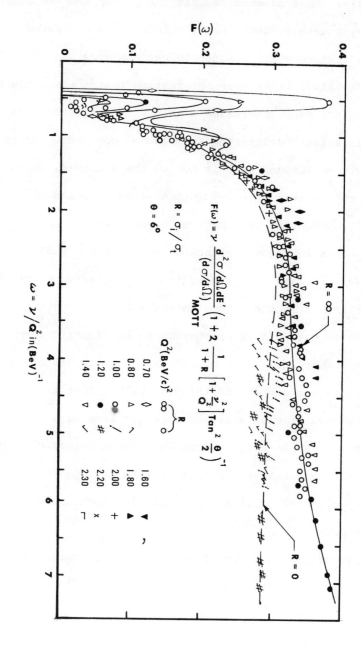

Figure 4. Scaling of the structure function.

momenta and energies as a function of ν/Q^2. What is plotted vertically is not W_2, but νW_2, and strictly speaking it is a little bit mixed up with W_1. The way that you plot it depends upon how much you assume W_1 is. If $W_1 = 0$, you get the upper curve, in which the dots do not fit on top of each other, while if W_1 is at its maximum value, then you get the curve marked R = 0 which is more or less a constant. What is interesting is that νW_2 plotted against ν/Q^2 seems to be approaching a universal curve so we should like to explain why that is a natural thing to expect.

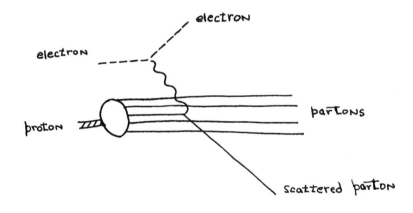

Figure 5

Parton Scattering Out of Incoming Proton

As before, in Figure 5 we represent the proton coming in

by a lot of partons. Now one of the partons is hit by an enormous momentum. The proton is moving along fast, but the sideways momentum from the photon is also very great and it knocks the parton off in a crazy direction, and leaves the other partons undisturbed. It might appear that the partons are all interacting with each other but they are not. When a thing is moving slowly, there is a lot of interaction between the parts. When the thing is moving very fast, the Lorentz time shift makes all the interaction go like a slow clock, and therefore the momentum increases so fast that it does not have much time to find out what happened. That is to say, we do not have to worry about the interactions ahead of time, nor do we have to worry about them very much afterwards, because when you sum over all the final states of the system there is a principle of completeness which says that the sum over all the final states is the same, whether they interact or not. (This is not absolutely true because there may be a shift in the energy. And when we insist experimentally that there be a given loss ν, it may not be exactly the same as the loss of a single parton because its energy may be somewhat changed by the interaction. At any rate, we can deal with these objects as free particles.)

The probability for this process involving the parton is

$$\text{Prob} = \int \frac{f(x)\,dx}{2\varepsilon}\, 2p_\mu 2p_\nu\, \delta((p+q)^2 - ?, \text{ say } p^2)$$

where $f(x)$ is the momentum distribution of the partons inside, each weighted by the square of the charge that it carries (in units of the electron charge), and x is the fraction of the z-momentum of the incoming proton that the parton has. The scattering probability for a photon on a parton is the "square" of the current, $2p_\mu 2p_\nu$, if the partons had spin zero; and for the other cases, you get more complicated things. But if you just look for W_2, you only need this factor. The delta function is over the square of the four-momentum before and afterwards. Let us suppose that afterwards it is equal to approximately p^2. So expanding,

Prob =

$$\int \frac{2f(x)\,dx}{Ex}\, P_\mu P_\nu x^2 \delta(q^2 + 2xP\cdot q + \text{transverse momenta, etc.}),$$

where we are taking $p_\mu = xP_\nu$ for the whole four-vector, so our notation for the p's is a little inconsistent. Inside the delta function we have a term $q^2 + 2xP \cdot q$, where P is in the z direction, plus some transverse momenta squared. There is an uncertainty as to the exact form for these extra terms and there may be little energies of interaction of the final object with the original object, so there are finite errors here. However, we are going to take the limit as Q^2 goes to infinity and as $2M\nu$ goes to infinity. As the energies and Q^2's get higher, we get deeper inelastic scattering, and the contribution of these finite terms to the delta function becomes less, and we have

$$\text{Prob} = \frac{2P_\mu P_\nu}{E} \int x \delta(-Q^2 + 2M\nu x) \, f(x) \, dx$$

$$= \frac{P_\mu P_\nu}{EM} \cdot \frac{1}{\nu} \, x \, f(x) \, ,$$

thereby getting,

$$\nu W_2 = x f(x), \quad x = \frac{Q^2}{2M\nu} \, ,$$

which shows that ultimately, as Q^2 and $2M\nu$ go to infinity, νW_2 is an invariant function independent of any variable except $Q^2/2M\nu$.

I can discuss the degree to which this should be correct. It turns out that it should be much more correct than we would guess. The errors are the order of something over ν, and then are only important if $xf(x)$ varies rapidly near the origin. But $xf(x)$ is constant at the origin, so if you misplace the value of x that you are looking at, it does not make much difference to νW_2, so I can say that this is a pretty good approximation. Actually, the energy that the things are measured at is not so terribly high, so it is rather nice that the universality of νW_2 is already showing up.

Now that I have explained the properties of the experiment in terms of the parton model, I would like to discuss the results of the experiment to find out what we know about the wave function for the proton. In the first place, we find that νW_2 is universal. What can we conclude from that? We can conclude that the charged partons are either spin 0 or spin 1/2. The coupling to a spin 1 particle increases with energy so rapidly that if there were spin 1 partons in there, the universality would be lost. We can even go a little bit further. If the spin were 0 for the partons, W_1 is zero. If the spin is 1/2, then W_1 reaches its maximum as expressed by the inequality we gave earlier. The question is, "What is W_1?" Now, here we go into a circle depending on how energetic a theorist we are. If we are sure that W_2 is universal, then we are also sure that the case corresponds to spin 1/2, because that is the case in which the curve is more universal when you plot it. But if, for some other reason, the curve is not universal, you would be hard pressed to argue which one is the right curve. There are going to be some more experiments at other angles and higher energies, that will make this a little bit clearer. At the present time, I would say that the experiment does favor the spin 1/2 and that, in principle, it could decide the question as to whether the charged objects which contribute to the current, the partons, are spin 1/2 or spin 0. I will conclude myself from what I have seen, that they are spin 1/2, and not spin 0.

Now it is interesting to turn these statements into a universal language that does not depend on our model. This is always a good thing to try to do. The model is just a scaffolding to discover something, but it is not the house. The first question is: What does it mean that νW_2 is universal? The $K_{\mu\nu}$ which we are measuring can be expressed directly in terms of things that do not involve the field theory

$$\int K_{\mu\nu}(q) \exp(iq \cdot x) d^4q = <P|[J_\nu(x), J_\nu(0)]|P> .$$

Therefore, in measuring $K_{\mu\nu}$ we are measuring the property of the commutator of two currents' expectation for a proton (minus expectation for a vacuum). Everybody knows that when two points in space-time are separated space-like the commutator shall be zero, and we also know that when the two points are time-like related, then the commutator is not zero, except in a trivial case. So what type of singularity is it as we sweep across from one region to another? What happens as we cross the light cone? According to the simplest theories, there is a delta function type singularity on the light cone for the above expression. It turns out that for a single particle in perturbation theory, particles of spin 0 and 1/2 have an ordinary delta function across the light cone, whereas particles of spin 1 have a gradient of a delta function. The proposition that νW_2 is a universal function is equivalent to the statement that the singularity on the light cone for the expectation value of the commutator is a simple delta function across the light cone and involves no gradient of the delta function. That is, if $<P|[J_\mu(x), J_\nu(0)]|P>$ varies as

$$P_\mu P_\nu g(t) \frac{1}{r} [\delta(r-t) - \delta(r+t)]$$

on the light cone, then

$$\nu W_2 \to G(x) = \int g(t) e^{-iMxt} dt \, .$$

We can therefore forget about partons if we want to, and state everything entirely in terms of commutators.

What does it mean that the parton has spin 1/2? Gell-

Mann, in writing his current commutation relationships assumed a kind of minimal electromagnetic coupling which is equivalent to the statement that the singularity across the light cone is the least possible. He also proposed that the commutation rules of the currents involve axial currents and that the axial currents and the regular currents commute, just as they would if they were coupled to a spin 1/2 object. So I think that with a little bit of luck and a little bit of fiddling around, I could conclude that the idea that the W_1 and W_2 are related in this limiting case for the spin 1/2, is related to the statement that the currents are part of a "thing" x "thing"; it does have to be SU(3), but could also be say SU(3) x SU(3). That means a coupling with γ_μ exists and that the opposite parity coupling, or the axial current, exists and has a coupling very similar to the normal coupling.

Let us go on to some other properties of this distribution $F(x)$. We have seen it plotted in the reciprocal variable, but I would like to plot x $F(x)$ in the regular variable $x = \frac{Q^2}{2M\nu}$.

Figure 6, when suitably normalized, yields

$$\int x\, f(x)\, dx = 0.16 \text{ or } f(x) \sim \frac{dx}{x} 0.16 \,.$$

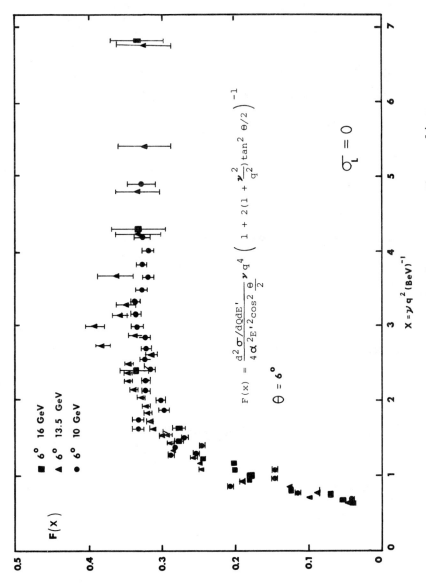

Figure 6. Structure function versus the scaling variable.

The first thing it shows is that there is an infinite number of partons because the mean number of charged partons goes as dx/x, and so the total number of them in an infinitely fast-moving proton would be logarithmically infinite at the low end. That is perfectly all right. However, it is very interesting to take a simple perturbation theory and make a model, say of proton going along that disintegrates into a parton of spin 1/2 and another parton of spin 0. If we then ask what is the distribution of this parton of spin 0, we will find that for spin 0, the distribution must always be xdx; for spin 1/2, dx; and for spin 1, $\frac{dx}{x}$. This is another representation of the angular momentum group and the angular momentum properties. For a single particle with nothing else around, perturbation theory gives the distribution $xdx/x^{2\alpha}$, for small x where α is the particle's spin, and a higher angular momentum would be more singular. Now, here we have a little peculiarity because dx/x is what you get from a vector particle, and dx/x is what we found; but a vector particle is the very one that we could not allow to couple. However, our arguments were based on the erroneous idea that a single particle could be in the small x region and that it is all right to do perturbation theory. There is an infinite number of particles for small x, and one cannot proceed by assuming a single particle and perturbation theory for the strong interactions.

However, very much like the particles in electrodynamics, when an electron is moving fast the field is condensed into a narrow splat by the Lorentz transformation, and if we analyze that in momentum space we get a uniform distribution in momentum space of the energy; the energy of each photon times the number of photons goes as dk/ω. So the reason for the dx/x is this dk/ω -- the field is squashed. (Incidentally, $\frac{dx}{x}$ looks like dk/ω, and if the photon had a finite mass as k goes to 0, it would have a finite limit. So does dx/x. It doesn't look like a finite limit, because the x is on a momentum scale which has been increased by the factor W to an enormous amount.)

Now the next thing of interest is the integral, $\int xf(x)\,dx = 0.16$. What is the normalization and what does 0.16 mean? I can express it this way: Suppose for a moment that all the partons were charged. If they all had unit charge, the square of their charges is also 1.0, and then this integral would be simply the sum of the momenta of all the particles. It is so normalized that it would be the total momentum of the original system and would be 1.0. It is remarkable that this comes out to be only 0.16, which is very small. For example, it seems to say that the number of neutrals in the system is five times the number with charges which is a little hard to accept. It could mean that the partons have a non-integral charge. There has been a lot of talk about quarks, and the proton may be made of two quarks of charge 2/3 and one of charge 1/3, in which case that number would come out 0.33, if all there was in the proton were three quarks. But, we just discovered that there must be an infinite number of partons in there. As Paschos has said, "Yes! Those are millions of pairs of quarks and antiquarks." So what we should do is take the statistical average of the 2/3 squared, the 2/3 squared, and the 1/3 squared, every kind of quark being equally likely. That comes out 0.22. But Paschos has disregarded the fact that the quarks will probably interact if there is going to be an infinite number of them. They are not going to be free, and there must be something which carries momentum back and forth between the quarks through interactions. This would be equivalent, I believe, to having a neutral quark around. So the interactors would lower it still further, and maybe it is quarks that are interacting. On the other hand, as long as the interactions are lowering it you can say that there is ordinary charge and an awful lot of interaction, such that there are indeed five neutrals for a single charged one. However, it is very interesting that these points of view are leading us into "semi-difficulties", or if you like, indications that the quarks and the partons are related.

One might ask: What is meant by the statement that the charge is an integer? It turns out that it can only be expressed as a property of four currents. Two current operators can make a commutator, but four currents can make commutators of commutators and things like that. The statement that the charge must be an integer on all the partons is a statement about a commutator with four currents in it. Experimentally, one can get a four current commutator by measuring the inelastic Compton effect from a proton at very high energy if the parton view is right. This is because in the Compton effect the amplitude goes as the charge squared; and when one works this model, one will be weighing charge to the fourth power. So by having another number on the charge, we can compare the charges. For example, if all are unit charges, then the same integral, where the weight is the charge to the fourth power, would give the same number. But, if the charges were less than a unit, then the integral for the Compton effect would give a smaller number. Of course, it is technically not quite feasible to do the Compton scattering. So I know of no direct way of testing it.

An attempt has been made to apply these ideas to hadronic scatterings, and I have made a number of speculations in print. For example, I can talk about the distribution of partons inside the hadrons and also about the distribution of hadrons which come out of a collision, and they are likely to be related but it is not obvious exactly how. One interesting feature, though, is this: Suppose for a moment that a proton moving to the right has a wave function to be partons and suppose that the exact wave function has virtually zero, or very low, amplitude to find any partons of very small x. That is, suppose all the partons are moving to the right and there is none standing still. Suppose also that another guy is moving the other way with exactly the same property. His partons are all moving to the left and nobody is nearly standing still. I am now talking about the wave function, not for the infinitely high speed, but for a very high speed where the limit has not quite been taken. Then each wave function is a solution for the whole Hamiltonian. The right moving solution involves only the right moving piece of the Hamiltonian, but it is a solution of the complete Hamiltonian. However, all the states in the right moving object are different from those in the left moving object, that is all the parton excitations are different. So if you will multiply these two wave functions together, it will still be a solution of the Hamiltonian equation. In short, there is no interaction!

There will be an interaction if one particle in the right moving proton can be thought of as being in the left moving proton. When we say that the distribution is dx/x, where xW is equal to the momentum of the parton in the z direction and W is the total momentum, it is the same as saying dp_z/p_z as long as p_x is large. But if x is of the order $1/W$, then p_z is no longer large, and I could not have taken out the square root. It is true that it is dx/x as long as x is bigger than a small quantity $1/W$; however, if x is of the order of $1/W$, then the argument that the partons cannot go backwards and that for dx/x distribution begin to fail. So in this coordinate system, there is a certain amplitude for finding partons at rest and that amplitude is approximately the integral of dx/x from 0 to $1/W$. The integral exactly is not of dx/x, but rather is of something like dk/ω. It is a cutoff, it is a function that looks like dx/x for large x, and then it cuts off somewhere of the order 1/W. So the area under this cutoff is approximately the cutoff height W times the width 1/W. In other words, the integral below 1/W is finite and independent of W. So a dx/x distribution of partons permits an interaction which leads to a cross section independent of the energy in the higher hadronic collisions. This may not be exactly right, there may be some errors, maybe a slight variation logarithmically or something.

If I suppose that the dx/x distribution extends also to the real particles which are produced in the collision, then the probability to have an exchange collision is the probability that there are no emitted particles. If a proton comes in, say, and exchanges a charge to become a neutron, it is the probability that the neutron is quiet. If I suppose that the probability that there is an emitted particle is also distributed as dx/x up to order 1/E, then

$$\text{Prob} = \int_{1/E}^{1} \frac{dx}{x} = c \, \ln E, \quad c = \text{constant},$$

and the mean number of emitted particles is $\bar{n} \sim c \, \ln E$, that is the mean number of pions and junk to be expected in an inelastic collision. But for charge exchange where no pions come out, the probability that none of them does come out is $e^{-\bar{n}} \sim E^{-c}$. This is the reason why when you have nice exchange reactions, where you ask for a specific exchange, things fall off with a power of the energy. They fall because the probability that they do not shatter the product into a lot of tiny partons, each with a mean number logarithmic will fall. I also expect the multiplicity in high energy collisions to go logarithmically, which apparently it does, as the number slowly increases in cosmic ray experiments.

In closing, the theoretical concept that I would like to emphasize is not so much the "partons", but that it is the wave functions of fast moving particles which might be useful for analyzing strong interactions, especially at high energies. By the way, there is no approximation made here and that is the complete wave function, so you could deal with any energy. For a different reason I would like, on the other hand, to emphasize for further study the interesting results of the experiments on the deep inelastic cross-sections. There are quite a few numbers associated with them, as well as simple relations, from which we ought to be able to learn something more, whether we do it by partons or by any other method.

DISCUSSION OF PROFESSOR FEYNMAN'S TALK

JULIAN ROSENMAN (University of Texas):

Can individual, free partons ever be seen?

FEYNMAN: No, I don't believe so and, at any rate, I'm supposing not. Just like the proton cannot have any amplitude to be a pure parton, the pure parton would itself disintegrate into real particles. This would be true, for example, for the parton kicked out by the photon in the deep inelastic scattering. There also might be some interaction with the remaining partons on the way out but I don't think that's as important as its disintegration. It could be that the parton's energy is higher than any of the fractions into which it can go.

ROSENMAN: Doesn't this mean that partons must have an integral charge?

FEYNMAN: Yes. So one worries if partons are quark-antiquark pairs, as Paschos suggests. What can we do, because certainly such a parton cannot be coming out? One of the ways that we can fix it is to assume that the interaction between them, say, a harmonic force to produce the linear masses, so when you pull a quark away, it snaps back by the harmonic force. Likewise, the kicked parton would be pulled back and all mixed up with the other ones, and the net result would still be pions and protons, not quarks. But it seemed to me at first that this large force would obviate the argument about the small energies. It doesn't. What matters is not the strength of the forces, but rather the spacing between the energy levels. The large force only means it takes longer to come around, and I've already summed over all the states and that's independent of things because of completeness. In fact, using the harmonic oscillator potential, I analyzed this and found that this theory should be right as $f(\frac{Q^2 \pm (300)^2}{2M\nu})$ for a spacing between the levels of 300 MeV, as for the proton excitation spectrum.

EDWARD M. MACKHOUSE (University of Texas):

If the partons are attracted with a quadratic force, do you postulate some other force to keep the self-energies from blowing up, to avoid the divergences?

FEYNMAN: What I postulate is that I can go as long as possible so long as I evade the question! It's true that any field theory has its divergences and it is also true that the proposition that the transverse momenta are limited is not quite right for a field theory. If you do it right field-theoretically, when you integrate over the transverse momenta, you get divergences, and so I think that there is something limiting the transverse momenta in a way that I do not understand, which makes convergence a lot easier. In other words, you leave your formulas with integrations over transverse momenta and just interpret them as numbers without actually carrying out the integration.

So I don't know how to cure the field theory and, therefore, I do not really propose that at the present time we should start with some specific field theory. I'm trying to meet both ways across the middle in order to have unitarity, and on the other hand to avoid divergences.

<u>ROBERT J. YAES (University of Texas)</u>:
Can particles interact by the exchange of a single parton which would give an appropriate peaking in the crossed channel?

<u>FEYNMAN</u>: No! The two things interact due to the overlap between the partons that have zero momentum, but that's not one parton; it's a characteristic of the system that the mean number of partons in there is infinite, not one. Perturbation theory is wrong. Here, diagrams of one, two, ... interacting partons, all going across, are to be added together to produce a resonance "gluck-glock" also known as a

Regge particle.

ROSENMAN: Doesn't the fact that partons decay imply that they have structure?

FEYNMAN: No! Such a decay only has to do with the energies. It is possible theoretically to have, say, a structure of three much lighter than just one, so you can't produce only one, for it would disintegrate into two three's or into some other combination of complicated objects. But these underlying objects must not be like quarks with a carrier quantum number of one-third, because then nothing can disintegrate. The quark picture is different, there you have to invent something to hold them together if you don't see them.

E. C. G. SUDARSHAN (University of Texas): Could it be that the partons are like the ether for light waves -- that the vibrations in the partons are like vibrations in the ether and may never be found?

FEYNMAN: Yes, and in the case of "the ether never being found", it was ultimately realized that there wasn't any ether at all, the ether was one of these scaffoldings to create a theory. It was later realized that the ether was an irrelevant complication and it may be that the partons are also nonexistent. The partons were only put in there to have a field theory but the world, in fact, is not field theory, rather it's something else. However, the world has unitarity, analyticity and relativistic invariance, ..., and so we use the field theory with partons. Thus, "this" and "that" is right but the real view is wrong, just like with the case of the ether.

SUBJECT INDEX

"PAST DECADE IN PARTICLE THEORY"

A.

Action-At-A-Distance
    E.C.G. Sudarshan   519

Advances in the Application of Symmetry
    Y. Ne'eman   55

Analytic S-Matrix Theory of Strong
Interactions, Exact Results In,
    V. Singh   151

B.

Behavior of Inelastic Nucleon-Nucleus
Cross Sections from 10^2 to 10^5 GeV
    G.B. Yodh, J. R. Wayland, and Yash Pal   233

Breaking of Conformal Symmetry and Renormalization of Non-Polynomial Theories
    A. Salam   1

C.

Complex J-Plane, Dispersion Theory and The,
    S. Mandelstam   295

Conformal Symmetry and Renormalization of
Non-Polynomial Theories, Breaking of,
    A. Salam   1

Construction of Crossing-Symmetric
Unitary Scattering Amplitudes
    R. L. Warnock   345

CP Violation
    L. Wolfenstein   719

Current x Current Theory,
Nonleptonic Decay and the
    S. Pakvasa and S. P. Rosen   437

D.

Development of Quantum Field Theory
    I. Segal   395

Dispersion Theroy and High Energy Scattering,
Experimental Verification of,
    S. J. Lindenbaum   255

Dispersion Theory and the Complex J-Plane
    S. Mandelstam   295

E.

Exact Results in the Analytic S-matrix
Theory of Strong Interactions for High Energies
    V. Singh   151

Experimental Verification of Dispersion Theory
and High Energy Scattering
    S. J. Lindenbaum   255

Experimental Verification of Weak Interactions
and the Muon Puzzle
    V. Telegdi   481

F.

Field Theory, Headlights of the Last Decade in,
    J. R. Klauder   561

Field Theory, Development of Quantum,
    I. Segal   395

G.

Gravitation and Particle Physics
    S. Deser   599

H.

High Energies, Exact Results in the Analytic
J-Matrix Theory of Strong Interactions for,
    V. Singh   151

High Energy Scattering, Experimental
Verification of Dispersion Theory and,
    S. J. Lindenbaum   255

High Energy Scattering, Phenomenology of,
    L. Durand III   171

I.

Inelastic Nucleon-Nucleus Cross Sections From
10^2 to 10^5 GeV, Behavior of,
    G. B. Yodh, J. R. Wayland, and Yash Pal   233

J.

J-Plane, Dispersion Theory and the Complex,
    S. Mandelstam   295

M.

Muon Puzzle, Experimental Verification of Weak
Interactions and the,
    V. Telegdi   481

N.

Nonleptonic Decay and the Current x Current Theory
    S. Pakvasa and S. P. Rosen   437

Non-Polynomial Theories, Breaking of Conformal
Symmetry and Renormalization of,
    A. Salam   1

Nucleon-Nucleus Cross Sections From 10 to 10
GeV, Behavior of Inelastic,
    G. B. Yodh, J. R. Wayland, and Yash Pal   233

P.

Particle Physics, Gravitation and,
S. Deser — 599

Particle Spectroscopy
N. P. Samios — 93

Partons
R. P. Feynman — 773

Phenomenology of High Energy Scattering
L. Durand III — 171

Q.

Quantum Field Theory, Development of,
I. Segal — 395

Quantum Mechanics, Relativity and,
P. A. M. Dirac — 747

R.

Relativity and Quantum Mechanics
P. A. M. Dirac — 747

Renormalization of Non-Polynomial Theories, Breaking of Conformal Symmetry and,
A. Salam — 1

S.

Scattering Amplitudes, The Construction of Crossing-Symmetric Unitary,
R. L. Warnock — 345

Small Mass Bosons, Symmetry Breakdown and,
Y. Nambu — 33

S-Matrix Theory of Strong Interactions For High Energies, Exact Results in the Analytic,
V. Singh — 151

Symmetry, Advances in the Application of,
    Y. Ne'eman   55

Symmetry and Renormalization of Non-Polynomial
Theories, Breaking of Conformal,
    A. Salam   1

Symmetry Breakdown and Small Mass Bosons
    Y. Nambu   33

Spectroscopy, Particle,
    N. Samios   93

V.

Violation, CP,
    L. Wolfenstein   719

Weak Interactions
    J. Nilsson   637

Weak Interactions and the Muon Puzzle,
Experimental Verification of,
    V. Telegdi   481